Photosynthesis-Assisted Energy Generation

Photosynthesis-Assisted Energy Generation

From Fundamentals to Lab Scale
and In-Field Applications

Edited by

Sathish-Kumar Kamaraj
Instituto Politécnico Nacional (IPN)-Centro de Investigación en Ciencia Aplicada y Tecnología Avanzada, Unidad Altamira (CICATA-Altamira), Altamira, Tamps., México

Iryna Rusyn
Lviv Polytechnic National University
Lviv, Ukraine

WILEY

Published by John Wiley & Sons, Inc., Hoboken, New Jersey.
Published simultaneously in Canada.

Library of Congress Cataloging-in-Publication Data

Names: Kamaraj, Sathish-Kumar, editor. | Rusyn, Iryna, editor.
Title: Photosynthesis-assisted energy generation : from fundamentals to lab scale and in-field applications / edited by Sathish-Kumar Kamaraj, Iryna Rusyn.
Description: Hoboken, New Jersey : Wiley, [2024] | Includes index.
Identifiers: LCCN 2023053436 (print) | LCCN 2023053437 (ebook) | ISBN 9781394172306 (hardback) | ISBN 9781394172313 (adobe pdf) | ISBN 9781394172320 (epub)
Subjects: LCSH: Photosynthesis. | Electric power production.
Classification: LCC QK882 .P37 2024 (print) | LCC QK882 (ebook) | DDC 572/.46—dc23/eng/20231219
LC record available at https://lccn.loc.gov/2023053436
LC ebook record available at https://lccn.loc.gov/2023053437

Cover Design: Wiley
Cover Image: Courtesy of author

Contents

List of Contributors

Wilgince Apollon
Department of Agricultural and Food Engineering, Faculty of Agronomy
Autonomous University of Nuevo León
General Escobedo, Nuevo León
México

Mohnish M. Borker
Department of Mechanical Engineering
Padre Conceicao College of Engineering
Affiliated to Goa University
Verna, Goa
India

L.A. Díaz-Colín
Division of Environmental Engineering
Technological of Higher Studies of Tianguistenco
Tianguistenco
México

Nicolas Flores-Álamo
Department of Chemical and Biochemical Engineering
Technological Institute of Toluca
National Technological of México
State of México
México

Selvaraj Gajalakshmi
Sustainable Fuel Cells Technology Lab
Centre for Pollution Control and Environmental Engineering
Pondicherry University
Puducherry
India

Nancy González Gamboa
Independent Researcher
Mérida, Yucatán, México

Soumya Ghosh
Department of Genetics, Faculty of Natural and Agricultural Sciences
University of the Free State
Bloemfontein
South Africa
and
Natural & Medical Science Research Center
University of Nizwa
Nizwa
Oman

Julio C. Gómora-Hernández
Division of Environmental Engineering
Technological of Higher Studies of Tianguistenco
Tianguistenco
México

Kuppurangan Gunaseelan
Sustainable Fuel Cells Technology Lab
Centre for Pollution Control and Environmental Engineering
Pondicherry University
Puducherry
India

Nguyen Trung Hiep
Research Institute for Sustainable Development
Ho Chi Minh University of Natural Resources and Environment
Ho Chi Minh City
Vietnam

Yolina Hubenova
Department of Biochemistry and Microbiology
Plovdiv University "Paisii Hilendarski"
Plovdiv
Bulgaria
and
Department of Electrocatalysis and Electrocrystallization
Institute of Electrochemistry and Energy Systems "Acad. E. Budevski" – Bulgarian Academy of Sciences
Sofia
Bulgaria

Pratima B. Jayarm
Department of Biochemistry and Biotechnology
Faculty of Science
Annamalai University
Chidambaram, Tamil Nadu
India

Nivedha Jayaseelan
Department of Biochemistry and Biotechnology
Faculty of Science
Annamalai University
Chidambaram, Tamil Nadu
India

Miriam J. Jiménez-Cedillo
Division of Environmental Engineering
Technological of Higher Studies of Tianguistenco
Tianguistenco
México

Kanimozhi Kaliyamoorthi
Department of Biochemistry and Biotechnology
Faculty of Science
Annamalai University
Chidambaram, Tamil Nadu
India

Sathish-Kumar Kamaraj
Instituto Politécnico Nacional (IPN)-Centro de Investigación en Ciencia Aplicada y Tecnología Avanzada
Unidad Altamira (CICATA-Altamira)
Altamira, Tamps.,
México

Vennila Lakshmanan
Department of Biochemistry and Biotechnology
Faculty of Science
Annamalai University
Chidambaram, Tamil Nadu
India

Victor A. Maldonado-Ruelas
Dirección de Posgrado e Investigación
Universidad Politécnica de Aguascalientes
Aguascalientes
México

Mario Mitov
Innovative Center for Eco Energy Technologies
South-West University "Neofit Rilski"
Blagoevgrad
Bulgaria

Anwesha Mukherjee
Department of Microbiology
Institute of Science
Nirma University
Ahmedabad, Gujarat
India

Lakshmipathy Muthukrishnan
Department of Conservative Dentistry & Endodontics
Saveetha Dental College and Hospitals
Saveetha Institute of Medical and Technical Sciences (SIMATS)
Saveetha University
Chennai, Tamil Nadu
India

Nalini Namasivayam
Department of Biochemistry and Biotechnology
Faculty of Science
Annamalai University
Chidambaram, Tamil Nadu
India

Vinh Nguyen
Thammasat University
School of Bio-Chemical Engineering and Technology
Sirindhorn International Institute of Technology
Thailand
and
Korea University
School of Civil, Environmental and Architectural Engineering
College of Engineering
Seoul
South Korea

Raúl A. Ortiz-Medina
Dirección de Posgrado e Investigación
Universidad Politécnica de Aguascalientes
Aguascalientes
México

Edith Osorio-de-la-Rosa
División de Ciencia, Ingeniería y Tecnología
CONAHCYT-Universidad Autónoma del Estado de Quintana Roo
Chetumal, Quintana Roo
México

Briska Jifrina Premnath
Department of Biochemistry and Biotechnology
Faculty of Science
Annamalai University
Chidambaram, Tamil Nadu
India

Ravichandiran Ragunath
Department of Biochemistry and Biotechnology
Faculty of Science
Annamalai University
Chidambaram, Tamil Nadu
India

Tani Carmel Raj
Department of Biochemistry and Biotechnology
Faculty of Science
Annamalai University
Chidambaram, Tamil Nadu
India

Roshan Regmi
Thammasat University
School of Bio-Chemical Engineering and Technology
Sirindhorn International Institute of Technology
Thailand
and
CSIRO
Microbiome for One Systems Health (MOSH)
Australia

Iryna Rusyn
Department of Ecology and Sustainable Environmental Management, Viacheslav Chornovil Institute of Sustainable Development
Lviv Polytechnic National University
Lviv
Ukraine

Manuel Sánchez-Cárdenas
Direccion de Postgrado e Investigación
Universidad Politécnica de Aguascalientes
Aguascalientes
México

Luis Antonio Sánchez-Olmos
Direccion de Postgrado e Investigación
Universidad Politécnica de Aguascalientes
Aguascalientes
México

Ranjita Sapkota
TERI School of Advance Studies
Department of Policy Studies
New Delhi
India
and
University of South Australia
Department of Business
Centre for Markets, Value and Inclusion, Adelaide, South Australia
Australia

Divya Shanmugavel
Programme of Nanoscience and Nanotechnology
CINVESTAV del IPN
Hydrogen and Fuel Cells Group
México
México

Omar Solorza-Feria
Department of Chemistry,
CINVESTAV del IPN
Hydrogen and Fuel Cells Group
México
México

Manoj K. Srinivasan
Department of Biochemistry and Biotechnology
Faculty of Science
Annamalai University
Chidambaram, Tamil Nadu
India

Olikkavi Subashchandrabose
Department of Biochemistry and Biotechnology
Faculty of Science
Annamalai University
Chidambaram, Tamil Nadu
India

Moogambigai Sugumar
Department of Medical Biotechnology and Integrative Physiology
Saveetha School of Engineering
Saveetha Institute of Medical and Technical Sciences
Thandalam, Chennai
India

Mirna Valdez-Hernández
Herbario, Departamento Conservación de la Biodiversidad
El Colegio de la Frontera Sur
Chetumal, Quintana Roo
México

Omar Francisco G. Vazquez
Department of Chemical and Biochemical Engineering
Instituto Tecnológico de Aguascalientes
Aguascalientes
México

Javier Vázquez-Castillo
División de Ciencia, Ingeniería y Tecnología
Universidad Autónoma del Estado de Quintana Roo
Chetumal, Quintana Roo
México

Marco A. Vázquez-Gutierrez
Tecnológico Nacional de México
Campus El Llano Aguascalientes
Aguascalientes
México

S. Ventura-Cruz
Division of Environmental Engineering
Technological of Higher Studies of Tianguistenco
Tianguistenco
México

Ma. Del Rosario M. Virgen
Department of Chemical and Biochemical Engineering
Instituto Tecnológico de Aguascalientes
Aguascalientes
México

Rosa M. Woo-García
Maestría en Ingeniería Aplicada, Facultad de Ingeniería de la Construcción y el Hábitat
Universidad Veracruzana
Boca del Río, Veracruz
México
and
Facultad de Ingeniería Eléctrica y Electrónica
Universidad Veracruzana
Boca del Río, Veracruz
México

Preface

Sustainable renewable energy technologies are gaining prominence. The biological basis for sustainable renewable energy is the most appealing priority on that list. Sustainable photosynthesis-assisted power generation is an intriguing renewable energy generation system, which generally exploits the light-harvesting systems of phototrophic prokaryotes (oxygenic phototrophic-cyanobacteria and anoxygenic phototrophic-purple, green, and heliobacteria) and higher plants in biological systems. This may effectively capture and transform solar energy, resulting in anabolism and catabolism activity and ensuring biological system growth and development. The conversion of organic material into electricity using biocatalysts in microbial energy-generating systems such as microbial fuel cells is an emerging renewable energy generation systems. The integration of these efficient energy generation systems with photosynthetic biological systems opens up new possibilities in the sustainable renewable energy industries. The first part of the book introduces the essence of getting bioelectricity through photosynthesis-assisted systems, describing their fundamentals and basic operating principles. Direct electronic transfer (DET) and indirect electronic transfer (IET), including abiotic factors, are highlighted as key links in power generation. The section provides insights into the diversity and importance of both technological and biological components of photosynthesis-assisted power-generating systems. Functional fundamental materials such as electrodes, photocatalytic and membranes, as well as their structural design, are recognized as having a significant impact on the efficiency and power production of photosynthesis-assisted systems. The possibilities, constraints, and significant sustainability factors for photosynthetic microbial fuel cells are addressed. The diversity of exploited plants, including C3, C4, and CAM plants, as well as wetland and drought-resistant plants, is assessed, as is their critical impact on the performance of photosynthesis-assisted power-generating systems. The second part of the book presents an overview of the diversity of photosynthetic species used in photosynthesis-assisted power generation, including anoxygenic photosynthetic bacteria as well as oxygenic photosynthetic organisms such as cyanobacteria, microalgae, bryophytes, and other plants. Electrogenic microalgae are studied, as well as the impact of biomass weight and light intensity, temperature, and pH on the performance of photosynthetic microbial fuel cells. Purple photosynthetic bacteria dry-surface biofilms are clarified. Recent research on bryophytes as a prospective object of photosynthesis-assisted power-generating technology is presented due to their great stress tolerance and survival in a wide range of temperature conditions. The effect of plant species on the efficiency of plant microbial fuel cells (PMFCs) is explored, as well as the potential of wetland and drought-resistant plants for photosynthesis-assisted power generation. The third part of the book highlights advancements and recent trends in photosynthesis-assisted microbial fuel cells to power various types of low-energy-consuming devices and IoT: as a power supply for real-time LED and digital clocks, IoT-based WSN sensor systems, powering batteries and partially loading a mobile phone, and so on. Because of their synergistic interrelationships, ammonium and phosphorus treatment with hybrid photoautotrophic algal-assisted MFCs presents an effective sustainable technology for the elimination of organic matter and the removal/recovery of nutrients from different wastes with simultaneous zero-waste bioenergy recovery. Photo-bioelectrochemical fuel cells for antibiotics and dye removal with simultaneous power generation have only recently begun to emerge, but they have made great progress. PBFC systems have a high power generation capacity and can eliminate up to 99–100% of antibiotics and dyes, which is the topic of this section. The progress of PMFC applications for bioremediation as well as plant health monitoring and biosensing is presented. The book also discusses the challenges of obtaining bioelectricity on green roofs in various climatic conditions, such as northern countries with frosty winters and southern countries with arid climates, which both provide biosensing and climatic benefits by reducing urban heat island temperatures using various types of photosynthesis-based microbial fuel cells. The prospect of agriculture-based crops in PMFC for futuristic sustainable protected agriculture is also considered. Finally, the most critical features of photosynthesis-assisted power generation sustainability are revealed. The topics encompassed in this discussion include renewable energy sources, the management of liquid effluents and solid residues in photosynthesis-assisted power generation, the environmental implications of bioelectrochemical systems, cost challenges, and the transition to a practical scale, as well as social considerations. With this in consideration, this book delves into the different distinct light-harvesting biological systems for energy generation, from principles to practical challenges ranging from the lab scale to in-field application.

October 2023 | Sathish-Kumar Kamaraj
Altamira, Tamaulipas, México | Iryna Rusyn

Acknowledgments

First, we want to thank God for blessing us with good health and the ability to edit this book. Our deepest gratitude goes to Wiley for believing in our work and accepting our book for publication. We thank all the authors who responded to the suggestion for cooperation and their excellent work. Thanks to everyone who helped make this book a reality – the authors, the reviewers, and everyone in between. We are grateful to the many publishers and authors who permitted us to use their work, especially the figures and tables.

Sathish-Kumar Kamaraj would like to express his gratitude to the Director General of Instituto Politécnico Nacional (IPN) and the Director of Centro de Investigación en Ciencia Aplicada y Tecnología Avanzada, Unidad Altamira (CICATA Altamira), for their constant support and facilities in promoting the research activities. Thanks for the project SIP:20231443, Secretaría de Investigación y Posgrado (SIP) – IPN. Further extensions to the funding agency of the National Council of Humanities, Sciences and Technologies (CONAHCyT – México) and the Secretary of Public Education (SEP – México). He extended his gratitude to Mrs. Mounika Kamaraj and Bbg. Aarudhraa for the family support.

Iryna Rusyn thanks God for the opportunity to work on this book during the difficult wartime in Ukraine and would like to express her gratitude to the family Mr. Oleksandr and Lukyan for their support.

Part I

The Basic Principle and Fundamentals of Photosynthesis-Assisted Power Generation

The part introduces the essence of obtaining bioelectricity by the photosynthesis-assisted systems presenting their fundamentals and basic principles of operation. Electron transfer mechanisms as an important link in power generation, the latest current data of their various types, direct electronic transfer (DET), and indirect electronic transfer (IET), including through abiotic elements, are discussed. Insights into the diversity and importance of both technological and biological components of photosynthesis-assisted power generation systems are revealed in this section. Functional basic materials such as electrodes, photocatalytic and membranes, and also their structural design that has some of the crucial effects on efficiency and power output of the photosynthesis-assisted systems are highlighted. Opportunities, limitations, and considerable sustainable parameters for photosynthetic microbial fuel cells are clarified. The diversity of exploited plants, C3, C4, and CAM plants, wetland and drought-resistant plants, and their essential impact on the performance of photosynthesis-assisted power generation systems are evaluated.

1

Introduction to Electron Transfer Mechanisms in Photosynthesis-Assisted Power Generation

Nancy González Gamboa

Independent Researcher, Mérida, Yucatán, México

1.1 Introduction

Electron transfer mechanisms (ETMs) are chains where microorganisms generate energy during their activities and this energy can be harnessed by bioelectrochemical systems. Microorganisms regulate their ETMs so that electrons from a donor are transferred to an available electron acceptor, which maximize their energy gain (Patil et al., 2012).

Bioelectrochemical systems, also known as plant microbial fuel cells (PMFC), are composed of two electrodes (anode and cathode) and an electrolyte. These systems utilize microorganisms that are either attached to one or both electrodes to catalyze the oxidation reaction at the anode and/or reduction reaction at the cathode (Hamelers et al., 2010).

Bioelectrochemical systems have the ability to transfer electrons bidirectionally between biotic and abiotic components, where redox-active microorganisms or biomacromolecules catalyze the exchange process. PMFCs work through the ETM that occurs in bioelectrochemical systems, where microorganisms transport electrons to a final acceptor, known as the anode. These electrons then pass through an external circuit to reach the cathode, and protons migrate through ions dissolved in the electrolyte to reach the cathode; this process closes the circuit and generates electrical energy and removes the organic matter. MFCs can be used to generate hydrogen, known as microbial electrolysis cells, for the

Photosynthesis-Assisted Energy Generation: From Fundamentals to Lab Scale and In-Field Applications, First Edition. Edited by Sathish-Kumar Kamaraj and Iryna Rusyn.

generation of solar energy, known as microbial solar cells, and microbial plant cells (photosynthesis) (Kumar et al., 2017).

The aim of this chapter is to describe the mechanism of electron transfer in PMFCs, occurring mainly at the anode and cathode of the systems, and the role of microorganisms working in the systems, as well as the various technologies that utilize the electron transport chain.

1.2 Electron Transfer Mechanism

A conventional microbial fuel cell (MFC) is composed of an anode and cathode chamber, which are separated by proton exchange membranes. The MFC generates its energy from the oxidation of organic matter by bacteria at the anode, and with the reduction of oxygen at the cathode (Logan et al., 2005).

The key advantages of biological fuel cells, in comparison to conventional fuel cells, are the mild operation conditions (ambient temperature and near-neutral pH) and the virtually unlimited range of potential fuels, for the oxidation of which we lack suitable electrocatalysts (Schröder, 2007).

Microorganisms are not evolutionarily designed to dispense energy to power a fuel cell—the majority of relevant redox processes take place within the microbial cells, and it is a great challenge and a major research issue to find means to efficiently divert electrons from the metabolism to the anode of the fuel cell. Various approaches have been proposed. They differ in the nature and the mechanism of the electron transfer from the microorganism to the fuel cell anode (Schröder, 2007).

Respiration in bacteria is a versatile process that involves multiple metabolic networks acting as conduits for electron flow to a terminal electron acceptor (TEA). Microorganisms harvest electrons from organic and/or inorganic molecules present in their environment and transfer these to specialized electron transport chains during cellular respiration. Electrons are ferried by the carrier proteins of the electron transport chain to the TEA for the purpose of generating energy via the creation of a transmembrane ion gradient for ATP synthesis. TEAs run the gamut from oxygen to organic carbon-based molecules in the case of fermentation, and inorganic molecules such as sulfate and nitrate. In the case of oxygen, its high redox potential, easy uptake into the cell, and abundance in the atmosphere makes it a natural TEA for the majority of living organisms. In anaerobic environments, the cells depend on alternative TEAs for energy generation. Molecules such as nitrate and sulfate in dissolved form can be ingested into the cell and used for the energy generation process. However, many anaerobic environments do not possess sufficient reserves of soluble TEAs. In such cases, microbes adapt by extending their redox circuitry across the cell membrane and accessing TEAs for the final discharge of electrons. This phenomenon, called the extracellular electron transfer (EET), requires special mechanisms and structures that allow selective transfer of electrons outside the cell membrane. Microbes capable of EET are variously termed electroactive bacteria, exoelectrogens, electricigens, or anode-respiring bacteria. This strategy, developed as a means to access the plentiful reserves of redox-active electron donors in their surroundings, forms the crux of the current generation in MFCs (Aiyer, 2020).

1.2.1 Electron Transfer at the Anode

The microorganisms present in the anodic chamber extract electrons and protons by oxidizing organic substrates. Electric current is generated by maintaining these microbes in anaerobic conditions provided in the anodic chamber, away from any TEA such as oxygen, besides anode. The ETMs can be broadly classified as direct electron transfer (DET) (via membrane-bound redox proteins or pilli without using any dissolved species), and mediated electron transfer (MET) (Nawaz et al., 2020).

The DET requires that microorganisms possess membrane-bound electron transport protein relays that transfer electrons from the inside of the bacterial cell to its outside, terminating in an outer-membrane (OM) redox protein that allows the electron transfer to an external, solid electron acceptor (a metal oxide or an MFC anode) (Schröder, 2007).

This process involves the direct transfer of electrons from the microbes to the electrode. For this to occur, physical contact between the bacterial cell membrane and the electrode is necessary. DET is accomplished via the outer membrane cytochromes (OMCs), conductive pili, or self-assembled appendages called nanowires (Figure 1.1a). C-type cytochromes, which are located on the outer surface of the bacterial membrane, are redox proteins that facilitate the transfer of electrons to an external TEA such as the MFC anode. The central role of *c-type cytochromes* resides in transferring electrons from the cell interior to an exterior TEA (Aiyer, 2020).

The formation of *conductive pili* (nanowires) in exoelectrogenic microbes that allow the microbes to transfer electrons to a solid electron acceptor is another way of establishing DET. These pili are linked to the bacterial membrane-bound cytochromes and allow electron transfer to TEAs that are physically distant from the microbes (Figure 1.1c). Another form of DET is through the formation of conductive biofilms. These *biofilms* have good electrical conductivity and are mainly associated with the *Geobacteraceae* family and related microorganisms enriched with sewage sludge (Figure 1.1c) (Aiyer, 2020; Schröder, 2007).

Many microorganisms lack the ability to transport electrons directly to the electrode; this mechanism is known as *indirect electron transfer* or mediated transfer (Figure 1.1b), because they are not in contact with the electrode surface. MET usually involves a redox carrier that acts as a shuttle to transfer electrons to the OER. In MFCs, this redox mediator gains electrons from the bacterial cells,

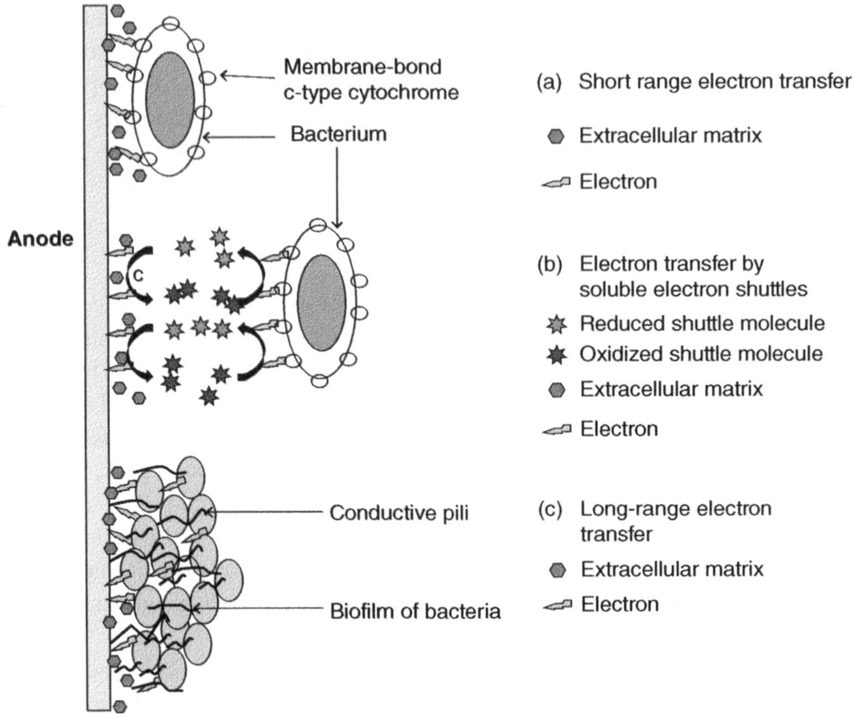

Figure 1.1 Different mechanisms of electron transfers from microorganisms to electrodes (Aiyer, 2020).

leaving them in a reduced state, and transfers electrons to the anode, oxidizing itself in the process (Aiyer, 2020).

1.2.2 Electron Transfer at the Cathode

Oxygen is the most popular TEA. Microorganisms directly transfer the electrons from the anode to oxygen, or they assist in the oxidation of transition metal compounds for electron delivery to oxygen (Song et al., 2019).

According to numerous investigations, the configurations of MFCs are quite varied, including air-cathode MFCs, aqueous cathodes using dissolved oxygen, and two-chamber reactors with soluble catholytes or potable potentials, tubular packed bed reactors, and so on. In these different configurations of MFCs, microbes exist only in the anode chamber. However, some research found that bacteria inevitably grow on the cathodes and could even increase energy production significantly. In early 1997, Hasvold et al. discovered that bacteria colonize the cathode and catalyze the reduction of oxygen (Song et al., 2019).

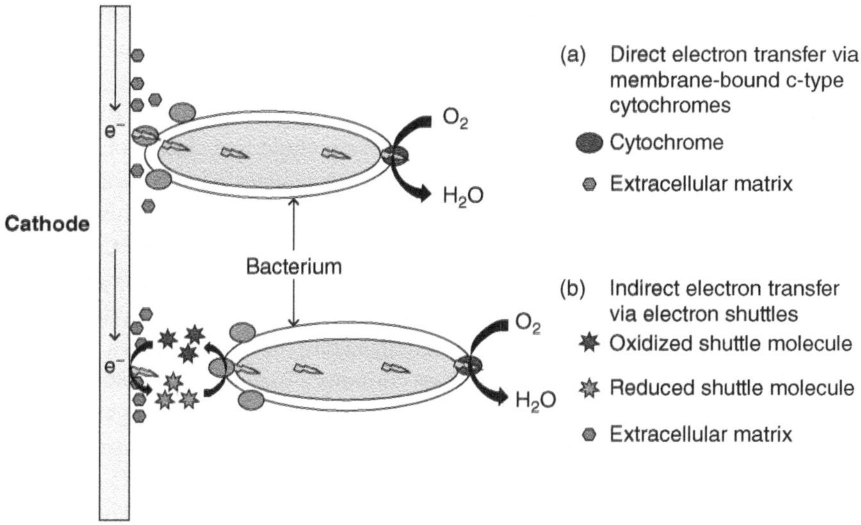

Figure 1.2 Mechanisms of electron transfers from electrode to microorganisms (Aiyer, 2020).

Two main mechanisms have been reported, namely direct and indirect electron transfer. *Direct electron transfer* requires physical contact between the bacterial cell membrane and the cathode electrode surface (Figure 1.2a), and electrons from the electrode are directly received by the outer membrane redox macromolecules such as cytochromes. This ETM was reported with *Geobacter* species or mixed cultures using fumarate, nitrate, tetrachloroethene, CO_2, O_2, Cr(VI), or U(VI) as an electron acceptor. *Shewanella putrefaciens*, which is versatile in anodic EET mechanisms such as excreted flavins and menaquinone-related redox mediators, as well as OMCs can utilize an outer membrane-bound redox compound for electron transfer in microbially cathodic oxygen reduction (Huang et al., 2011).

Bergel reported that the presence of the biofilm on the MFCs' cathode surface leads to efficient electron density. These findings encouraged the development of bio-cathode MFC, where bacteria serve as biocatalysts to accept electrons from the cathode electrode. Up to now, bio-cathode MFCs have attracted much attention and have been considered the promising MFCs (Song et al., 2019). Marine sediment MFC is one of the earlier applications of MFCs, where bacteria colonize on the cathode surface forming slimes, which catalyze the reduction of oxygen. This results in increase in catalytic activity of the cathode and an increase in the on-load cell voltage from typically 1.2 to 1.6 V (Huang et al., 2011).

The *indirect electron transfer* mechanism (Figure 1.2b) does not need physical contact between the bacterial cell membrane and the cathode electrode surface.

It has been suggested that in MFCs with naturally occurring microbial communities, extracellular substances are always involved in the electron transfer between microbes and electrodes (Huang et al., 2011).

Several phylogenetically diverse microbes (mostly bacteria) have been reported to generate electricity in mediator-less MFCs. These include five classes of microorganisms such as *Proteobacteria*, *Firmicutes*, and *Acidobacteria* phyla, and some microalgae, yeast, and fungi, which have shown current generation in this technology. The prevalent bacterial species known to produce electricity in MFCs include dissimilatory iron-reducing *Geobacter spp.*, *Shewanella spp.*, *Rhodoferax ferrireducens*, *Aeromonas hydrophila*, *Pseudomonas aeruginosa*, *Clostridium butyricum*, and *Enterococcus gallinarum*. Alternatively, microalgae have been utilized as a substrate or biocathode in MFCs. Microalgae biomass contains a high level of proteins (32%) and carbohydrates (51%), which are readily degradable by the exoelectrogens to produce electricity. Velasquez-Orta et al. obtained a power density of 277 W/m^3 in MFC using *Chlorella vulgaris* (microalgae) powder as a substrate (Kumar et al., 2015).

Oxygen can function as an indirect electron acceptor in MFCs. This occurs when manganese oxides and iron salts are first reduced by the cathode (abiotically) and then re-oxidized by bacteria. To achieve high electron transfer efficiency, manganese and iron are used as mediators of oxygen transfer under aerobic conditions (Song et al., 2019).

1.3 Photosynthesis in the Electron Transfer Mechanism

Major limitations of MFC are environmental toxicity, cost of electrode materials, and the availability of the final electron acceptor at the cathode. These obstacles could be overcome by using bio-electrolytes, especially if the biocathode chamber had an oxygenator organism to reduce the costs and work as a final electron acceptor (Elshobary et al., 2021). There is a current trend to utilize photosynthetic organisms such as plants and algae in MFCs to improve the performance of the technology and reduce costs. The PMFC, a biological cell that converts solar energy into bioelectricity with the aid of the microbes in the rhizosphere region of a plant, seems to be an emergent source of sustainable energy (Nitisoravut and Regmi, 2017). Other microorganisms that can rapidly generate O_2 normally on this planet are microalgae that include eukaryotic algae and prokaryotic cyanobacteria. Microalgae are renowned photosynthetic organisms, which have the potential to produce bioelectricity by incorporating with MFCs (Elshobary et al., 2021).

Photosynthesis is dependent on two factors: carbon dioxide concentration and photosynthetic active radiation (PAR) light intensity (Maddalwar et al., 2021).

Recently, the utilization of the photosynthesis process as a source of renewable energy by generating hydrogen, bioelectricity, and other biofuels, has been considered an efficient alternative to fossil fuels. The photosynthesis process depends on the conversion of solar energy into chemical energy for biomass production, and the conversion efficiency varies according to the species of photosynthetic organisms (Elshobary et al., 2021).

Similar to plants, the photosynthesis process is initiated by photon induction in algal cells, resulting in the fixation of carbon into various storage compounds such as carbohydrates, lipids, and proteins. Bioelectricity production using algae seems to be a wise approach to extract energy from sunlight in an economical and sustainable manner. This is achieved through the integration of photosynthesis with an MFC. Algae have been used commonly in MFCs to reduce oxygen at the cathode or as a substrate for bacteria. However, sufficient electric current can also be generated at the anode, where cytochromes help indirect shuttling of electrons generated in photosystem II of the algal cells and can be called photosynthetic algal MFC (Shukla and Kumar, 2018).

Concurrent bioelectricity and biomass production make PMFC an appealing choice forfuture green energy. Although wind, solar, geothermal, and hydroelectric power undoubtedly decrease CO_2 footprints, they also have some disadvantages such as landscape transformation, energy-intensive processes, and geographic limitations. In contrast, PMFCs can generate continuous energy without competition for food and can be operated at any location. Mild operating conditions make PMFC more attractive than these traditionally considered alternative sources of energy (Nitisoravut and Regmi, 2017).

The selection of plant species also depends on the type of application such as power generation. It might be expected to have other applications such as heavy metal removal, phytoremediation, or sewage treatment plants (Maddalwar et al., 2021).

Microalgae convert solar energy into different forms of chemical energy using CO_2 directly from the atmosphere. Sunlight is captured via the antenna of the chlorophyll and transferred to special chlorophyll dimers (P680 of photosystem I and P700 of photosystem II). Once photons excite these chlorophyll dimers, electron transportation is initiated through various transporter (*Plastoquinone (PQ), Cytochrome b6F (Cyt b6F), Plactocyanine (PC)*) and across membranes to produce ATP. The ferredoxin enzyme receives electrons from Photosystem I and transfers them to ferredoxin NADP+ oxidoreductase (FNR) to generate NADPH, which can be utilized in the dark reaction for CO2 fixation. Deviation from normal electrical conditions can split two water molecules into four photons to produce one oxygen molecule, four protons, and four exciting electrons, where two photons are used to reduce each NADP+ molecule (Elshobary et al., 2021).

1.3.1 Anodic Electrode

Electrons are transferred from the anode to the cathode to reduce the oxidized mediator, which then enters the microalgal cells to further oxidize, and release its electrons. The microalgal cells release the oxidized mediator into the medium, and the cycle repeats in this way. These electron mediators can be added externally or produced naturally by the microalgae themselves. This demonstrates that microalgae could be used as electron donors on the anode side in MFCs to generate electricity. In the anode compartment, different biodegradable microorganisms, mostly bacteria, are used to degrade organic matter of external origin or produced by microalgae. However, this oxidation process relies on the release of electrons from the microbial cell to the anode, unlike other biochemical and cellular processes, where electrons are transferred through a redox reaction (Elshobary et al., 2021).

Electrogenic microalgae are used at the anode as electron donors to generate bioelectricity through the process of photosynthesis in autotrophic algae or the degradation of organic compounds in heterotrophic algae. However, the generation of oxygen by algae on the anode side poses a serious problem for microalgae-assisted MFC (MA-MFC), as oxygen is an electron acceptor that could significantly reduce the electron flow. Most anode configurations for microalgae are restricted to the design of single-chamber fuel cells (Elshobary et al., 2021).

PMFCs are an innovative, promising, and environmentally friendly way of generating renewable bioelectricity. The method of PMFCs consists of installation of electrode systems into the substrate with developing plants through the collection of bioelectricity produced by the electroactive microorganisms in the rhizosphere as a result of consuming products excreted via the roots by photosynthesis, and also feeding by plant decomposition products and substrate compounds (Rusyn, 2021).

Plants with C3 or C4 pathways are preferred for PMFC applications because the biomass production and the rate of photosynthesis are highest in C4 plants than C3 plants, and it is much lesser in plants with *Crassulacean* acid metabolism (CAM) pathway (C4 > C3 > CAM) (Maddalwar et al., 2021).

1.3.2 Cathodic Electrode

The O_2 level in the cathodic chamber determines the performance of the cathode. Microalgae serve as a renewable source of oxygen via the photosynthesis process, which can increase the level of O_2 on the cathode side to increase MFC cathode potentials and power densities and to decrease cathode overpotentials, as most of the active cathodic sites are involved in the reaction (Elshobary et al., 2021).

There are different MFC configurations in which microalgae can be used in the cathode compartment. In the wet cathodic configuration, microalgae photosynthesize under light and produce electrons and protons depending on the utilization of CO_2 and light. In this process, organic substances, algal biomass, and oxygen are produced as end products. The oxygen released during photosynthesis is consumed by the microalgae in the dark respiration process to degrade the organic matter and continue producing electrons and protons during the dark period. In the wet cathode fuel cell, the cathode and anode are immersed in the same fluid as the fuel cell, where the protons travel to the wet cathode without any mediator or ion exchange membrane to be oxidized within the fuel cell (Elshobary et al., 2021). *Air cathode* MFCs differ from the above by the use of an ion exchange membrane to transport protons to the cathode, which is located outside the fuel cell exposed to air for utilizing atmospheric oxygen as the final acceptor. In this design, pure algal cultures or a mix with some degradable bacteria can be used to improve the electron generation and the performance of the MFC by using organic materials produced by the microalgae as a substrate for bacterial degradation (Elshobary et al., 2021). *Two-chamber MFC* in the absence of an ion exchange membrane, where algal cells are grown in one chamber under light conditions and fermenting microorganisms, are grown in the other chamber under dark conditions. This configuration relies on the degradation of organic matter (e.g. wastewater) by the fermenting microorganisms to produce electrons that flow through the external circuits, in addition to the CO_2 used by the photosynthetic microalgae to reduce the electrons in the cathode chamber. *Dual-chamber MFC with proton exchange membrane* is the most common for using microalgae as oxygen generators on the cathode side of the MFC. Activated sludge is used in the dark anode compartment and microalgae culture on the cathode side. The CO_2 produced by the anode bacteria is transferred to the cathode compartment for use by the microalgae. There are also *sediment MFC* configurations, which rely on natural differences in electropotential created by using an anode buried in the sediment and a cathode submerged on top of the sediment. Electron flow is achieved through the decomposition of organic or inorganic compounds in the sediment by the microorganisms, while the generated CO_2 is consumed by the microalgae cells in the upper cathode compartment, and the O_2 produced by the microalgae acts as the final electron acceptor (Elshobary et al., 2021).

There are some studies in which plant species are grown in the cathodic chamber because the plant provides oxygen as a byproduct of photosynthesis in the cathodic chamber, which reacts with the electrons received through the MFC circuit to form water (Maddalwar et al., 2021).

Organic substrates are mainly supplied by the root deposits of aquatic plants in PMFCs and are degraded by bacteria on the anode surface to generate energy. The classical configuration of a PMFC is described by placing the anode immersed in a support matrix close to the plant roots to obtain the exudates as fuel, and the cathode position is enclosed to the plant root surface (Goto et al., 2015).

In recent years, plant-based catalysts have been identified as producing high yields due to their large surface area, and research into these catalysts is expected to continue to grow (Martín-Betancor et al., 2017).

1.4 Technologies In Which the Photosynthesis Process Can Be Applied for Energy Generation

Despite MA-MFC would be more sustainable than using MFC alone, further developments in such systems are still necessary for improving its efficiency and achieving a real-world application on a large scale. In this context, understanding the bio-electrochemical mechanism of MA-MFC, including electron shuttling and oxygen generation, is very important. Emissions of CO_2 due to the consumption of fossil fuels have been estimated to increase from 388.5 in 2009 to 409.95 ppm by 2019, within only 10 years. To ensure energy security and reduce CO_2 emissions, it is of utmost importance to develop alternative clean energy sources (Elshobary et al., 2021).

Algae and MFC integration has been found promising with respect to efficiency. Yagishita et al. (1997) established a system that converted light energy into electrical power with a conversion efficiency of only 3% (Elshobary et al., 2021; Yagishita et al., 1997).

An MFC assembled with an algae cathode and a *Rhodospirillum rubrum* suspension anode (sandblasted platinum electrodes) achieved, after 21 hours of continuous illumination, an open circuit voltage of 0.96 V and a short circuit density of 75 $\mu A/cm^2$. A cell free of organisms, operated for comparison, gave a decrease in open-circuit potential over 7 hours, from 0.19 to 0.03 V. The short circuit current density decreased from 7.0 to 6.1 $\mu A/cm^2$. These data demonstrate the ability of specific microorganisms to convert light energy to electrical energy. The electrical power derived from these cells is estimated to be approximately 0.1–0.2% of that available from the incident radiation. This low efficiency is attributed to both biological and electrochemical inefficiencies (Song et al., 2019)

In a study by Xu et al. (2015), *Chlorella pyrenoidosa* was introduced at the anode of air cathodic CBM, which reported a maximum power density of 6.03 W/m^2 anodically, more than 200 times higher than that obtained with bacterial MFC (0.03 W/m^2) (Zhou et al., 2022; Xu et al., 2015). The highest electrical power (1.64 mW/m^2) was recorded in the dark with the highest voltage (390 mV) compared to the daytime electrical power, which did not exceed 0.132 mW/m^2. Lin et al. (2013) used the same MA-MFC design but with *Spirulina platensis* biofilm in the anode electrolyte. The results showed that the maximum open circuit voltage (OCV) of the MA-MFC (490 mV) and the maximum power density reached 10 mW/m^2 when connected to an external 1 kΩ resistor (Elshobary et al., 2021; Lin et al., 2013).

Two different algae have been used as dry feed powder in the MFCs of an anode chamber, namely *C. vulgaris* (microalgae) and *Ulva lactuca* (macroalgae). The *C. vulgaris* feedstock generated a higher power density (980 mW/m^2) than that produced by *U. lactuca* (760 mW m) with a power generation of 2.5 kWh/kg, while *U. lactuca* showed a high degradability rate higher than that of *C. vulgaris* (Velasquez-Orta et al., 2009). On the other hand, a full MFC can be performed by microalgae in both electrolytic chambers using algal biomass in the bioanode chamber and live microalgae on the cathode side. This MFC relies on the use of CO_2 released in the anode chamber to grow live microalgae in the cathode chamber. In this context, dead *Scenedesmus sp.* microalgae biomass (collected from polluted streams) was used as a substrate on the anode side and live *C. vulgaris* was used in the cathodic chamber of the MFC (Elshobary et al., 2021; Cui et al., 2014).

Wang et al. (2014) demonstrated a maximum power density of 38 mW/m^3 by the MFC system equipped with a carbon nanotube (CNT)-coated cathode inoculated with microalga (Allegue and Hinge, 2014) (Wang et al., 2014). Sun et al. used a microalgal MFC that completed superior removal effect of nitrogen (94.05% NH_4^{3+}-N and 77.35% inorganic nitrogen) and organic matter (76.66%) in nitrogenous maricultural wastewater, which ascribed to the integrated performance of microalgae, electrodes, and anodic bacteria (Jiaqi et al., 2020). The most exciting investigations of microalgal MFC are to make it a self-sustaining system, for example, the nutrient required for algae growth can originate from photosynthesis products themselves during the reduction of CO_2 (Xu et al., 2021).

Lan et al. (2013) studied the impact of red and blue monochromatic light of wavelengths 620–750 and 450–495 nm, respectively, on the anode-assisted *Chlamydomonas reinhardtii* and reported that red LED light increased the power density to 12.95 mW/m^2, which was more than 60% higher than the power density provided by blue LED light at the same intensity. This phenomenon was attributed to the increased growth rate when grown under red light wavelength, ranging from 650 to 750 nm, thus enhancing the photosynthesis process, electron flow rate, and energy production (Elshobary et al., 2021; Lan et al., 2013). Table 1.1 summarizes different types of MFCs with different algae and their performance.

The study carried out by Sekar et al. (2014) focused on the production of direct electricity using a cyanobacterium called *Nostoc sp.* (NOS) as a photo-biocatalyst immobilized on CNTs on the anode of the NOS immobilized on a CNT-modified electrode generated a photocurrent density of 30 mA/m^2 in photo-bioelectrochemical cells.

Table 1.1 Maximum power density using microalgae at anode, cathode, or substrate for power generation (Elshobary et al., 2021; Shukla and Kumar, 2018).

Species	Function	Type of MFC	P_{max} (mW/m^2)	P_{max} (mW/m^3)
Spirulina platensis	Bioanode	Single chamber		1.64
S. platensis	Bioanode	Single chamber	10	
Chlamydomonas reinhardtii	Bioanode	Single chamber	75	2100
Cyanobacteria	Bioanode	Single chamber	72	72
Synechocystis sp.	Bioanode	Two-chambered		6.7
Mixed cyanobacteria	Bioanode	Two-chambered	72	
Mixed microalgae	Bioanode	Single chamber	76	1060
Mixed microalgae	Bioanode	Single chamber	0.003	0.005
Chlorella pyrenoidosa	Bioanode	Dual chamber	30.2	120
C. pyrenoidosa	Bioanode	Dual chamber	2.5	450
Microcystis aeruginosa	Biocathode	Dual-chamber		58
Chlorella vulgaris	Biocathode	Dual-chamber	13.5	
C. vulgaris	Biocathode	Dual-chamber	18.7	
C. vulgaris	Biocathode	Dual-chamber		2485.35
C. pyrenoidosa	Biocathode	Dual-chamber	2.5	450
Desmodesmus sp.	Biocathode	Dual-chamber	64.2	820
Scenedesmus acutus	Biocathode	Dual-chamber		400
Mixed microalgae	Biocathode	Dual-chamber		2800
Cyanobacterial blooms	Biocathode	Sediment	16.7	
Mixed microalgae cyanobacteria	Biocathode	Sediment	11	
C. vulgaris	Biocathode	Sediment	68	
C. vulgaris	Biocathode	Sediment	38	
C. vulgaris	Biocathode	Sediment	48.5	
Mixed microalgae	Biocathode	Sediment	22.19	
Mixed microalgae	Biocathode	Dual-chamber		5700
C. vulgaris	Biocathode	Dual-chamber		600
C. vulgaris	Biocathode	Three-chambered		151
Microcystis aeruginosa	Substrate	Two-chambered		4140
Arthrospira maxima	Substrate	Two-chambered		100
Synechococcus leopoliensis	Substrate	Stack of 9MFCS		4250
Cyanobacteria	Substrate	Single chamber	114	2570
C. vulgaris	Substrate	Single chamber	980	2770
C. vulgaris	Substrate	Two-chambered		3700

Table 1.1 (Continued)

Species	Function	Type of MFC	P_{max} (mW/m^2)	P_{max} (mW/m^3)
C. vulgaris	Substrate	Two-chambered	15	370
Dunaliella tertiolecta	Substrate	Two-chambered	5.3	672
Scenedesmus obliquus	Substrate	Two-chambered	102	950
Scenedesmus sp.	Substrate	Two-chambered	1780	
Mixed algae	Substrate	Two-chambered	207	1035
Mixed algae	Substrate	Two-chambered	110	
Laminaria saccharina	Substrate	Two-chambered	218	
Ulva lactuca	Substrate	Single chamber	760	2150

The cell generated a maximum current density of 250 mA/m^2 and a peak power density of 35 mW/m^2 without any mediator. With the addition of 1,4-benzoquinone as a redox mediator, the electricity generation capability was significantly enhanced, with a current density of 2300 mA/m^2 and a power density of 100 mW/m^2 (Sekar et al., 2014).

Ramanathan et al. (2011) conducted a study on nine marine microalgae in a fuel cell and reported maximum electricity production (0.792 V) by *Tetraselmis gracilis*. Similarly, other microalgae, such as *Mougeotia*, *Chlorella sp.*, *Scenedesmus*, and blue-green algae have been successfully used by many researchers for bioelectricity generation. Mixed cultures have also been tested for electricity production. Extracting electrons from algae and thereby, generating electricity is a novel idea (Shukla and Kumar, 2018; Ramanathan et al., 2011).

Zhao et al. (2012) utilized four different feedstocks, namely cyanobacterial acid fermentation broth, crude cyanobacteria, sodium acetate, and glucose, with a power of almost 3 W/m^3. The acid fermentation broth with cyanobacteria showed the maximum power density of 0.072 W/m^2, with a substrate degradation rate of 0.607 kg COD/m^3d. This concludes that the presence of mediator compounds in the cyanobacterial acid fermentation broth increased the possibility of electron transfer to the anode electrolyte (Elshobary et al., 2021; Zhao et al., 2012).

1.5 Future Vision of the Use of Photosynthesis in Energy Generation

The chemical energy stored in algal biomass can be harnessed for the sustainable production of fuels and other value-added products. The production of bioelectricity from algae appears to be a successful approach for extracting energy from sunlight in a cost-effective and sustainable way. Electric current can also be generated at the anode, where cytochromes assist in the indirect transfer of electrons generated in the algal cell photosystem and can be referred to as a photosynthetic algal MFC (Shukla and Kumar, 2018).

The photosystem has attracted a great deal of attention and research interest. The favorable superiorities lie in higher treatment effects, improved energy generation, as well as broader potential applications (Xu et al., 2021).

In recent years, laboratory-scale studies on MFCs have received considerable attention; however, successful commercial-scale applications have not yet been adequately reported. To improve performance, a large number of anode modifications have been proposed, while nanocomposites have now been extensively investigated as coating materials for the anode surface. A large number of studies have reported nanocomposites (metal/metal oxides) embedded on the anode surface, which increase bacterial adhesion, conductivity, electron transfer, and reduction of ohmic losses. The simultaneous wastewater treatment and electricity production is currently a widely explored area and a large number of research works have been carried out in this field. Wet-earth coupled MFCs and MFCs operated in continuous flow mode have been investigated to be efficient systems at the laboratory scale; however, their performance under real conditions has not been satisfactorily reported. The incorporation of photosynthetic microorganisms in MFCs has been extensively studied. These systems are proficient in capturing CO_2 with minimal risk of CO_2 emissions, as well as in producing oxygen in the photosynthetic light phase, which eliminates the need of external oxygen supply, making it beneficial in reducing the cost of aeration (Nawaz et al., 2020).

There are still some constraints that limit the commercial application of microalgae MFCs (MA-MFCs) and need to be improved. General limitations include the slow activity of autotrophic microalgae compared to heterotrophic species, which affects the whole MA-MFC performance; energy loss caused by high potential; open-air cathode increases the charge transfer resistance; the inferior performance of biocathodes compared to the chemical cathode in terms of energy generation; the influence of sunlight intensity in an open-air operated cell; the problems of harvesting algae biomass in large-scale production, and so on, which make it difficult to apply this technology at scale. On the other hand, the long-term operation of the MA-MFC results in a voltage decrease due to the basic microenvironment on the cathode side. The passage of organic materials from the anode to the cathode can reduce the performance of the MFC and the biocathode due to their toxic effect on the microalgae on the cathode side and convert autotrophic biocathodes into heterotrophs. Suspended organic compounds also tend to increase the turbidity of the algae culture and reduce light penetration into the culture (Elshobary et al., 2021).

Photosynthetic efficiency can be increased in different ways, such as regulating the light-harvesting antenna, manipulating growth conditions, and reducing the size of the antenna per reaction center complex. Increasing light intensity can greatly increase photosynthetic efficiency by reducing cell shading and allowing light to penetrate deeper. Reducing pigment content per algal cell and decreasing antenna size are targets for optimizing photosynthetic efficiency in mass culture (Elshobary et al., 2021).

Continuous and sustainable electricity production is a challenge in photosynthetic systems, as performance is affected by illumination and dissolved oxygen. The relationship between the amount of electricity and pollutant removal, optimization of system configuration and operations are also important in assessing the sustainability of these systems. The use of algae for energy generation can easily overcome the obstacles faced in hydrogen production. The enzymes responsible for hydrogen evolution are [Fe-Fe]-hydrogenase (in green algae) and [NiFe]-hydrogenase (in cyanobacteria), which are sensitive to the oxygen produced in the photosynthesis process generated in the cells. The use of algae to generate electricity is a great innovation, but there are also certain problems that restrict its large-scale use for now (Shukla and Kumar, 2018).

1.6 Conclusion

Bacterial respiration involves different metabolic mechanisms acting at the same time, leading electrons toward a final acceptor. Electrons can be transported directly and indirectly, through the protein chain present in C-type cytochrome bacteria. Where the bacteria are close to the solid acceptor and deliver their electrons, they can be transferred from one bacterium to another until they reach the solid surface, and where they are transferred to conductive pili locks.

It has been found that there are transfer mechanisms in photosynthesis that can work in MFC systems using algae and microalgae in the cathode and anode compartments. Algae have proven to be a more economical and better performing catalytic material than chemicals, and as a substrate have proven to have a high capacity for organic matter removal and CO_2 reduction in the environment, which is one of the focuses of international efforts in reducing tons emitted into the environment. It is therefore considered that by bringing algal and microalgal bioelectrochemical systems to full scale, a new industrial technology will be available to improve the quality of human life. It is expected that in the near future, algal CBMs will have a wider application on a real scale, due to their large capacity and energy generation potential.

References

Aiyer, K.S. (2020). How does electron transfer occur in microbial fuel cells? *World Journal of Microbiology and Biotechnology 36* (2): 19. https://doi.org/10.1007/s11274-020-2801-z.

Allegue, L.B. and Hinge, J. (2014). *Biogas Upgrading Evaluation of Methods for H2S Removal*. Danish Technological Institute December 31.

Cui, Y., Rashid, N., Hu, N. et al. (2014). Electricity generation and microalgae cultivation in microbial fuel cell using microalgae-enriched anode and bio-cathode. *Energy Conversion and Management 79*: 674–680. https://doi.org/10.1016/j.enconman.2013.12.032.

Elshobary, M.E., Zabed, H.M., Yun, J. et al. (2021). Recent insights into microalgae-assisted microbial fuel cells for generating sustainable bioelectricity. *International Journal of Hydrogen Energy 46* (4): 3135–3159. https://doi.org/10.1016/j.ijhydene.2020.06.251.

Goto, Y., Yoshida, N., Umeyama, Y. et al. (2015). Enhancement of electricity production by graphene oxide in soil microbial fuel cells and plant microbial fuel cells. *Frontiers in Bioengineering and Biotechnology* 3: https://doi.org/10.3389/fbioe.2015.00042.

Hamelers, H.V.M., Ter Heijne, A., Sleutels, T.H.J.A. et al. (2010). New applications and performance of bioelectrochemical systems. *Applied Microbiology and Biotechnology 85* (6): 1673–1685. https://doi.org/10.1007/s00253-009-2357-1.

Huang, L., Regan, J.M., and Quan, X. (2011). Electron transfer mechanisms, new applications, and performance of biocathode microbial fuel cells. *Bioresource Technology 102* (1): 316–323. https://doi.org/10.1016/j.biortech.2010.06.096.

Jiaqi, S., Lifen, L., and Fenglin, Y. (2020). Successful bio-electrochemical treatment of nitrogenous mariculture wastewater by enhancing nitrogen removal via synergy of algae and cathodic photo-electro-catalysis. *Science of the Total Environment 743*: 140738. https://doi.org/10.1016/j.scitotenv.2020.140738.

Kumar, A., Hsu, L.H.-H., Kavanagh, P. et al. (2017). The ins and outs of microorganism–electrode electron transfer reactions. *Nature Reviews Chemistry 1* (3): 24. https://doi.org/10.1038/s41570-017-0024.

Kumar, R., Singh, L., Wahid, Z.A., and Din, M.F.M. (2015). Exoelectrogens in microbial fuel cells toward bioelectricity generation: a review. *International Journal of Energy Research 39* (8): 1048–1067. https://doi.org/10.1002/er.3305.

Lan, J.C.-W., Raman, K., Huang, C.-M., and Chang, C.-M. (2013). The impact of monochromatic blue and red LED light upon performance of photo microbial fuel cells (PMFCs) using Chlamydomonas reinhardtii transformation F5 as biocatalyst. *Biochemical Engineering Journal 78*: 39–43. https://doi.org/10.1016/j.bej.2013.02.007.

Lin, C.-C., Wei, C.-H., Chen, C.-I. et al. (2013). Characteristics of the photosynthesis microbial fuel cell with a Spirulina platensis biofilm. *Bioresource Technology 135*: 640–643. https://doi.org/10.1016/j.biortech.2012.09.138.

Logan, B.E., Murano, C., Scott, K. et al. (2005). Electricity generation from cysteine in a microbial fuel cell. *Water Research 39* (5): 942–952. https://doi.org/10.1016/j.watres.2004.11.019.

Maddalwar, S., Kumar Nayak, K., Kumar, M., and Singh, L. (2021). Plant microbial fuel cell: opportunities, challenges, and prospects. *Bioresource Technology 341*: 125772. https://doi.org/10.1016/j.biortech.2021.125772.

Martín-Betancor, K., Aguado, S., Rodea-Palomares, I. et al. (2017). Co, Zn and Ag-MOFs evaluation as biocidal materials towards photosynthetic organisms. *Science of the Total Environment 595*: 547–555. https://doi.org/10.1016/j.scitotenv.2017.03.250.

Nawaz, A., Hafeez, A., Abbas, S.Z. et al. (2020). A state of the art review on electron transfer mechanisms, characteristics, applications and recent advancements in microbial fuel cells technology. *Green Chemistry Letters and Reviews 13* (4): 365–381. https://doi.org/10.1080/17518253.2020.1854871.

Nitisoravut, R. and Regmi, R. (2017). Plant microbial fuel cells: a promising biosystems engineering. *Renewable and Sustainable Energy Reviews 76*: 81–89. https://doi.org/10.1016/j.rser.2017.03.064.

Patil, S.A., Hägerhäll, C., and Gorton, L. (2012). Electron transfer mechanisms between microorganisms and electrodes in bioelectrochemical systems. *Bioanalytical Reviews 4* (2): 159–192. https://doi.org/10.1007/s12566-012-0033-x.

Ramanathan, G., Birthous, R.S., Abirami, D., and Highcourt, D. (2011). Efficacy of marine microalgae as exoelectrogen in microbial fuel cell system for bioelectricity generation. *World Journal of Fish and Marine Sciences 3* (1): 79–87.

Rusyn, I. (2021). Role of microbial community and plant species in performance of plant microbial fuel cells. *Renewable and Sustainable Energy Reviews 152*: 111697. https://doi.org/10.1016/j.rser.2021.111697.

Schröder, U. (2007). Anodic electron transfer mechanisms in microbial fuel cells and their energy efficiency. *Physical Chemistry Chemical Physics 9* (21): 2619–2629. https://doi.org/10.1039/B703627M.

Sekar, N., Umasankar, Y., and Ramasamy, R.P. (2014). Photocurrent generation by immobilized cyanobacteria via direct electron transport in photo-bioelectrochemical cells. *Physical Chemistry Chemical Physics 16* (17): 7862–7871. https://doi.org/10.1039/C4CP00494A.

Shukla, M. and Kumar, S. (2018). Algal growth in photosynthetic algal microbial fuel cell and its subsequent utilization for biofuels. *Renewable and Sustainable Energy Reviews 82*: 402–414. https://doi.org/10.1016/j.rser.2017.09.067.

Song, H.-L., Zhu, Y., and Li, J. (2019). Electron transfer mechanisms, characteristics and applications of biological cathode microbial fuel cells – a mini review. *Arabian Journal of Chemistry 12* (8): 2236–2243. https://doi.org/10.1016/j.arabjc.2015.01.008.

Velasquez-Orta, S.B., Curtis, T.P., and Logan, B.E. (2009). Energy from algae using microbial fuel cells. *Biotechnology and Bioengineering 103* (6): 1068–1076. https://doi.org/10.1002/bit.22346.

Wang, D.-B., Song, T.-S., Guo, T. et al. (2014). Electricity generation from sediment microbial fuel cells with algae-assisted cathodes. *International Journal of Hydrogen Energy 39* (25): 13224–13230. https://doi.org/10.1016/j.ijhydene.2014.06.141.

Xu, C., Poon, K., Choi, M.M.F., and Wang, R. (2015). Using live algae at the anode of a microbial fuel cell to generate electricity. *Environmental Science and Pollution Research 22* (20): 15621–15635. https://doi.org/10.1007/s11356-015-4744-8.

Xu, Z., Chen, S., Guo, S. et al. (2021). New insights in light-assisted microbial fuel cells for wastewater treatment and power generation: a win-win cooperation. *Journal of Power Sources 501*: 230000.

Yagishita, T., Sawayama, S., Tsukahara, K., and Ogi, T. (1997). Effects of intensity of incident light and concentrations of Synechococcus sp. and 2-hydroxy-1,4-naphthoquinone on the current output of photosynthetic electrochemical cell. *Solar Energy 61* (5): 347–353. https://doi.org/10.1016/S0038-092X(97)00069-8.

Zhao, J., Li, X.-F., Ren, Y.-P. et al. (2012). Electricity generation from Taihu Lake cyanobacteria by sediment microbial fuel cells. *Journal of Chemical Technology & Biotechnology 87* (11): 1567–1573. https://doi.org/10.1002/jctb.3794.

Zhou, E., Lekbach, Y., Gu, T., and Xu, D. (2022). Bioenergetics and extracellular electron transfer in microbial fuel cells and microbial corrosion. *Current Opinion in Electrochemistry 31*: 100830. https://doi.org/10.1016/j.coelec.2021.100830.

2

Role of Functional Materials Involved in the Photosynthesis-Assisted Power Generation

Manoj K. Srinivasan[1], Pratima B. Jayarm[1], Ravichandiran Ragunath[1], Briska Jifrina Premnath[1], Nalini Namasivayam[1,] and Sathish-Kumar Kamaraj[2]*

[1] *Department of Biochemistry and Biotechnology, Faculty of Science, Annamalai University, Chidambaram, Tamil Nadu, India*
[2] *Instituto Politécnico Nacional (IPN)-Centro de Investigación en Ciencia Aplicada y Tecnología Avanzada, Unidad Altamira (CICATA-Altamira), Altamira, Tamps., México*

2.1 Introduction

MFCs (microbial fuel cells) are BES (bioelectrochemical systems) designed to produce environmentally friendly power. Electrochemically active bacteria use their metabolic processes to produce electrical energy in MFCs (Domínguez-Garay and Esteve-Núñez, 2018). This technology is not new; research into using microbes as fuel cell catalysts dates back to the 1970s. In addition, the concept of producing energy via microbial cultures has been mentioned since the early 20th century. According to Lovely, the oxidation of renewable biomass and organic waste by MFC technologies represents a novel method for using bacteria to produce bioelectricity (Lovely, 2006). Researchers improved MFCs power output, making real-time practical applications possible (Schröder et al., 2003). Enzymes and microorganisms comprise the first and second groups of biological reactions used in an MFC to create a biomass oxidation reaction. The two-electron oxidation system using GOx (glucose oxidase) or GDH (glucose dehydrogenase) is the most extensively researched mechanism for utilizing enzymes (Azuma and Ojima, 2018).

*Corresponding author

Photosynthesis-Assisted Energy Generation: From Fundamentals to Lab Scale and In-Field Applications, First Edition. Edited by Sathish-Kumar Kamaraj and Iryna Rusyn.

Compared to the methods currently employed to generate energy from organic matter (OM), MFCs also have various operational and functional advantages: (i) In contrast to all current bioenergetic processes, MFC technologies work proficiently in an open environment (such as environmental temperatures); (ii) they do not require gas treatment; (iii) they can be widely applied in areas without electrical infrastructure; (iv) they do not require power input, according to Liu et al., and (v) they increase the efficiency of the system by directly converting the energy of the substrate into electricity (Rabaey, 2009).

A byproduct of the MFC is the PMFC, which produces bioelectricity by feeding the electroactive bacteria (EAB) at the anode with exudates from plant roots (De Schamphelaire et al., 2010; Powell et al., 2014; Strik et al., 2008; Tapia et al., 2017). PMFC technologies were created to increase the power outputs' effectiveness (Helder et al., 2013). For the first time, Strik et al. (2008) planned, developed, and tested the concept of a PSMFC (plant sediment MFC), which capitalizes on the stream of OM at the anode during SMFC operation. Reed manna grass, Glyceria maxima, helped the SMFC by producing a maximum power density output of 67 mW/m^2 per PGA (plant growth area). Later, Zou et al. (2009). created a photosynthetic MFC to compare the power generation improvement in response to light to two conventional MFCs (polypyrrole-coated anodes MFC and polyaniline-coated MFC). Correspondingly, the power density attained in the untreated PMFC was 0.36 mW/m^2, as opposed to conventional MFC, which showed values of 1.3 mW/m^2 (polypyrrole-coated anodes) and 0.95 mW/m^2 (for polyaniline coating). The electrons supplied to the anode electrode during bioelectricity production by PMFC are subsequently spent in the cathode compartment by a typical reduction of O_2 to H_2O (Strik et al., 2011). The primary goal of the PMFC is to have plants store energy in chemical form via photosynthetic activity, which is then converted to electrical power using this system. The following equation depicts a P-chemical MFC reaction:

$$6CO_2 + 6H_2O \rightarrow C_6H_{12}O_6 + 6O_2 + \sim 38\,ATP$$

Furthermore, PMFCs can operate anywhere and continuously produce electricity, even during food deprivation. Because of the favorable operating conditions, PMFCs have drawn more consideration than conventional power sources. Without a doubt, adding plants to MFCs might boost biomass, current, and electricity production without killing or harvesting the plants (Moqsud et al., 2015). PMFC technologies can produce a sustainable energy level 18 times higher than a standard SMFC. The improvement in OM obtainability at the anode chamber for microbial oxidation gives this power generation performance. Before this technology's commercial application, more research is necessary to address the difficulties that come with it. In PMFCs, the moisture content of the soil has also been researched as a factor that affects electricity production, permitting the transport of electrons to the electrode cathode (Moqsud et al., 2015; Chiranjeevi et al., 2012). It has been established that there is a direct correlation between soil moisture and electrical output. A significant drop in electrical energy following irrigation cessation in earlier PMFC studies (Li et al., 2016). The performance and underlying principles of PMFCs and MSCs (microbial solar cells) and the associated difficulties for applying these systems in the future were discovered (Strik et al., 2011). Moqsud et al. (2015) also examined the variables influencing PMFC performance. PMFCs have also been verified in C3 (Calvin cycle) and C4 (Hatch-Stack cycle) plants modified to saturate soil circumstances with H_2O. Yet, this technology has received less research in plants utilizing crassulacean acid metabolism (CAM). Over the past 10 years, an in-depth study of PMFC configuration and potential applications has been done. Thus, PMFC has evolved into a bryophyte MFC system, an ecological engineering system, an erected wetland system coupled with MFC (CW-MFC), and a macrophyte MFC system, among other things (Bombelli et al., 2016; Chiranjeevi et al., 2013; Doherty et al., 2015; Kabutey et al., 2020; Piyare et al., 2017; Venkata Mohan et al., 2010). PMFCs have been designed to produce electricity in semi-arid regions (with CAM plants), paddy fields, a wetland, floating water bodies, and green roofs (Tapia et al., 2017; Helder et al., 2013; Apollon et al., 2020; Kiran Kumar et al., 2020; Sudirjo et al., 2018; Sudirjo et al., 2019; Wetser et al., 2015). As a beneficial alternative to creating ecosystems and maintaining air quality, plants grown indoors, on green roofs, and in rooftop gardens can also be used in PMFCs to produce electricity, claim (Tapia et al., 2017). The living plant is a continuous source of OM, which makes PMFCs more appealing as a potential low-cost source of renewable bioenergy. Consuming plants in MFC is a practical approach to progress the cohort of bioelectricity (Nitisoravut and Regmi, 2017).

This chapter aims to present up-to-date data on significant PMFC components, such as cathode anodes and membranes, that are used to enhance bioelectricity generation. In the end, the PMFC's challenges and applications and the factors that impact how well they work were discussed. New concepts regarding the critical features of BES (i.e. PMFC) were created due to this chapter; these concepts must now be researched to be developed for use in a large-scale, sustainable, and economical manner.

2.2 Plant-Mediated Microbial Fuel Cells

2.2.1 Basic Concept of PMFCs

A relatively new technology called plant-microbial fuel cells (PMFCs) uses living plants to produce electricity from solar energy. In this system, the oxidation of organic compounds by bacteria to produce electricity is combined with plants'

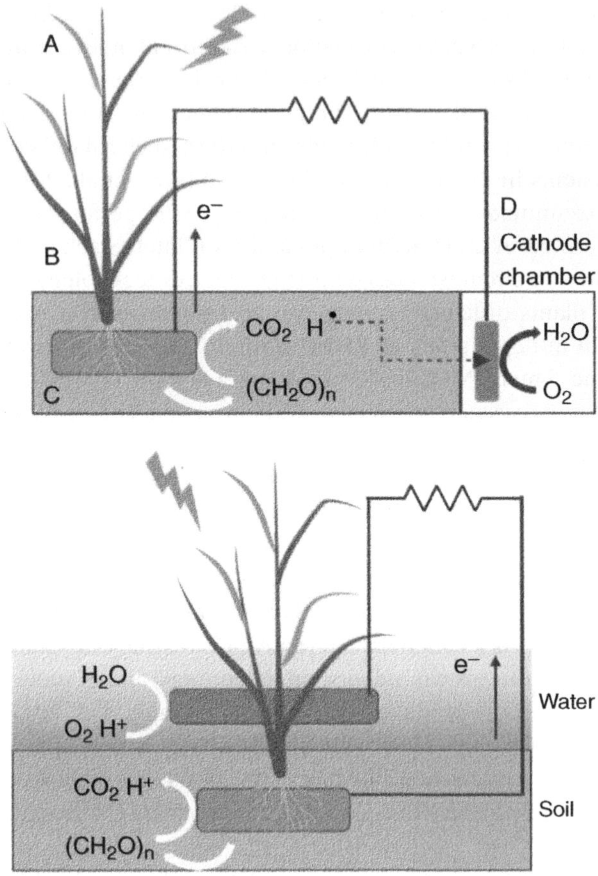

Figure 2.1 Rice paddy field MFC (left) and plants MFC (right).

deposition of organic compounds into the rhizosphere. Figure 2.1 depicts a PMFC configuration. The four basic PMFCs developed so far are the flat-plate PMFC, tubular plant fuel cell, sediment types MFC, and PMFC (Helder et al., 2012; Wetser et al., 2017). PMFCs are an alternative technology that utilizes living plants and microbial communities in soil and plant roots to break down organic compounds into simpler ones. In theory, photosynthesis is a process that living plants use to create organic compounds like glucose and sucrose so they can grow and reproduce. The majority of the organic substances are then released into the soil by the roots. Organic compounds will naturally be converted into energy by microorganisms near the sources. The electrons released into the environment are the process' excesses. Some microbes, including Geobacter species, Rhodobacter species, and Shewanella species, can transfer electrons out of the body.

The function of electron transfer is to obtain energy from the variance between the number of electrons outside and inside the cell and maintain the stability of biochemical processes (diffusion). A direct electric current is produced when the electrode traps the free electrons (Doherty et al., 2015; Watanabe et al., 2010).

Description of their action that occurs in the PMFC process:

A. Photosynthesis
$CO_2 + 6H_2O > C_6H_{12}O_6 + 6O_2$
B. Transport of organic compounds in the anode chamber
C. Oxidation of organic compounds by microorganisms in the anode chamber
$C_6H_{12}O_6 + 6H_2O > 6HCO_3 + 30H^+ + 24e^-$
D. Reduction in the cathode chamber
$6O_2 + 24H^+ + 24e^- > 12H_2O$

2.2.2 Plants and Their Bioelectricity Generation Capabilities

The production of bioelectricity in various vascular and nonvascular plants, as well as plants grown in paddies, swamps, and marshes, has been tested in multiple plants. The simultaneous production of biomass and bioelectricity is a dual benefit of PMFC technology. Figure 2.2 shows the schematic representation of the bioelectricity production process of PMFC.

The rice plant (*Oryza sativa*), which can grow in sopping conditions and thus promote the oxidation reaction of carbon compounds, is one of the most extensively

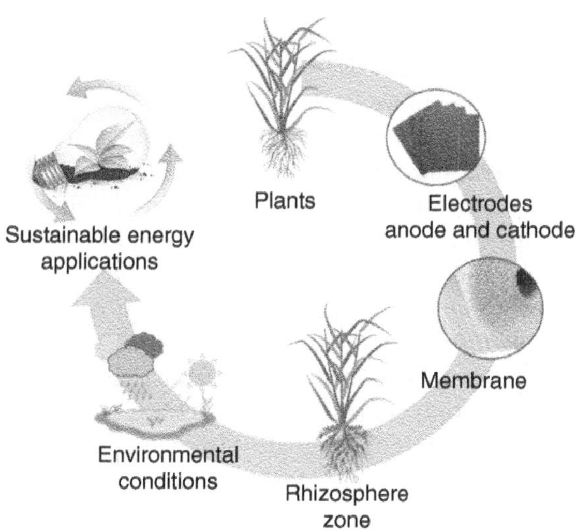

Figure 2.2 Schematic representation of the bioelectricity production process of PMFC.

studied plant species in PMFC research. However, since rice fields are significant sources of greenhouse gas methane, including rice fields in PMFC studies will give researchers plenty of chances to make a study on methane extenuation in the paddy field. Numerous types of research on *O. sativa* were conducted, expanding the potential for installing PMFCs in paddy fields to produce food with electricity while avoiding any conflict with food production. Moqsud et al. (2015) conducted studies using *O. sativa* to examine the impact of soil and compost on bioelectricity and biomass growth. Kaku et al. (2008) studied *O. sa*livate to comprehend the relationship between plants and microbes by enhancing plant growth with phosphorus, potassium, nitrogen fertilizers, and acetate solution. Kouzuma et al. (2013) investigated how rice plants formed a biofilm on a graphite anode surface beneath the impact of the growth medium. Bombelli et al. (2013) conducted a comparison study on a photovoltaic system using *O. sativa* and a related weed called *Echinochloa glabrescent*. Arends et al. (2014) investigated the reduction of methane gas in rice PMFC, including Hoagland solution as the nutrient medium. Rice plants were examined for their capacity to produce bioelectricity, the effects of electrode size and placement underneath the impact of fertilizers and soil, and the influence of Hoagland solution as the nutrient medium in the anode chamber (Takanezawa et al., 2010; Ueoka et al., 2016). Studying PMFCs with rice plants may offer important insights into the co-production of large-scale crops and energy.

A saltwater grass called *Spartina anglica*, which can rise in environments with high water content and is known for its effortlessness of maintenance and salinity tolerance, is one of the most extensively researched plant species in the PMFC study, second only to the rice plant. *S. anglica* has been used to generate bioelectricity and biomass and assess its long-term performance (Helder et al., 2010; Wetser et al., 2017; Wetser et al., 2015). It was discovered to be the perfect plant for future large-scale bioelectricity generation in wetlands.

To produce bioelectricity using the *Canna indica* plant in oligotrophic circumstances without the need for peripheral substrates, Lu et al. (2015) explored the impacts of a microbial community. Numerous other grass species have been studied for various applications, primarily for bioelectricity production and wastewater treatment within PMFCs. These include *E. glabrescens*, *Cyprus involucrate*, *Lolium perennial*, *Pennisetum setaceum*, *Ipomoea aquaticaAcorus calamus*, *Echinorriea crassipes*, and *Typha latifolia*. In addition, indoor plants with the potential to reduce air pollution, such as *Dracaena braunii* and *Epipremnum aureum*, were also studied in the PMFC study (Bombelli et al., 2013; Klaisongkram and Holasut, 2014; Chiranjeevi et al., 2012; Habibul et al., 2016; Liu et al., 2013; Yan et al., 2015; Oon et al., 2015; Sarma and Mohanty, 2018). Azri et al. (2018) also investigated indoor ornamental plants such as *Papyrus diffuses, Chlorophytum comosum*, and *Chasmanthe floribunda*.

Nonvascular plants and commonly employed vascular plants were also looked into for their potential to generate bioelectricity. Nonvascular plants with distinct bryophytes and physiology can store water and nutrients and have a higher tolerance for dehydration than vascular plants. They feature a network of rhizoids, or hairy roots that adhere to the external environment on which they cultivate, stabilizing the soil and reducing erosion-related nutrient loss. Bryophytes regulate the C and N_2 cycles in the atmosphere and serve as a food supply for microbial consortiums. The first bryo-MFC (bryophyte MFC), based on the forest moss *Dicranum montanum*, was announced during the First International Plant Power Symposium in Ghent, Belgium, in 2011. In their study of the model moss species *Physcomitrella patens*, Bombelli et al. (2016) discovered that the electrical output from a group of bryo-MFCs connected in series was enough to power a commercial radio.

2.3 Applications of PMFC technology

Dispersing a green rooftop PMFC system is practical and may be used in underdeveloped and isolated locations to produce power and food (Helder et al., 2013; Skinner, 2006; Deng et al., 2014). Rice fields may assist farmers in rural places in generating income from power generation during cultivation procedures while reducing greenhouse gas emissions and improving environmental quality. Another potential expansion for PMFC investigations is fitting a PMFC system in rice cultivation with fish farming in a hydroponics system. The control of plant disease, biosensors, and soil remediation are further cutting-edge applications for PMFC technology.

Heavy metal (chromium) removal units and low-cost wastewater treatment units use PMFC technology (Habibul et al., 2016; Lu et al., 2015). Maximum COD (chemical oxygen demand), turbidity, and color removal were shown using a simple, affordable small ecological system integrated with a PMFC (Chiranjeevi et al., 2013; Venkata Mohan et al., 2011). Compared to an SMFC, benzo[a]pyrene and pyrene degradation were increased in a PMFC (Habibul et al., 2016). Additionally, the plants used in an SMFC demonstrated effective nitrate, ammonium, and COD elimination.

PMFCs may be employed in EES (ecological engineering system) because eco-electrogenic engineered systems use wastewater treatment and recycling in wetlands while generating electricity simultaneously to decrease or alleviate pollution issues (Chiranjeevi et al., 2013; Habibul et al., 2016; Moulin et al., 2001; Venkata Mohan et al., 2011). FTSs (flow-through systems) entrenched with PMFCs replicating natural ecosystems may be utilized to recreate wetlands by including fishponds. FTSs may be used to repair hazardous waste sites, rehabilitate lakes and rivers, and rebuild the environment damaged by extensive mining activity. EES (one of the FTS) treatments targeting the environmental challenges may be used to alter degrading existing ecosystems ecologically soundly.

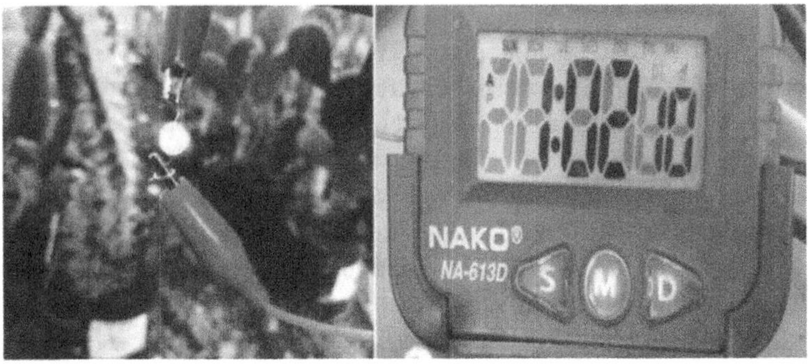

Figure 2.3 *Opuntia* PMFCs prototype powered LED and digital clock (Apollon et al., 2020).

PMFC usage on a green roof adds visual worth to the building's contemporary design and improves its image as sustainable. Rather than spending additional money on non-sustainable solutions, more than 50% of businesses worldwide are open to considering sustainable alternatives for their systems. Through decentralized implementation, PMFCs will have extremely high social acceptance in underdeveloped nations. Farmers may make money in addition to the regular income from their produce by creating power while their crops are grown. In total, 1.2 billion people worldwide may access power generation using PMFC as a decentralized system (Helder et al., 2012).

In 2013, "Plant-e," a spin-off company founded by David Strik, a scientific researcher, and CEO Marjole in Helder of Wageningen University, began emerging new tubular designs for spread over PMFCs in flood plains, river deltas, and wetland systems for power generation that enables multifunctional land use. In addition, plant-e has created rooftop PMFC systems and 1 m^2 modular ornamental systems for residential or commercial settings.

For the first time, Apollon et al. (2020) developed PMFC-based prototype-powered LED and digital clock using a plant such as *Opuntia ficus* with the h power density of 489.02 mW/m^2 (Figure 2.3).

2.4 Development of Electrodes and Membranes for Plant Microbial Fuel Cells

The assembly, architecture, and general shape of the membrane and electrodes are primarily responsible for unit cell cost and PMFC performance. PMFC is a cutting-edge technology that uses direct bioconversion to produce electricity and has been proposed as a solution to several environmental and energy problems. In a PMFC, bioelectricity is produced by the oxidation of a substrate in an anode

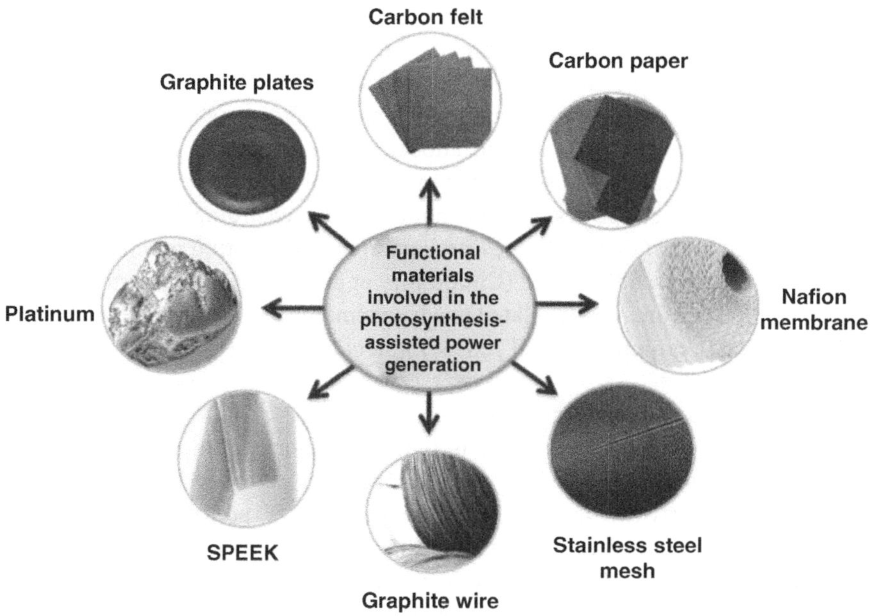

Figure 2.4 Materials used in PMFC for bioelectricity generation.

chamber, which results in the creation and movement of proton and electron (He et al., 2005; Rabaey and Verstraete, 2005). Industrial, municipal, and agricultural wastes may provide the biomass-rich organic substrates the bacteria eat (Ismail and Jaeel, 2013; Mansoorian et al., 2016; Samsudeen et al., 2015). As it is a direct conversion of chemical energy without partial heat losses, the Carnot cycle does not affect the amount of electricity produced in a PMFC. As a result, it may offer a high energy conversion yield of up to 75%, similar to traditional chemical fuel cells (Mathuriya and Yakhmi, 2016). The functionality and efficacy of PMFCs vary according to the substrates available for microbial metabolism (Pant et al., 2010). Other electron acceptors, such as oxygen, can affect PMFC performance and microbial impact. The advancement in electrode materials and membranes is thoroughly detailed to improve PMFC performance and scale up PMFC technology. Some of the important functional materials used in PMFC for bioelectricity generation are shown in Figure 2.4.

2.4.1 Progress in Electrode Materials

To enhance power generation and reduce expenses during the growth of a PMFC, it is necessary to use particularly efficient electrode materials (Chiranjeevi et al., 2019). For efficient electron transfer with the aid of the bacteria from the rhizosphere to the electrode in the PMFC, many electrode materials have been employed, including carbon fiber, graphite felt, graphite rods/discs/plates, carbon cloth (CC), graphite granules, and stainless-steel mesh. In PMFC, electron transfer efficiency and power output improvement are shown when identical electrodes are used for the cathode and anode. According to reports, graphite felt works as a cathode and anode electrode material in PMFC. For cost-effective and scalable PMFCs, electrode design is the main obstacle, according to (Rabaey, 2009). For the development of PMFCs, researchers are concentrating on configuration and designing electrode material (Choudhury et al., 2017). The performance and viability of PMFCs are mostly related to advancements in cathode and anode electrode materials (Rahimnejad et al., 2015; Zhou et al., 2011).

2.4.1.1 Progress in Anode Materials

Through their impact on bacterial adhesion, the efficiency of interfacial substrate oxidation, and electron transfer from microbes to electrodes, the primary and structural characteristics of anode directly impact PMFC performances (Li et al., 2014). These characteristics are related to EET efficiency via interactions between the microbe–electrode–electrolyte system and an enhanced surface area that is specifically designed for the creation of biofilms. Anode material, which offers anaerobic terminal electron acceptors for oxidizing organic materials in wastes, has also increased the metabolic rates of anaerobic bacteria (Li et al., 2012; Srivastava et al., 2015). To boost PMFC power output, suitable anode material selection and modification are thus necessary.

Carbon-based Anode Materials Carbon-based anode materials have recently been used in PMFC systems, and carbon-structured anode materials include graphite plates, carbon fiber, carbon rods, carbon felt, and carbon brushes.

Power densities of 69.32 and 6.12 mW/m^2 have been achieved using carbon-based anode materials, CC, and carbon felt (Oon et al., 2015; Sophia and Sreeja, 2017). Additionally, the treatment of red soil has been utilized to generate a maximum power density of 163 mW/m^2 using graphite plate electrodes in a PMFC (Chiranjeevi et al., 2012). The wastewater treatment in continuous-flow Cassette electrode-PMFC was improved using a graphite-felt microstructured anode with fins. This resulted from the vertical orientation of anode fins, which improved the interaction of microorganisms with wastewater organics and horizontal water flow dispersion (Ishii et al., 2017). Nitrogen-doped porous carbons attained the most excellent power density of 679 mW/m^2 as an anode compared to commercial CC. This increase in PMFC performance was primarily made possible by these materials' improved electrical conductivity and very porous structure (Wetser et al., 2015). Due to their stability, increased electrical conductivity, cost-effectiveness, and better specific surface area, carbon-based electrode materials are potential PMFC choices. Exoelectrogenic bacteria that can produce bioelectricity from household effluent may thrive on graphite fiber brush anodes because of their enormous porosities and surface area (Logan et al., 2007).

Metal-based Anode Materials Owing to their durability, increased electrical conductivity, and enthusiastic encouragement of microbial adherence, metal-based anodes such as stainless steel, aluminum, nickel, silver, molybdenum, copper, gold, and titanium are suitable for a productive act in PMFCs.

In PMFC applications, for instance, molybdenum has produced power densities of 1296 mW/m^2. Metal anode materials are advantageous due to their low cost and resistance to corrosion (Iannaci et al., 2017; Yamashita and Yokoyama, 2018). For example, the upright access of a 3-dimensional porous structure constructed of titanium oxide nanosheets may create a high surface area, producing a bigger lively surface that improves electron transport and molecular diffusion (Yin et al., 2015). In contrast, when bare metals and carbon were utilized as electrodes, it could lower electron transfer efficiency and produce lesser adhesion qualities by bacteria in PMFCs. According to Wu et al. (2013), using platinum as the anode, sediment/rhizodeposits generated a maximum power density of 14.0 mW/m^2.

Composite Anode Materials Alternately, it has been determined that the addition or alteration of nanomaterials to or on typical carbon-structured electrode materials, including polyaniline (PANI), carbon nanotubes (CNT), 3D carbon black, graphene, and porous carbon, may be advantageous. Modifying anodes with metals affects improved bacterial cell adherence and reduced ohmic loss.

The newly created anode with a composite made of CNT and iron (II, III) oxide (Fe_3O_4) displayed an 830 mW/m^2 power density. Fe_3O_4 is adhered to carbon nanotubes, creating a multilayered network that endorses electron transport (Park et al., 2014). As an alternative, C/Hematite has been applied to CC using fast and affordable pyrolysis of ferrocene at room temperature. Compared to a clean CC electrode in PMFC, this procedure significantly increased the EET efficacy of *S. oneidensis* and gave a current density of around six times greater (Zhou et al., 2015). Recently, rGO (reduced graphene oxide) and carbon nanotubes-based monohybrids with SnO_2 (tin(IV) oxide) have been created utilizing sonochemical synthesis and hydrothermal-assisted microwave dispersion, respectively (Garino et al., 2016; Mehdinia et al., 2014). A power density of 1624 mW/m^2 was attained utilizing a hybrid nanocomposite electrode made of SnO_2/rGO-CC as the anode, which was five times that achieved using a bare electrode (Mehdinia et al., 2014).

The high specific surface area was primarily responsible for better performance, while the hybrid composites improved biocompatibility, resulting in more excellent bacterial adhesion and electron transfer efficiency (Guo et al., 2015; Herrero-Hernández et al., 2013). By using a novel technique for distributing Ni-NPs (nickel nanoparticles) onto ACFs/CNF (carbon micro-nanofibers), an effective electrode for double-chamber plant MFCs (DCPMFCs) has been created. Ni-ACF/CNF electrodes generated the finest power density in DCMFC, 1145 mW/m^2 (Singh and Verma, 2015). With a power density of 2065 mW/m^2, a different composite electrode consisting of MgO (manganese oxide) and rGO coated on carbon felt has been employed to improve the adhesion and EET efficiency of exoelectrogenic bacteria (Zhang et al., 2016). The extreme power density of 2143 mW/m^2 is produced by combining thin graphene films with stainless steel fiber felt (SSFF), which is higher than the power density of 1018 mW/m^2 produced by graphene-modified CC-PMFC (Hou et al., 2015). These surface alterations have significantly increased the specific surface area and the biocompatibility characteristics, which leads to better communication between the electrode and the biofilm. Therefore, bacterial colonization on the electrode macropores causes the anode's reactive surface area to increase (Hou et al., 2014).

In a different method, a 3D stainless-steel mesh anode with carbon black (SSM/CB) has produced a higher power density of $3215 \pm 80\,mW/m^2$. A substantial aspect of the improved PMFC efficacy of SSM/CB electrodes was their effective electrical and microbial adhesion capabilities; moreover, these electrodes are constructed of inexpensive constituents and are reasonably simple to synthesize (Zhang et al., 2015). In a PMFC design, a unique composite anode made of fiber brush and carbon textile produced electricity while treating wastewater. With 1140 mW/m^2 of power density, the maximum COD removal value of 0.330 h was attained. A unique composite anode was necessary for better electricity production and wastewater treatment since it promoted electrochemical activity, mass transfer processes, and microbial growth in the PMFC. An effective electrode for PMFC was recently reported to be a 3D laminated FZ (Zinc coated metallic wire/carbon composite), with an electrical potential of 291 mV produced and 80% COD elimination. The FZ laminated composite electrodes' hydrophilicity, wet ability, and larger surface area affected how much power was produced from sewage treatment (An et al., 2011). To produce energy and remediate wastewater, Xu et al. (2018) used a special filtration composite anode (composed of carbon textile and fiber brush) in PMFC. By facilitating the mass transfer, the filtering feature of composite electrodes increased microbial population and improved electricity production and COD removal rates (Xu et al., 2018). Table 2.1 summarizes PMFC performance with various anodes.

Table 2.1 Classifications of anode materials used in PMFC for bioelectricity generation.

Anode	Substrates	Type of bioelectrical system	Power density (mW/m^2)	References
Carbon felt	Mixed culture sludge	CW-MFC	6.12	Oon et al. (2015)
Carbon cloth	Compost soil mix	PMFC	69.32	Sophia and Sreeja (2017)
Graphite plates	Red soil	R-MFC	163	Chiranjeevi et al. (2012)
Nitrogen-doped porous carbon on carbon cloth	Plant growth medium	PMFC	679	Wetser et al. (2015)
Platinum	Sediment/Rhizodeposits	PSMFC	14.0	Wu et al. (2013)
Fe_3O_4/CNT composite	Wastewater	PMFC	830	Park et al. (2014)
rGO/SnO_2-CC	Sludge	PMFC	1624	Mehdinia et al. (2014)
Ni-ACFs/CNFs	Anaerobic sludge	CW-MFC	1145 ± 20	Singh and Verma (2015)
Reduced graphene oxide(rGO) and manganese oxide (MnO_2) coated on carbon felt	Microbes	PMFC	2065	Zhang et al. (2016)
3D-Carbonblack/stainless steel mesh	Wastewater	PMFC	3215 ± 80	Zheng et al. (2015)
Carbon fiber brush with carbon textile	Wastewater	PMFC	1140	Xu et al. (2018)
Graphite felt	Speelzand and ammonium rich 1/2 Hoagland solution	PMFC	60	Zhou et al. (2015)
Ferric cyanide	Microorganisms in the soil	PMFC	469	Li et al. (2012)
Stainless steel mesh	Granular activated carbon (GAC)	CW-MFC	0.149	Pu et al. (2017)
Graphite wire	Garden soil	PMFC	242 ± 10.5	Logan et al. (2007)
Oxygen (Pt)	Paper and carbon fiber	BryoMFC	6.7 ± 0.6	Bombelli et al. (2016)
Graphite felt	Cow urine	PMFC	42.79 ± 14.62	Apollon et al. (2020)
Graphite rods	Potato waste	BMFC	36.84	Yaqoob et al. (2022)

Note: PMFC, plant microbial fuel cell; pSMFC, plant sediment microbial fuel cell; R-MFC, rhizosphere-microbial fuel cell; CW-MFC, constructed wetland–microbial fuel cell; BryoMFC, bryophyte microbial fuel cell; BMFC, benthic microbial fuel cell.

2.4.1.2 Progress in Cathode Materials

Due to their possible effect on ORR (oxygen reduction reaction), cathodes are another component of PMFCs that may affect power production and wastewater treatment in addition to the anode. Ferricyanide and permanganate have been utilized as oxidants (electron acceptors) to simulate the ORR process, despite oxygen being thought to be the utmost appropriate oxidant in the cathode compartment to improve PMFC action during oxygen reduction reaction (Pandit et al., 2011; Lu and Li, 2012; Kodali et al., 2017; Noori et al., 2016; Tang et al., 2015). Intermolecular bridging and biofouling often cause ORR kinetics to be reduced in PMFCs (An et al., 2011). As an ORR catalyst, platinum (Pt)-based electrodes are often utilized (Pei et al., 2016; Rozendal et al., 2008). These electrode drawbacks are their high cost, the tendency for biofouling, and surface microbial poisoning. These factors have led to the exploration of several low-priced, highly catalytically active, and biofouling-resistant materials to increase power production from PMFCs. These materials include N-doped carbon, conducting polymers, metal complexes, metal oxides, and carbon structures. By changing the cathode surface to improve the microbial features, affordable chemical activation of the electrode materials also enhances the electrochemical and adsorption effects of PMFC. Correspondingly, this increased ORR characteristic in PMFC enabled chemically treated graphite to provide greater power densities of 7.9 and 6.5 W/m^3 (Zhang et al., 2016).

Carbon structures have shown promise among the different electrode materials. For instance, heteroatom-doped carbon obtained from alfalfa leaves produced more significant cathodic catalyst activity and performance (1328.9 mW/m^2), comparable to the standard Pt/C cathodic catalyst (1337.7 mW/m^2) (Deng et al., 2017), for wastewater treatment in PMFC, a cathode constructed using polyvinylidene fluoride, carbon fiber cloth, and catalytic components C, Mn, Fe, and O proved successful. The exclusion efficacy of ammonium, COD, and total phosphate was considered 80, 90, and 65%, respectively. Future wastewater treatment with this kind of cathode membrane has been proposed with higher commercialization values (Li et al., 2015).

In a cathode single chamber PMFC with a 930 ml capacity, Tremouli et al. (2016) employed MnO_2 as a cathode catalyst and graphite as an anode to treat wastewater continuously. From then, owing to catalyst non-electrogenic and degradation microbe competition during long-standing operation, they removed COD from synthetic and municipal wastewater by 85 and 45%, respectively (Tremouli et al., 2016). In a 48-hour fed-batch process, a 72 l stacked PMFC with packed bed electrodes of GAC (granular activated carbon) produced 50.9 1.7 W/m^3 of power density with 90% COD removal rates from wastewater (Wu et al., 2016). Wastewater was treated using an affordable membrane bioreactor-PMFC coupling system with a cathode membrane made of carbon fiber cloth covered with poly (vinylidene fluoride, or PVDF). GAC was added to the cathode membrane to boost electrical output and reduce biofouling. The elimination effectiveness of ammonium and COD was, therefore, determined to be 80% and 90%, respectively. ROS (reactive oxidizing species), precisely OH, utilizing doped GAC on the cathode, produced 90 percent oxidative elimination of tetracycline hydrochloride (Li et al., 2014).

With a power density of 1771 mW/m^2, graphite plates used as the cathode and anode have produced exceptional PMFC performance (Kaewkannetra et al., 2011). Besides this, Fe–N–G (iron- and nitrogen-functionalized graphene) has been successfully working as a cathode catalyst replacement in PMFCs, yielding a power density of 1149.8 mW/m^2 (Li et al., 2012). Additionally, the carbonaceous design of the CC cathodes and carbon brush anodes in PMFCs produced a power density of 2110 mW/m^2 (Feng et al., 2011). An O_2-reduction process in PMFC-mediated wastewater treatment was made possible using a GO/MgO (GO hybridized MgO) nanocomposite on CC at the cathode. Compared to Pt/C catalysts with identical power output, the power density of 755.63 mW/m^2 was reached with a 79.5% COD elimination efficiency, saving 93.3% of the cost (Li et al., 2017). To improve the PMFC performance and operating stability, new anode and cathode materials have been developed as alternatives. The electrode materials for the cathode and anode have been applied as Ni nanoparticles distributed across ACFs/CNFs, with a power density of 1145,20 mW/m^2 obtained (Singh and Verma, 2015). $NiCo_2S_4$/AC, a transition metal chalcogenide supported by activated carbon, has been employed as a cathode material in PMFC to provide power densities that are 2.25 times more than bare AC (888.56 mW/m^2) at 2000 $\pm$ 59 mW/m^2. The occurrence of sulfur and the large micropore surface area of $NiCo_2S_4$ created an enhanced number of active sites to boost ORR activity producing an effective ORR activity because of the presence of pyridinic nitrogen, quaternary nitrogen, and iron species (Iannaci et al., 2017). However, owing to their high nitrogen content, significant precise area of a substantial number of active sites, porous structure, O_2 reduction, MOF (metal–organic frameworks), and nitrogen-doped carbon structures have also been recognized as suitable cathode materials for PMFC (Singh and Verma, 2015; Pu et al., 2017). With a maximum power density of 3118.9 mW/m^2, a 3D Fe–N–C (Fe-entrenched of N-doped carbon) framework was also reported to permit good oxygen reduction reaction activity in PMFC. A greater ORR performance was considered according to their substantial surface area, dispersion of Fe–N active species, and well-ordered mesoporous structure in the 3D carbon network. After removing the silica template, the carbon adopts a 3D-linked network with mesoporous walls (Zhang et al., 2014). Significant attention is paid to novel pairings and various dimensional architectures of carbon and metal nanoparticles (Tang et al., 2017; Wang et al., 2018; Zhang et al., 2017). A biocathode

Table 2.2 Classifications of cathode materials used in PMFC for bioelectricity generation.

Cathode	Substrates	Type of bioelectrical system	Power density (mW/m^2)	References
Iron and nitrogen functionalized grapheme (Fe–N–G)	Wastewater	PMFC	1149.8	Li et al. (2012)
Carbon brush	Compost soil mix	PMFC	69.32	Sophia and Sreeja (2017)
Go/MgO	Plant growth medium	PMFC	755.63	Li et al. (2017)
ACFs/CNFs	Sediment/ Rhizodeposits	PMFC	1145.20	Singh and Verma (2015)
$NiCo_2S_4$/AC	Wastewater	PMFC	2000 ± 59	Pu et al. (2017)
3D porous Fe–N C on carbon paper	Sludge	PMFC	3118.9	Tang et al. (2017)
Granular activated carbon	Anaerobic sludge and nutrient solution	CW-MFC	55.05	Liu et al. (2014)
Stainless steel mesh	Anaerobic sludge	CW-MFC	12.42	Liu et al. (2013)
Graphite	Wastewater	PMFC	120	Gilani et al. (2016)
Carbon-painted	Human urine	MFC	754 ± 16	Pasternak et al. (2018)
Pt-coated carbon cloth	Potato wastewater	MFC	576	Sato et al. (2021)
Stainless-steel mesh	Goat urine	PMFC	46.97 ± 0.67	Apollon et al. (2022)
Graphite rods	Mango waster	BMFC	87.71	Yaqoob et al. (2022)

Note: MFC, Microbial Fuel Cell; PMFC, Plant Microbial Fuel Cell; CW-MFC, Constructed Wetland–Microbial Fuel Cell; BMFC, Benthic Microbial Fuel Cell.

(anammox bacteria) for removing carbon and nitrogen while generating energy was described by Kokabian et al. (2018). This resulted in an ammonium removal rate of 90% and a power density of $0.092\,W/m^3$. For instance, the air-cathode PMFC was employed with the laccase-producing white rot fungus *Ganoderma lucidum* BCRC 36123 as the biocathode to simultaneously generate $207.74\,mW/m^2$ of electricity and decolorize the azo dye AO7 (Acid Orange 7) (Lai et al., 2017). Table 2.2 summarizes PMFC performance with various cathodes.

2.4.2 Development of Membranes for PMFC Performance

The PMFC membrane separates the negative and positive electrodes while mediating ion transport. In addition, the membrane plays a crucial role in preventing crossover between the cathode and anode chambers, particularly regarding CO_2, O_2, and effluent (Leong et al., 2013; Li et al., 2011). Figure 2.5 depicts the PMFC membrane function, which has been described (Mohan et al., 2014; Nor et al., 2015). Considerations for membrane selection include substrate loss, internal resistance, oxygen diffusion, and biofouling (Asensio et al., 2018; Ghasemi et al., 2012; Hernández-Fernández et al., 2015). Numerous PEMs (polymer electrolyte membranes) have been used in devices with various energy levels and proton transportation processes in the PMFC (Di Palma et al., 2018; Tiwari and Ghangrekar, 2018). Perfluorinated polymer membranes, composite membranes (organic–inorganic and organic–organic hybrid polymer membranes), and hydrocarbon polymer membranes are the categories under which PEMs fall.

Perfluorinated sulfonic acid membranes, such as DuPontTM Nafion® membranes, are effective PEMs for PMFC applications because of their high proton conductivity; however, their non-biocompatible and high-cost properties are drawbacks (Leong et al., 2013; Angioni et al., 2017; Ghasemi et al., 2013). To manufacture an effective Nafion-based PEM, several inorganic additives and polymers (ZrO_2, TiO_2, SiO_2, etc.) have been added to the Nafion ionomer to address the cost problem (Angioni et al., 2016). As a low-priced PEM in PMFCs for energy production, a poly (vinyl alcohol) (PVA)-Nafion-borosilicate membrane has recently been created. Additionally, OM degradation rates in wastewater ranged from 55% to 65% (Tiwari and Ghangrekar, 2018). Another strategy involves creating a brand-new semifluorinated sulfonated polytriazole and cross-linking it to polyvinyl alcohol for PMFCs. In this arrangement, the cross-linking led to enhanced oxidative stability of the PMFC, more regulated water intake, and dimensional change (Saha et al., 2018). Different substitute constituents, particularly sulfonated polymer polymers, have been employed as separators to substitute the Nafion membrane in PMFCs. In PMFC applications, ceramics and other fibers are also considered ion exchangers. As an alternative PEM, a low-cost nanocomposite membrane made of natural cotton fabric and covered with polyvinyl acetate-grafted PVAc-g-PVDF (polyvinylidene fluoride) has recently been created. It improves PMFC capacity for proton transfer, effectiveness in removing COD, and power density (Wang et al., 2018). The PEM in PMFC has also been modified using inexpensive PVA-H (PVA hydrogel) and light-extended clay aggregate; the clay and PVA-H high water holding capacity have enhanced proton conductivity and durability. PVA-H has been utilized as a PEM in a tubular plant air-cathode PMFC to produce bioelectricity with a toluol elimination proficiency of more than 99% (Liu et al., 2018).

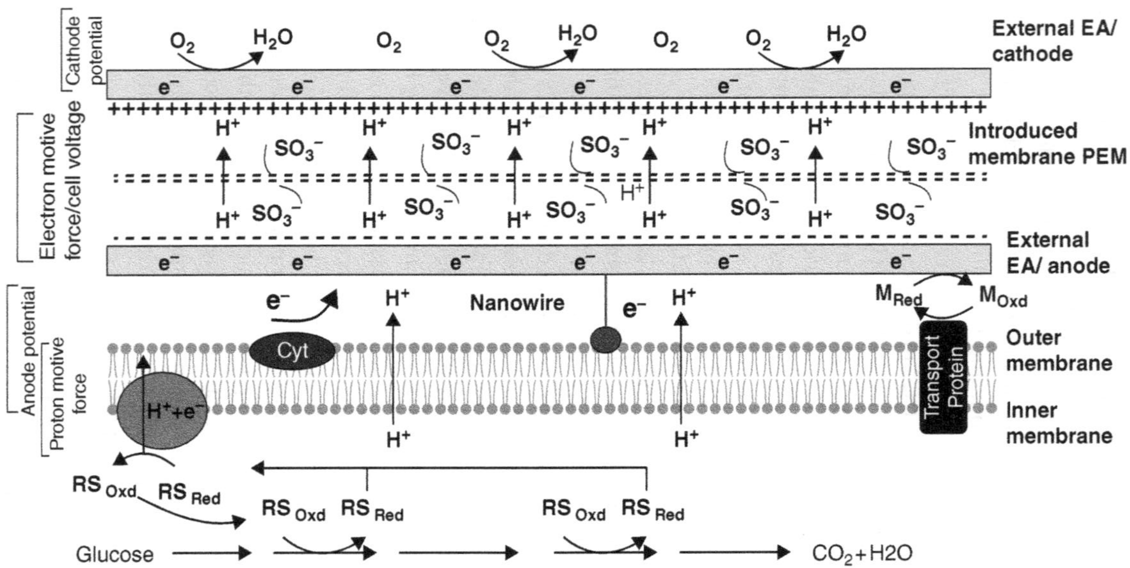

Figure 2.5 Diagrammatic representation of proton transport across the membrane in PMCF.

Recently, multiple inorganic–organic polymer composite membranes have been created to improve the performance of the PMFC and membrane characteristics (Park et al., 2017; Sivasankaran et al., 2016; Venkatesan and Dharmalingam, 2015). Polymers of various kinds, such as polybenzimidazole (PBI), polyimide (PI), polyether sulfone, and polyether ether ketone (PEEK), performed better in PMFCs and were more reasonably priced. Polyether ketone-based polymers are viable membrane options compared to other materials because of their superior mechanical and thermal stability, affordable price, and strong proton conductivity (Leong et al., 2015). Polymer membrane sulfonation ($-SO_3H$) also allegedly increased membrane process ability, hydrophilicity, and proton conductivity (Sedesheva et al., 2016). To evaluate PMFC performance, membranes with various DS (sulfonation) degrees in SPEEK (sulfonated PEEK), such as DS-20.8, 41, 63.6, and 76%, were manufactured. The 63.6% variant of the DS membrane outperformed the others while operating the PMFC. The CE, COD removal percentage, and power density can all be improved with more DS in SPEEK polymers. At the same time, performance in certain areas was hindered by the membrane's structural alterations and the emergence of useless cations during PMFC cell operation (Ghasemi et al., 2016). Using a PMFC with a SPEEK membrane, phenol and acetone have also been taken out of wastewater while producing energy. This energy-efficient technique removed acetone and phenol from wastewater while producing electricity using PMFC (Wu et al., 2018). Nanofiller-infused composite membranes made from non-fluorinated SPEEK have outperformed Nafion series (Nafion® 112 and Nafion® 117) membranes in terms of performance. PMFC using GO-SPEEK and SPEEK as separators achieved COD removal effectiveness of 83.01 and 82.65% (Leong et al., 2015). Ayyaru and Dharmalingam (2011) used SPEEK as PEM in PMFC to treat home and dairy wastewater and produce electricity. With a 15–80% COD removal rate for household wastewater and a 7–75% COD deletion rate for dairy wastewater, the resulting power density was 55% higher than Nafion 117. An SCMFC membrane with a better power density and less oxygen crossing was created using a SPEEK/magnetite (Fe_3O_4) nanocomposite. Particularly in PMFCs, the Fe_3O_4 nanoparticles increased the proton conductivity, water-mediated, and proton-hopping process (Ghasemi et al., 2013). Another instance is the solvent evaporation process to incorporate inorganic filler made of H-faujasite (zeolite) onto the SPEEK membranes to create an improved PEM. In contrast to Nafion membranes, zeolite-covered SPEEK membranes demonstrated increased proton conductivity, stability, surface area, consistent porosity, and power density (Venkatesan and Dharmalingam, 2015). To enhance the SPEEK membrane's capacity to retain water and its proton conductivity, silico tungstic acid (STA) particles have also been trapped in addition to inorganic fillers (Venkatesan and Dharmalingam, 2017). Another method involves using a low-cost polyether sulfone (PES) as a PEM for PMFCs. This PES has superior chemical and thermal characteristics and mechanical stability owing to reduced chain mobility. Through the use of nonsolvent-induced phase separation methods, a unique SPES-PES blended membrane has been created. The resultant sulfonated PES significantly improved the anti-biofouling capabilities of PEM and decreased its roughness (Zinadini et al., 2017). Fe_3O_4 was recently added to PES to create a superior organic–inorganic hybrid nanocomposite membrane (Di Palma et al., 2018). In contrast, sulfonated PVDF and sulfonated polystyrene (SPS) have been developed and have shown improved power density generation (Behera et al., 2018; Kim et al., 2014). As a result of the greater sulfonation in the porous structure, which increased their ionic selectivity and ion exchange capacity while lowering oxygen permeability and electrical resistance, the porous sulfonated PVDF membranes functioned similarly to Nafion membranes (Kim et al., 2014). S-SiO_2 (sulfonated silica) is regarded as an effective PEM for usage in PMFCs when added to SSEBS (sulfonated polystyrene ethylene butylene polystyrene). After adding inorganic nanofillers, the composite membrane surfaces remain uniform, demonstrating the sulfonated SiO_2 homogeneous dispersion over the membrane polymer matrix. By doing so, the composite PEM's proton conductivity may be efficiently increased, and the system's internal resistance can be decreased (Sivasankaran et al., 2016). Additionally, an environmentally benign biopolymer has been effectively used as a substitute for PEM in PMFCs (Srinophakun et al., 2017; Yue et al., 2017). Bio-based composite membranes include bacterial cellulose-poly (4-styrene sulfonic acid) hybrid membranes and phosphorylated-sorbitol-chitosan membranes (Holder et al., 2016). A novel combined PMFC and electric membrane bioreactor was created for effective wastewater treatment using a quartz sand chamber as PEM. Additionally, a cohesive trickling filter and an air contact oxidation bed with filled volcanic rock were created and employed to lower the cost of aeration and improve the amount of dissolved oxygen in the cathode. This system was shown to have >97% COD removal proficiency, >93% ammonium nitrogen removal, and 50% phosphorous removal (Guo et al., 2015). A sulfonated bacterial cellulose membrane (SBC/PANI) was cross-linked with polyaniline to create a composite membrane (Yue et al., 2017). In a PMFC design, the produced SBC/PANI membrane has been used as a gel polymer electrolyte. Bacterial cellulose is sulfonated to increase ionic conductivity and boost biocompatibility. The SBC/PANI membranes stated ionic conductivity and ion-exchange capacity are 3.92 mequiv/g and 5.2×10^{-3} S/cm, respectively (Apollon et al., 2022). Changes were made to the proton conductivity, thickness, ion exchange capacity, and water uptake based on the characteristics of the membrane. The hybrid composite structures may be compared to useful membranes for the new PMFC technologies. The characteristics of the membranes utilized in PMFCs are shown in Table 2.3.

Table 2.3 Properties of different membranes for PMFC applications.

Membrane	Thickness	Proton conductivity (S/cm)	Water uptake (%)	IEC (mequiv/g)	References
PVA-Nafion-borosilicate (MPN)	167 μm	0.07	105%	—	Tiwari and Ghangrekar (2018)
Sulfonated-oxy-poly benzimidazole	20–30 μm	0.0783	29%	1.0	—
Sulfonated polystyrene-ethylene-butylene-polystyrene (SPSEBS)	180 μm	0.382	164%	—	Ayyaru and Dharmalingam (2011)
SSEBS-S-$SiO_2$7.5%	180 μm	0.321	210 ± 8	3.015	Sivasankaran et al. (2016)
7.5%Fe_3O_4/SPEK	0.018(±0.002)	0.165×10^{-2}	20.63	1.88	Ghasemi et al. (2013)
SPEAK	0.018(±0.002)	0.163×10^{-2}	15.85	1.47	Sivasankaran et al. (2016)
SPEEK+ 7.5%TiO_2	0.018(±0.002)	0.187×10^{-2}	21.83	1.98	Sivasankaran et al. (2016)

2.5 Challenges and Future Perspective

PMFCs are still developing their products and extensively studying the upscaling procedure. In the Netherlands, PMFCs are utilized with green roofs because they may slow down the pace of rainwater flows, provide aesthetic value, increase biodiversity, enhance air quality, and maintain a constant inside temperature. To lessen reliance on fossil fuels for power, electricity generation was anticipated to be 3.2 W/m^2. PMFCs may be combined with a specific wetland, such as a field, leisure area, or swimming pool. Additionally, PMFCs have the potential to be developed in nations like Indonesia since they can lower greenhouse gas emissions like CO_2 and CH_4 (Oon et al., 2015; Srivastava et al., 2015). PMFCs may be a solution for distant water quality monitoring about the clean water problem. As a constant supply of reliable electricity, PMFCs may be employed as power for in-situ measuring equipment. Convenience, cost, and field usability are the obstacles to using PMFCs as biosensors.

Studies on the method's effectiveness are still established to get the lowest price and make an item tradeable (Xu et al., 2015). PMFCs still have several restrictions, mainly in the electrical power produced. Redox conditions, organic loading, microbes surrounding the rhizosphere, roots/wetland plants, electrode material, and configurations impact wastewater treatment's electrical output and efficacy. The effectiveness of substrate transformed into energy by microbes depends on the organic component. The number of electrons produced by microbes increases with COD. However, an appropriate COD must be identified for the best processing outcomes. The voltage gradient between the two compartments is impacted by redox conditions (anode–cathode). The differential in oxygen content between the two compartments is the cause of this. The ideal redox conditions may be maintained by maintaining the DO. It is also important to consider the significance of wetland flora. Good root systems are essential for wetlands plants because they increase metabolic efficiency. The root is where oxygen and organic chemicals produced by plants are supplied. Microbes have a significant impact on the whole process. Maintaining and regulating the consortium microorganisms' symbiotic mutualism in the PMFCs system is essential. Increasing the distance between the electrodes results in greater electrical circuit impedance, thereby affecting the overall reactor design.

The electrode material and water flow rate can also affect how effectively power is generated (Doherty et al., 2015; Watanabe et al., 2010). In Indonesia, studies on PMFC are still very few. As a result, there is an opportunity to identify the ideal configuration and modify this technology for Indonesia. PMFCs must be combined with other technologies to treat wastewater more effectively or act as a hybrid. Proton exchange membranes and cation exchange membranes are two components of PMFCs that are still relatively costly and difficult to locate in Indonesia.

Another significant problem that requires further research is the MFC system's stability and long-term performance. To get the best circumstances for PMFCs, media fouling issues and PMFCs must be evaluated. Future energy and water supply issues need interdisciplinary collaboration to be solved.

2.6 Conclusion

To address the current energy crisis in the world and lessen the overuse of other harmful energy sources, PMFC technologies, which are BES, must be further investigated. These systems are bearable, renewable, and environmental replacements for traditional energy sources. Furthermore, compared to conventional MFCs, PMFCs are a promising choice for producing bioelectricity. A promising increase of 90.13% has been made in the initial bioelectricity production (67–679 Mw/m^2 PGA). Improvements in power density generation in PMFCs

(PMFC performance) depend on several variables, including electrode material types, membrane material types, and suitable plant species (high rhizodeposition). Because of the transfer of charged hydroxide and phosphate ions, the anion membrane's potency is also increased compared to the cation membrane. Additionally, chemical products, such as sodium hydroxide and hydrogen peroxide in PMFCs, are generated and retrieved using a cation exchange or anion exchange membrane. Lastly, if this system is to be used in places without enough energy, plants that release the most organic chemicals will need to be made, and technology will need to improve so that electrons can be moved more quickly.

References

An, J., Jeon, H., Lee, J., and Chang, I.S. (2011). Bifunctional silver nanoparticle cathode in plant microbial fuel cells for microbial growth inhibition with comparable oxygen reduction reaction activity. *Environmental Science & Technology* 45: 5441–5446. https://doi.org/10.1021/es2000326.

Angioni, S., Millia, L., Bruni, G. et al. (2016). Improving the performances of Nafion™-based membranes for plant microbial fuel cells with silica-based, organically-functionalized mesostructured fillers. *Journal of Power Sources* 334: 120–127. https://doi.org/10.1016/j.jpowsour.2016.10.014.

Angioni, S., Millia, L., Bruni, G. et al. (2017). Novel composite polybenzimidazole-based proton exchange membranes as efficient and sustainable separators for plant microbial fuel cells. *Journal of Power Sources* 348: 57–65. https://doi.org/10.1016/j.jpowsour.2017.02.084.

Apollon, W., Kamaraj, S.-K., Silos-Espino, H. et al. (2020). Impact of Opuntia species plant biobattery in a semi-arid environment: demonstration of their applications. *Applied Energy* 279: 115788.

Apollon, W., Rusyn, I., González-Gamboa, N. et al. (2022). A comprehensive review of the improvement of zero waste sustainable recovery using microbial energy generation systems. *Science of the Total Environment* 817: 153055. https://doi.org/10.1016/j.scitotenv.2022.153055.

Arends, J.B., Speeckaert, J., Blondeel, E. et al. (2014). Greenhouse gas emissions from rice microcosms amended with a plant microbial fuel cell. *Applied Microbiology and Biotechnology* 98: 3205–3217. https://doi.org/10.1007/s00253-013-5328-5.

Asensio, Y., Fernandez-Marchante, C.M., Lobato, J. et al. (2018). Influence of the ion-exchange membrane on the performance of double-compartment plant microbial fuel cells. *Journal of Electroanalytical Chemistry* 808: 427–432. https://doi.org/10.1016/j.jelechem.2017.06.018.

Ayyaru, S. and Dharmalingam, S. (2011). Development of MFC using sulfonated polyether ether ketone (SPEEK) membrane for electricity generation from wastewater. *Bioresource Technology* 102: 11167–11171. https://doi.org/10.1016/j.biortech.2011.09.021.

Azri, Y.M., Tou, I., Sadi, M. et al. (2018). Bioelectricity generation from three ornamental plants: *Chlorophytum comosum*, *Chasmanthe floribunda*, and *Papyrus diffuses*. *International Journal of Green Energy* 15: 254–263. https://doi.org/10.1080/15435075.2018.1432487.

Azuma, M. and Ojima, Y. (2018). Catalyst development of microbial fuel cells for renewable-energy production. In: *Current Topics in Biochemical Engineering*. IntechOpen https://doi.org/10.5772/intechopen.81442.

Behera, S.D., Kumari, U., Shankar, R., and Mondal, P. (2018). Performance analysis of a double-chambered plant microbial fuel cell employing a low-cost sulfonated polystyrene proton exchange membrane. *Ionics* 24: 3573–3590. https://doi.org/10.1007/s11581-018-2480-z.

Bombelli, P., Iyer, D., Covshoff, S. et al. (2013). Comparison of power output by rice (*Oryza sativa*) and an associated weed (*Echinochloa glabrescens*) in vascular plant bio-photovoltaic (VP-BPV) systems. *Applied Microbiology and Biotechnology* 97: 429–438. https://doi.org/10.1007/s00253-012-4473-6.

Bombelli, P., Dennis, R.J., Felder, F. et al. (2016). The electrical output of bryophyte microbial fuel cell systems is sufficient to power a radio or an environmental sensor. *Royal Society Open Science* 3 (10): 160249. https://doi.org/10.1098/rsos.160249.

Chiranjeevi, P., Mohanakrishna, G., and Mohan, S.V. (2012). Rhizosphere-mediated electrogenesis with the function of anode placement for harnessing bioenergy through CO_2 sequestration. *Bioresource Technology* 124: 364–370. https://doi.org/10.1016/j.biortech.2012.08.020.

Chiranjeevi, P., Chandra, R., and Venkata Mohan, S. (2013). Ecologically engineered submerged and emergent macrophyte based system: an integrated eco-electrogenic design for harnessing power with simultaneous wastewater treatment. *Ecological Engineering* 51: 181–190.

Chiranjeevi, P., Yeruva, D.K., Kumar, A.K. et al. (2019). Plant-microbial fuel cell technology. *Microbial Electrochemical Technology* 549–564.

Choudhury, P., Prasad Uday, U.S., Bandyopadhyay, T.K. et al. (2017). Performance improvement of the microbial fuel cell (MFC) using a suitable electrode and bioengineered organisms: a review. *Bioengineered* 8 (5): 471–487. https://doi.org/10.1080/21655979.2016.1267883.

De Schamphelaire, L., Cabezas, A., Marzorati, M. et al. (2010). Microbial community analysis of anodes from sediment microbial fuel cells powered by rhizodeposits of living rice plants. *Applied and Environmental Microbiology* 76 (6): 2002–2008. https://doi.org/10.1128/AEM.02432-09.

Deng, H., Wu, Y.C., Zhang, F. et al. (2014). Factors affecting the performance of single-chamber soil microbial fuel cells for power generation. *Pedosphere* 24: 330e338.

Deng, L., Yuan, Y., Zhang, Y. et al. (2017). Alfalfa leaf-derived porous heteroatom-doped carbon materials as efficient cathodic catalysts in plant microbial fuel cells. *ACS Sustainable Chemistry & Engineering* 5: 9766–9773. https://doi.org/10.1021/acssuschemeng.7b01585.

Di Palma, L., Bavasso, I., Sarasini, F. et al. (2018). Synthesis, characterization and performance evaluation of Fe_3O_4/PES nanocomposite membranes for plant microbial fuel cell. *European Polymer Journal* 99: 222–229. https://doi.org/10.1016/j.eurpolymj.2017.12.037.

Doherty, L., Zhao, Y., Zhao, X. et al. (2015). A review of a recently emerged technology: constructed wetland—microbial fuel cells. *Water Research* 85: 38–45. https://doi.org/10.1016/j.watres.2015.08.016.

Domínguez-Garay, A. and Esteve-Núñez, A. (2018). Designing strategies for operating microbial electrochemical systems to clean up polluted soils under non-flooded conditions. *Bioelectrochemistry* 124: 142–148. https://doi.org/10.1016/j.bioelechem.2018.03.006.

Feng, Y., Yang, Q., Wang, X. et al. (2011). Treatment of biodiesel production wastes with simultaneous electricity generation using a single-chamber plant microbial fuel cell. *Bioresource Technology* 102: 411–415. https://doi.org/10.1016/j.biortech.2010.05.059.

Garino, N., Sacco, A., Castellino, M. et al. (2016). Microwave-assisted synthesis of reduced graphene oxide/SnO_2 nanocomposite for oxygen reduction reaction in plant microbial fuel cells. *ACS Applied Materials & Interfaces* 8: 4633–4643. https://doi.org/10.1021/acsami.5b11198.

Ghasemi, M., Shahgaldi, S., Ismail, M. et al. (2012). New generation of carbon nanocomposite proton exchange membranes in microbial fuel cell systems. *Chemical Engineering Journal* 184: 82–89. https://doi.org/10.1016/j.cej.2012.01.001.

Ghasemi, M., Daud, W.R.W., Ismail, A.F. et al. (2013). Simultaneous wastewater treatment and electricity generation by plant microbial fuel cell: performance comparison and cost investigation of using Nafion 117 and SPEEK as separators. *Desalination* 325: 1–6. https://doi.org/10.1016/j.desal.2013.06.013.

Ghasemi, M., Daud, W.R.W., Alam, J. et al. (2016). Sulfonated poly ether ketone with different degree of sulphonation in plant microbial fuel cell: application study and economical analysis. *International Journal of Hydrogen Energy* 41: 4862–4871. https://doi.org/10.1016/j.ijhydene.2015.10.029.

Gilani, S.R., Yaseen, A., Zaidi, S.R.A. et al. (2016). Photocurrent generation through plant microbial fuel cell by varying electrode materials. *Journal of the Chemical Society of Pakistan* 38 (1): 17–27.

Guo, K., Prévoteau, A., Patil, S.A., and Rabaey, K. (2015). Engineering electrodes for microbial electrocatalysis. *Current Opinion in Biotechnology* 33: 149–156. https://doi.org/10.1016/j.copbio.2015.02.014.

Habibul, N., Hu, Y., Wang, Y.K. et al. (2016). Bioelectrochemical Chromium(VI) removal in plant-microbial fuel cells. *Environmental Science & Technology* 50 (7): 3882–3889. https://doi.org/10.1021/acs.est.5b06376.

He, Z., Minteer, S.D., and Angenent, L.T. (2005). Electricity generation from artificial wastewater using an up-flow plant microbial fuel cell. *Environmental Science & Technology* 39: 5262–5267. https://doi.org/10.1021/es0502876.

Helder, M., Strik, D.P., Hamelers, H.V. et al. (2010). Concurrent bio-electricity and biomass production in three plant-microbial fuel cells using *Spartina anglica*, *Arundinella anomala* and *Arundo donax*. *Bioresource Technology* 101 (10): 3541–3547. https://doi.org/10.1016/j.biortech.2009.12.124.

Helder, M., Strik, D.P., Hamelers, H.V., and Buisman, C.J. (2012). The flat-plate plant-microbial fuel cell: the effect of a new design on internal resistances. *Biotechnology for Biofuels* 5, Published 2012 Sep 21 (1): 70. https://doi.org/10.1186/1754-6834-5-70.

Helder, M., Strik, D.P., Hamelers, H.V. et al. (2012). New plant-growth medium for increased power output of the plant-microbial fuel cell. *Bioresource Technology* 104: 417–423. https://doi.org/10.1016/j.biortech.2011.11.005.

Helder, M., Strik, D.P., Timmers, R.A. et al. (2013). Resilience of rooftop plant-microbial fuel cells during Dutch winter. *Biomass and Bioenergy* 51: 1–7.

Helder, M., Chen, W.S., Van der Harst, E.J.M. et al. (2013). Electricity production with living plants on a green roof: environmental performance of the plant-microbial fuel cell. *Biofuels, Bioproducts and Biorefining* 7: 52–64.

Hernández-Fernández, F.J., De Los Ríos, A.P., Salar-García, M.J. et al. (2015). Recent progress and perspectives in plant microbial fuel cells for bioenergy generation and wastewater treatment. *Fuel Processing Technology* 138: 284–297. https://doi.org/10.1016/j.fuproc.2015.05.022.

Herrero-Hernández, E., Smith, T.J., and Akid, R. (2013). Electricity generation from wastewater with starch as a carbon source using a mediatorless plant microbial fuel cell. *Biosensors & Bioelectronics* 39: 194–198. https://doi.org/10.1016/j.bios.2012.07.037.

Holder, S.L., Lee, C.H., Popuri, S.R., and Zhuang, M.X. (2016). Enhanced surface functionality and plant microbial fuel cell performance of chitosan membranes through phosphorylation. *Carbohydrate Polymers* 149: 251–262. https://doi.org/10.1016/j.carbpol.2016.04.118.

Hou, J., Liu, Z., Yang, S., and Zhou, Y. (2014). Three-dimensional macroporous anodes based on stainless steel fibre felt for high-performance plant microbial fuel cells. *Journal of Power Sources* 258: 204–209. https://doi.org/10.1016/j.jpowsour.2014.02.035.

Hou, J., Liu, Z., Li, Y. et al. (2015). A comparative study of graphene-coated stainless steel fibre felt and carbon cloth as anodes in PMFCs. *Bioprocess and Biosystems Engineering* 38: 881–888. https://doi.org/10.1007/s00449-014-1332-0.

Iannaci, A., Sciarria, T.P., Mecheri, B. et al. (2017). Power generation using a low-cost sulfated zirconium oxide-based cathode in single chamber plant microbial fuel cells. *Journal of Alloys and Compounds* 693: 170–176. https://doi.org/10.1016/j.jallcom.2016.09.159.

Ishii, Y., Miyahara, M., Watanabe, K., 2017. Anode macrostructures influence electricity generation in plant microbial fuel cells for wastewater treatment. Journal of Bioscience and Bioengineering 123, 91-95. https://doi.org/10.1016/j.jbiosc.2016.07.014

Ismail, Z.Z. and Jaeel, A.J. (2013). Sustainable power generation in continuous flow plant microbial fuel cell treating actual wastewater: influence of biocatalyst type on electricity production. *The Scientific World Journal* 2013: 713515. https://doi.org/10.1155/2013/713515.

Kabutey, F.T., Ding, J., Zhao, Q. et al. (2020). Electrical current generation from a continuous flow macrophyte biocathode sediment microbial fuel cell (mSMFC) during the degradation of pollutants in urban river sediment. *Environmental Science and Pollution Research International* 27 (28): 35364–35380. https://doi.org/10.1007/s11356-020-09812-y.

Kaewkannetra, P., Chiwes, W., and Chiu, T.Y. (2011). Treatment of cassava mill wastewater and production of electricity through plant microbial fuel cell technology. *Fuel* 90: 2746–2750. https://doi.org/10.1016/j.fuel.2011.03.031.

Kaku, N., Yonezawa, N., Kodama, Y. et al. (2008). Plant/microbe cooperation for electricity generation in a rice paddy field. *Applied Microbiology and Biotechnology* 79 (1): 43–49. https://doi.org/10.1007/s00253-008-1410-9.

Kim, Y., Shin, S.H., Chang, I.S., and Moon, S.H. (2014). Characterization of uncharged and sulfonated porous poly (vinylidene fluoride) membranes and their performance in plant microbial fuel cells. *Journal of Membrane Science* 463: 205–214. https://doi.org/10.1016/j.memsci.2014.03.061.

Kiran Kumar, V., Man Mohan, K., Manangath Sreelakshmi, P. et al. (2020). Resource recovery from paddy field using plant microbial fuel cell. *Process Biochemistry* 99: 270–281.

Klaisongkram, N. and Holasut, K. (2014). Electricity generation of plant microbial fuel cell (PMFC) using Cyperus involucratus R. *KKU Engineering Journal* 42: 117–124. https://doi.org/10.14456/kkuenj.2015.2.

Kodali, M., Santoro, C., Herrera, S. et al. (2017). Bimetallic platinum group metal-free catalysts for high power generating plant microbial fuel cells. *Journal of Power Sources* 366: 18–26. https://doi.org/10.1016/j.jpowsour.2017.18.110.

Kokabian, B., Gude, V.G., Smith, R., and Brooks, J.P. (2018). Evaluation of anammox biocathode in microbial desalination and wastewater treatment. *Chemical Engineering Journal* 342: 410–419. https://doi.org/10.1016/j.cej.2018.02.088.

Kouzuma, A., Kasai, T., Nakagawa, G. et al. (2013). Comparative metagenomics of anode-associated microbiomes developed in rice paddy-field microbial fuel cells. *PLoS One* 8 (11): e77443. https://doi.org/10.1371/journal.pone.0077443.

Lai, C.Y., Liu, S.H., Wu, G.P., and Lin, C.W. (2017). Enhanced bio-decolorization of acid orange 7 and electricity generation in plant microbial fuel cells with superabsorbent-containing membrane and laccase-based bio-cathode. *Journal of Cleaner Production* 166: 381–386. https://doi.org/10.1016/j.jclepro.2017.08.047.

Leong, J.X., Daud, W.R.W., Ghasemi, M. et al. (2013). Ion exchange membranes as separators in plant microbial fuel cells for bioenergy conversion: a comprehensive review. *Renewable and Sustainable Energy Reviews* 28: 575–587. https://doi.org/10.1016/j.rser.2013.08.052.

Leong, J.X., Daud, W.R.W., Ghasemi, M. et al. (2015). Composite membrane containing graphene oxide in sulfonated polyether ether ketone in plant microbial fuel cell applications. *International Journal of Hydrogen Energy* 40: 11604–11614. https://doi.org/10.1016/j.ijhydene.2015.04.082.

Li, W.W., Sheng, G.P., Liu, X.W., and Yu, H.Q. (2011). Recent advances in the separators for plant microbial fuel cells. *Bioresource Technology* 102: 244–252. https://doi.org/10.1016/j.biortech.2010.03.090.

Li, S., Hu, Y., Xu, Q. et al. (2012). Iron-and nitrogen-functionalized graphene as a non-precious metal catalyst for enhanced oxygen reduction in an air-cathode plant microbial fuel cell. *Journal of Power Sources* 213: 265–269. https://doi.org/10.1016/j.jpowsour.2012.04.002.

Li, B., Zhou, J., Zhou, X. et al. (2014). Surface modification of plant microbial fuel cells anodes: approaches to practical design. *Electrochimica Acta* 134: 116–126. https://doi.org/10.1016/j.electacta.2014.04.136.

Li, J., Ge, Z., and He, Z. (2014). A fluidized bed membrane bioelectrochemical reactor for energy-efficient wastewater treatment. *Bioresource Technology* 167: 310–315. https://www.sciencedirect.com/science/article/pii/S0960852414008785.

Li, Y., Liu, L., Yang, F., and Ren, N. (2015). Performance of carbon fibre cathode membrane with C–Mn–Fe–O catalyst in MBR–MFC for wastewater treatment. *Journal of Membrane Science* 484: 27–34. https://doi.org/10.1016/j.memsci.2015.03.006.

Li, X., Wang, X., Zhao, Q. et al. (2016). Carbon fibre enhanced bioelectricity generation in soil microbial fuel cells. *Biosensors & Bioelectronics* 85: 135–141. https://doi.org/10.1016/j.bios.2016.05.001.

Li, M., Zhou, S., and Xu, M. (2017). Graphene oxide supported magnesium oxide as an efficient cathode catalyst for power generation and wastewater treatment in single-chamber plant microbial fuel cells. *Chemical Engineering Journal* 328: 106–116. https://doi.org/10.1016/j.cej.2017.07.031.

Liu, S., Song, H., Li, X., and Yang, F. (2013). Power generation enhancement by utilizing plant photosynthate in microbial fuel cell coupled constructed wetland system. *International Journal of Photoenergy* 172010. https://doi.org/10.1155/2013/172010.

Liu, S., Song, H., Wei, S. et al. (2014). Bio-cathode materials evaluation and configuration optimization for power output of vertical subsurface flow constructed wetland-Microbial fuel cell systems. *Bioresource Technology* 166: 575–583.

Liu, S.H., Chang, C.M., and Lin, C.W. (2018). Modifying proton exchange membrane in a plant microbial fuel cell by adding clay mineral to improve electricity generation without reducing removal of toluene. *Biochemical Engineering Journal* 134: 101–107. https://doi.org/10.1016/j.bej.2018.03.013.

Logan, B., Cheng, S., Watson, V., and Estadt, G. (2007). Graphite fibre brush anodes for increased power production in air-cathode plant microbial fuel cells. *Environmental Science & Technology* 41: 3341–3346. https://doi.org/10.1021/es062644y.

Lovely, D.R. (2006). Bug juice: harvesting electricity with microorganisms [published correction appears in Nat Rev Microbiol. 2006 Oct;4(10):797]. *Nature Reviews. Microbiology* 4 (7): 497–508. https://doi.org/10.1038/nrmicro1442.

Lu, M. and Li, S.F.Y. (2012). Cathode reactions and applications in plant microbial fuel cells: a review. *Critical Reviews in Environmental Science and Technology* 42: 2504–2525. https://doi.org/10.1080/10643389.2011.592744.

Lu, L., Xing, D., and Ren, Z.J. (2015). Microbial community structure accompanied with electricity production in a constructed wetland plant microbial fuel cell. *Bioresource Technology* 195: 115–121. https://doi.org/10.1016/j.biortech.2015.05.098.

Mansoorian, H.J., Mahvi, A.H., Jafari, A.J., and Khanjani, N. (2016). Evaluation of dairy industry wastewater treatment and simultaneous bioelectricity generation in a catalyst-less and mediator-less membrane plant microbial fuel cell. *Journal of Saudi Chemical Society* 20: 88–100. https://doi.org/10.1016/j.jscs.2014.08.002.

Mathuriya, A.S. and Yakhmi, J.V. (2016). Plant microbial fuel cells—applications for generation of electrical power and beyond. *Critical Reviews in Microbiology* 42: 127–143.

Mehdinia, A., Ziaei, E., and Jabbari, A. (2014). Multi-walled carbon nanotube/SnO_2 nanocomposite: a novel anode material for plant microbial fuel cells. *Electrochimica Acta* 130: 512–518. https://doi.org/10.1016/j.electacta.2014.03.011.

Mehdinia, A., Ziaei, E., and Jabbari, A. (2014). Facile microwave-assisted synthesized reduced graphene oxide/tin oxide nanocomposite and used as anode material of plant microbial fuel cell to improve power generation. *International Journal of Hydrogen Energy* 39: 10724–10730. https://doi.org/10.1016/j.ijhydene.2014.05.008.

Mohan, S.V., Velvizhi, G., Modestra, J.A., and Srikanth, S. (2014). Plant microbial fuel cell: critical factors regulating bio-catalyzed electrochemical process and recent advancements. *Renewable and Sustainable Energy Reviews* 40: 779–797. https://doi.org/10.1016/j.rser.2014.07.109.

Moqsud, M., Azizul Hannan, M.A., and Omine, K. (2015). Assessment of factors influencing bioelectricity generation in paddy plant microbial fuel cells. *Global Advanced Research Journal of Agricultural Science* 4 (12): 840–850.

Moqsud, M.A., Yoshitake, J., Bushra, Q.S. et al. (2015). Compost in plant microbial fuel cell for bioelectricity generation. *Waste Management* 36: 63–69. https://doi.org/10.1016/j.wasman.2014.11.004.

Moulin, L., Munive, A., Dreyfus, B., and Boivin-Masson, C. (2001). Nodulation of legumes by members of the beta-subclass of Proteobacteria [published correction appears in Nature 2001 Aug 30;412(6850):926]. *Nature* 411 (6840): 948–950. https://doi.org/10.1038/35082070.

Nitisoravut, R. and Regmi, R. (2017). Plant microbial fuel cells: a promising biosystems engineering. *Renewable and Sustainable Energy Reviews* 76: 81–89.

Noori, M.T., Ghangrekar, M.M., and Mukherjee, C.K. (2016). V_2O_5 microflower decorated cathode for enhancing power generation in air-cathode plant microbial fuel cell treating fish market wastewater. *International Journal of Hydrogen Energy* 41: 3638–3645.

Nor, M.H.M., Mubarak, M.F.M., Elmi, H.S.A. et al. (2015). Bioelectricity generation in plant microbial fuel cell using natural microflora and isolated pure culture bacteria from anaerobic palm oil mill effluent sludge. *Bioresource Technology* 190: 458–465. https://doi.org/10.1016/j.biortech.2015.02.103.

Oon, Y.L., Ong, S.A., Ho, L.N. et al. (2015). Hybrid system up-flow constructed wetland integrated with microbial fuel cell for simultaneous wastewater treatment and electricity generation. *Bioresource Technology* 186: 270–275. https://doi.org/10.1016/j.biortech.2015.03.014.

Pandit, S., Sengupta, A., Kale, S., and Das, D. (2011). Performance of electron acceptors in catholyte of a two-chambered plant microbial fuel cell using anion exchange membrane. *Bioresource Technology* 102: 2736–2744. https://doi.org/10.1016/j.biortech.2010.11.038.

Pant, D., Van Bogaert, G., Diels, L., and Vanbroekhoven, K. (2010). A review of the substrates used in plant microbial fuel cells (PMFCs) for sustainable energy production. *Bioresource Technology* 101: 1533–1543. https://doi.org/10.1016/j.biortech.2009.10.017.

Park, I.H., Christy, M., Kim, P., and Nahm, K.S. (2014). Enhanced electrical contact of microbes using Fe_3O_4/CNT nanocomposite anode in mediator-less plant microbial fuel cell. *Biosensors & Bioelectronics* 58: 75–80. https://doi.org/10.1016/j.bios.2014.02.044.

Park, S.G., Chae, K.J., and Lee, M. (2017). A sulfonated poly (arylene ether sulfone)/polyimide nanofiber composite proton exchange membrane for microbial electrolysis cell application under the coexistence of diverse competitive cations and protons. *Journal of Membrane Science* 540: 165–173. https://doi.org/10.1016/j.memsci.2017.06.048.

Pasternak, G., Greenman, J., and Ieropoulos, I. (2018). Dynamic evolution of anodic biofilm when maturing under different external resistive loads in microbial fuel cells. Electrochemical perspective. *Journal of Power Sources* 400: 392–401.

Pei, S., Zhou, Z., Chen, X. et al. (2016). Co/CoO nanoparticles/Ag nanowires/nitrogen codoped electrospun carbon nanofibers as efficient electrocatalysts for oxygen reduction. *International Journal of Electrochemical Science* 11: 8994–9006.

Piyare, R., Murphy, A.L., Tosato, P., and Brunelli, D. (2017). Plug into a plant: using a plant microbial fuel cell and a wake-up radio for an energy neutral sensing system. In: *IEEE Proceedings of 2017 IEEE 42nd Conference on Local Computer Networks Workshops (LCN Workshops) Singapore*, 18–25. IEEE Press October 9.

Powell, R.J., White, R., and Hill, R.T. (2014). Merging metabolism and power: development of a novel photoelectric device driven by photosynthesis and respiration. *PLoS One* 9 (1): e86518. Published 2014 Jan 22. https://doi.org/10.1371/journal.pone.0086518.

Pu, L., Liu, D., Li, K. et al. (2017). Carbon-supported binary transition metal chalcogenide used as cathode catalyst for oxygen reduction in plant microbial fuel cells. *International Journal of Hydrogen Energy* 42: 14253–14263. https://doi.org/10.1016/j.ijhydene.2017.04.074.

Rabaey, K. (2009). *Bioelectrochemical Systems: From Extracellular Electron Transfer to Biotechnological Application*. Water Intelligence Online. IWA Publishing.

Rabaey, K. and Verstraete, W. (2005). Microbial fuel cells: novel biotechnology for energy generation. *Trends in Biotechnology* 23 (6): 291–298. https://doi.org/10.1016/j.tibtech.2005.04.008.

Rahimnejad, M., Adhami, A., Darvari, S. et al. (2015). Plant microbial fuel cell as new technology for bioelectricity generation: a review. *Alexandria Engineering Journal* 54: 745–756. https://doi.org/10.1016/j.aej.2015.03.031.

Rozendal, R.A., Hamelers, H.V., Rabaey, K. et al. (2008). Towards practical implementation of bioelectrochemical wastewater treatment. *Trends in Biotechnology* 26: 450–459. https://doi.org/10.1016/j.tibtech.2008.04.008.

Saha, S., Kumar, A.G., Noori, M.T. et al. (2018). New crosslinked sulfonated polytriazoles: proton exchange properties and plant microbial fuel cell performance. *European Polymer Journal* 103: 322–334. https://doi.org/10.1016/j.eurpolymj.2018.04.023.

Samsudeen, N., Radhakrishnan, T.K., and Matheswaran, M. (2015). Bioelectricity production from plant microbial fuel cell using mixed bacterial culture isolated from distillery wastewater. *Bioresource Technology* 195: 242–247. https://doi.org/10.1016/j.biortech.2015.07.023.

Sarma, P.J. and Mohanty, K. (2018). *Epipremnum aureum* and *Dracaena braunii* as indoor plants for enhanced bio-electricity generation in a plant microbial fuel cell with electrochemically modified carbon fibre brush anode. *Journal of Bioscience and Bioengineering* 126 (3): 404–410. https://doi.org/10.1016/j.jbiosc.2018.03.009.

Sato, C., Paucar, N.E., Chiu, S. et al. (2021). Single-chamber microbial fuel cell with multiple plates of bamboo charcoal anode: performance evaluation. *Processes* 9 (12): 2194.

Schröder, U., Niessen, J., and Scholz, F. (2003). A generation of microbial fuel cells with current outputs boosted by more than one order of magnitude. *Angewandte Chemie (International Ed. in English)* 42 (25): 2880–2883. https://doi.org/10.1002/anie.200350918.

Sedesheva, Y.S., Ivanov, V.S., Wozniak, A.I., and Yegorov, A.S. (2016). Proton-exchange membranes based on sulfonated polymers. *Oriental Journal of Chemistry* 32: 2283–2296. https://doi.org/10.13005/ojc/320501.

Singh, S. and Verma, N. (2015). Fabrication of Ni nanoparticles-dispersed carbon micro-nanofibers as the electrodes of a plant microbial fuel cell for bio-energy production. *International Journal of Hydrogen Energy* 40: 1145–1153. https://doi.org/10.1016/j.ijhydene.2014.11.073.

Sivasankaran, A., Sangeetha, D., and Ahn, Y.H. (2016). Nanocomposite membranes based on sulfonated polystyrene ethylene butylene polystyrene (SSEBS) and sulfonated SiO_2 for plant microbial fuel cell application. *Chemical Engineering Journal* 289: 442–451. https://doi.org/10.1016/j.cej.2015.12.095.

Skinner, C.J. (2006). Urban density, meteorology and rooftops. *Urban Policy and Research* 24: 355–367.

Sophia, A.C. and Sreeja, S. (2017). Green energy generation from plant microbial fuel cells (PMFC) using compost and a novel clay separator. *Sustainable Energy Technologies and Assessments* 21: 59–66.

Srinophakun, P., Thanapimmetha, A., Plangsri, S. et al. (2017). Application of modified chitosan membrane for plant microbial fuel cell: roles of proton carrier site and positive charge. *Journal of Cleaner Production* 142: 1274–1282. https://doi.org/10.1016/j.jclepro.2016.06.153.

Srivastava, P., Yadav, A.K., and Mishra, B.K. (2015). The effects of microbial fuel cell integration into constructed wetland on the performance of constructed wetland. *Bioresource Technology* 195: 223–230. https://doi.org/10.1016/j.biortech.2015.05.072.

Strik, D.P.B.T.B., Snel, J.F.H., and Buisman, C.J.N. (2008). Green electricity production with living plants and bacteria in a fuel cell. *International Journal of Energy Research* 32: 870–876.

Strik, D.P., Timmers, R.A., Helder, M. et al. (2011). Microbial solar cells: applying photosynthetic and electrochemically active organisms. *Trends in Biotechnology* 29 (1): 41–49. https://doi.org/10.1016/j.tibtech.2010.10.001.

Sudirjo, E., Buisman, C.J.N., and Strik, D.P.B.T.B. (2018). Electricity generation from wetlands with activated carbon bioanode. In: *IOP Institute of Physics Publishing Conference Series: Earth and Environmental Science; 2018 March*. Bristol, UK: IOP Publishing 012046.

Sudirjo, E., de Jager, P., Buisman, C.J.N., and Strik, D.P.B.T.B. (2019). Performance and long distance data acquisition via LoRa technology of a tubular plant microbial fuel cell located in a paddy field in West Kalimantan, Indonesia. *Sensors (Basel)* 19 (21): 4647. Published 2019 Oct 25. https://doi.org/10.3390/s19214647.

Takanezawa, K., Nishio, K., Kato, S. et al. (2010). Factors affecting electric output from rice-paddy microbial fuel cells. *Bioscience, Biotechnology, and Biochemistry* 74: 1271–1273. https://doi.org/10.1271/bbb.90852.

Tang, H., Cai, S., Xie, S. et al. (2015). Metal–organic-framework-derived dual metal-and nitrogen-doped carbon as efficient and robust oxygen reduction reaction catalysts for plant microbial fuel cells. *Advancement of Science* 3: 1500265. https://doi.org/10.1002/advs.201500265.

Tang, H., Zeng, Y., Zeng, Y. et al. (2017). Iron embedded nitrogen-doped carbon frameworks as a robust catalyst for oxygen reduction reaction in plant microbial fuel cells. *Applied Catalysis B: Environmental* 202: 550–556. https://doi.org/10.1016/j.apcatb.2016.09.062.

Tapia, N., Claudia, F.R., Carlos, A.B., and Ignacio, T.V. (2017). Evaluation of Sedum as a driver for plant microbial fuel cells in a semi-arid green roof ecosystem. *Ecological Engineering* 108: 203–2010.

Timmers, R.A., Strik, D.P., Hamelers, H.V., and Buisman, C.J. (2013). Electricity generation by a novel design tubular plant microbial fuel cell. *Biomass and Bioenergy* 51: 60–67.

Tiwari, B.R. and Ghangrekar, M.M. (2018). Electricity production during distillery wastewater treatment in a plant microbial fuel cell equipped with low-cost PVA-Nafion-borosilicate membrane. *Journal of Clean Energy Technologies* 6: 155–158. https://doi.org/10.18178/jocet.2018.6.2.452.

Tremouli, A., Martinos, M., Bebelis, S., and Lyberatos, G. (2016). Performance assessment of a four-air cathode single-chamber plant microbial fuel cell under conditions of synthetic and municipal wastewater treatments. *Journal of Applied Electrochemistry* 46: 515–525. https://doi.org/10.1007/s10800-016-0935-3.

Ueoka, N., Sese, N., Sue, M. et al. (2016). Sizes of anode and cathode affect electricity generation in rice paddy-field microbial fuel cells. *Journal of Sustainable Bioenergy Systems* 06: 10–15. https://doi.org/10.4236/jsbs.2016.61002.

Venkata Mohan, S., Mohanakrishna, G., Chiranjeevi, P. et al. (2010). Ecologically engineered system (EES) designed to integrate floating, emergent and submerged macrophytes for the treatment of domestic sewage and acid-rich fermented-distillery wastewater: evaluation of long term performance. *Bioresource Technology* 101 (10): 3363–3370. https://doi.org/10.1016/j.biortech.2009.12.027.

Venkata Mohan, S., Mohanakrishna, G., and Chiranjeevi, P. (2011). Sustainable power generation from floating macrophytes based ecological microenvironment through embedded fuel cells along with simultaneous wastewater treatment. *Bioresource Technology* 102 (14): 7036–7042. https://doi.org/10.1016/j.biortech.2011.04.033.

Venkatesan, P.N. and Dharmalingam, S. (2015). Development of cation exchange resin-polymer electrolyte membranes for plant microbial fuel cell application. *Journal of Materials Science* 50: 6302–6312. https://doi.org/10.1007/s10853-015-9167.

Venkatesan, P.N. and Dharmalingam, S. (2015). Effect of cation transport of SPEEK–Rutile TiO_2 electrolyte on plant microbial fuel cell performance. *Journal of Membrane Science* 492: 518–527. https://doi.org/10.1016/j.memsci.2015.06.025.

Venkatesan, P.N. and Dharmalingam, S. (2017). Characterization and performance study of phase-inversed sulfonated poly ether ether ketone–silico tungstic composite membrane as an electrolyte for plant microbial fuel cell applications. *Renewable Energy* 102: 77–86. https://doi.org/10.1016/j.renene.2016.10.027.

Wang, C., Kaneti, Y.V., Bando, Y. et al. (2018). Metal–organic framework-derived one-dimensional porous or hollow carbon-based nanofibers for energy storage and conversion. *Materials Horizons* 5: 394–407. https://doi.org/10.1039/C8MH00133B.

Watanabe, K., Dan K., Nishio. 2010. Book edited by Jatin Nathwani and Artie Ng, ISBN 978-953-307-401-6.

Wetser, K., Sudirjo, E., Buisman, C.J. et al. (2015). Electricity generation by a plant microbial fuel cell with an integrated oxygen-reducing biocathode. *Applied Energy* 137: 151–157. https://doi.org/10.1016/j.apenergy.2014.10.006.

Wetser, K., Dieleman, K., Buisman, C.J. et al. (2017). Electricity from wetlands: tubular plant microbial fuels with silicone gas-diffusion biocathodes. *Applied Energy* 185: 642–649. https://doi.org/10.1016/j.apenergy.2016.10.122.

Wu, X., Song, T., Zhu, X. et al. (2013). Research on different wetland plants to construct the plant-sediment microbial fuel cell. *Renewable Energy* 31 (9): 78–82.

Wu, S., Li, H., Zhou, X. et al. (2016). A novel pilot-scale stacked plant microbial fuel cell for efficient electricity generation and wastewater treatment. *Water Research* 98: 396–403. https://doi.org/10.1016/j.watres.2016.04.043.

Wu, H., Fu, Y., Guo, C. et al. (2018). Electricity generation and removal performance of a plant microbial fuel cell using sulfonated poly (ether ether ketone) as proton exchange membrane to treat phenol/acetone wastewater. *Bioresource Technology* 260: 130–134. https://doi.org/10.1016/j.biortech.2018.03.133.

Xu, L., Zhao, Y., Doherty, L. et al. (2015). The integrated processes for wastewater treatment based on the principle of microbial fuel cells: a review. *Critical Reviews in Environmental Science and Technology* 1–32. https://doi.org/10.1080/10643389.2015.1061884.

Xu, T., Wang, Q., Wu, S. et al. (2018). A novel filtration composite anode configuration of plant microbial fuel cell for efficient wastewater treatment and enhanced power generation. *Journal of Cleaner Production* 178: 305–313. https://doi.org/10.1016/j.jclepro 2017.12.259.

Yamashita, T. and Yokoyama, H. (2018). Molybdenum anode: a novel electrode for enhanced power generation in plant microbial fuel cells, identified via extensive screening of metal electrodes. *Biotechnology for Biofuels* 11: 39. https://doi.org/10.1186/s13068-018-1046-7.

Yan, Z., Jiang, H., Cai, H. et al. (2015). Complex interactions between the macrophyte Acorus calamus and microbial fuel cells during pyrene and benzo[a] pyrene degradation in sediments. *Scientific Reports* 5: 10709. https://doi.org/10.1038/srep10709.

Yaqoob, A.A., Al-Zaqri, N., Yaakop, A.S., and Umar, K. (2022). Potato waste as an effective source of electron generation and bioremediation of pollutant through benthic microbial fuel cell. *Sustainable Energy Technologies and Assessments* 53: 102560.

Yaqoob, A.A., Guerrero-Barajas, C., Ibrahim, M.N.M. et al. (2022). Local fruit wastes driven benthic microbial fuel cell: a sustainable approach to toxic metal removal and bioelectricity generation. *Environmental Science and Pollution Research* 29: 32913–32928.

Yin, T., Lin, Z., Su, L. et al. (2015). Preparation of vertically oriented TiO_2 nanosheets modified carbon paper electrode and its enhancement to the performance of PMFCs. *ACS Applied Materials & Interfaces* 7: 400–408. https://doi.org/10.1021/am506360x.

Yue, L., Xie, Y., Zheng, Y. et al. (2017). Sulfonated bacterial cellulose/polyaniline composite membrane for use as gel polymer electrolyte. *Composites Science and Technology* 145: 122–131. https://doi.org/10.1016/j.compscitech.2017.04.002.

Zhang, B., Wen, Z., Ci, S. et al. (2014). Synthesizing nitrogen-doped activated carbon and probing its active sites for oxygen reduction reaction in plant microbial fuel cells. *ACS Applied Materials & Interfaces* 6: 7464–7470. https://doi.org/10.1021/am5008547.

Zhang, X., Ren, H., Pyo, S. et al. (2015). A high-efficiency DC–DC boost converter for a miniaturized plant microbial fuel cell. *IEEE Transactions on Power Electronics* 30: 2041–2049. https://doi.org/10.1109/TPEL.2014.2323075.

Zhang, C., Liang, P., Yang, X. et al. (2016). Binder free graphene and manganese oxide coated carbon felt anode for high-performance plant microbial fuel cell. *Biosensors & Bioelectronics* 81: 32–38. https://doi.org/10.1016/j.bios.2016.02.051.

Zhang, B., Jiang, Y., and Han, J. (2017). Facile fabrication of PVAc-g-PVDF coating on surface-modified cotton fabric for applications in oil/water separation and heavy metal ions removal. *Fibers and Polymers* 18: 1754–1762. https://doi.org/10.1007/s12221-017-1224-4.

Zheng, S., Yang, F., Chen, S. et al. (2015). Binder-free carbon black/stainless steel mesh composite electrode for high-performance anode in microbial fuel cells. *Journal of Power Sources* 284: 252–257.

Zhou, M., Chi, M., Luo, J. et al. (2011). An overview of electrode materials in plant microbial fuel cells. *Journal of Power Sources* 196: 4427–4435. https://doi.org/10.1016/j.jpowsour.2011.01.012.

Zhou, S., Tang, J., and Yuan, Y. (2015). Conduction-band edge dependence of carbon-coated hematite stimulated extracellular electron transfer of Shewanella oneidensis in bioelectrochemical systems. *Bioelectrochemistry* 102: 29–34. https://doi.org/10.1016/j.bioelechem.2014.11.005.

Zinadini, S., Zinatizadeh, A.A., Rahimi, M. et al. (2017). High power generation and COD removal in a plant microbial fuel cell operated by a novel sulfonated PES/PES blend proton exchange membrane. *Energy* 125: 427–438. https://doi.org/10.1016/j.energy.2017.02.146.

Zou, Y., Pisciotta, J., Billmyre, R.B., and Baskakov, I.V. (2009). Photosynthetic microbial fuel cells with a positive light response. *Biotechnology and Bioengineering* 104 (5): 939–946. https://doi.org/10.1002/bit.22466.

3

An Overview of the Non-noble Electrocatalysts as Air Cathodes in Biocells

Omar Francisco G. Vazquez and Ma. Del Rosario M. Virgen

Department of Chemical and Biochemical Engineering, Instituto Tecnológico de Aguascalientes, Aguascalientes, México

3.1 Introduction

Within the multiple fundamental parameters that govern both the efficiency and the construction costs of fuel biocells to produce sustainable electrical energy, ion-capturing electrodes play a preponderant role (Wei et al., 2011). The electrodes are the devices in charge of generating oxidation–reduction reactions on their surface, as well as capturing, releasing, and transporting the ions within the biocell. Based on this, these electrodes must meet different characteristics, such as resistance to corrosion, mechanical resistance, high electrical conductivity, and even bioaffinity for microorganisms (Kaur et al., 2020), all this under the focus of being as efficient as possible. In this sense, the choice of the materials that will make up the appropriate electrodes is of the utmost importance.

Biofuel cell systems can be divided according to the number of chambers that comprise it, existing biocells with one, two, or even three chambers (Wang et al., 2020); however, they all work under the same principle, in which at least two electrodes are required, given the polarity characteristics of the generated current (direct current) (Figure 3.1). The electrodes are named after the type of reaction that happens in them. The electrode at which oxidation reactions take place is called the anode, while the electrode at which reduction reactions take place is called the cathode.

Photosynthesis-Assisted Energy Generation: From Fundamentals to Lab Scale and In-Field Applications, First Edition. Edited by Sathish-Kumar Kamaraj and Iryna Rusyn.

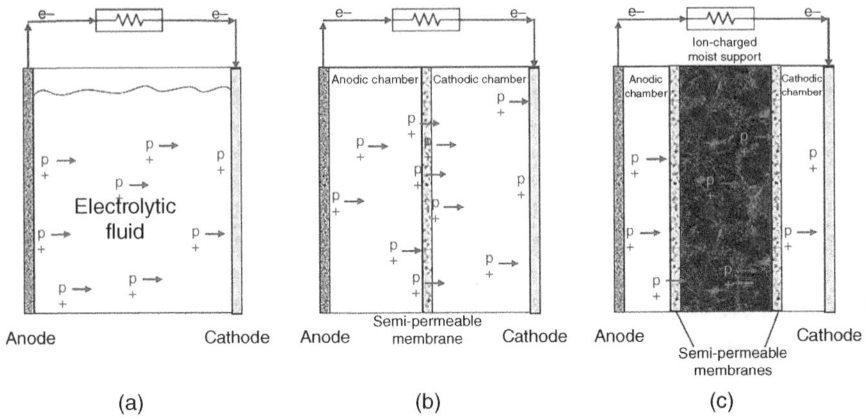

Figure 3.1 Examples of biocells with one chamber (a), two chambers (b), and three chambers (c).

The cathodes themselves play a crucial role in the operation of biofuel cells, since they are responsible for the reduction of the oxidizing agent (Zhang et al., 2016), which captures the protons and recovers electrons from the external circuit, forming water in this process (González Vázquez et al., 2022). The oxidizing agent is usually diatomic oxygen due to its high electronegativity and relatively high availability, in this sense, the medium in which the oxygen is available defines some characteristics of the cathode; if the oxygen is taken in gaseous form from the air, it is called an aerated cathode, while if it is taken from a liquid phase, which is usually water, it is called an unaerated or aqueous cathode. This presupposes a great difference between them, since to carry out a series of successful reactions it is crucial that the oxygen can be easily captured and the reaction must be efficient, so for this to happen, normally it must have a catalytic agent that promotes said reaction, since at pH close to neutral and at room temperature said reaction does not usually happen (HaoYu et al., 2007). This determines that the cathode must normally have some catalytic material that promotes the reduction reaction, additionally, said material must have both desirable electrical and mechanical properties, such as those mentioned above. This greatly limits the materials available for that purpose. In this sense, electrodes made from noble metals such as Pt, Pd, or Rh (Gao et al., 2021), they allow to catalyze diverse reactions with great success, among them those of oxygen reduction; however, the use of these materials presupposes a challenge in the economic sense and difficulty in the synthesis methods. Therefore, various researchers have focused on obtaining various materials that could replace these materials, trying to reduce costs and match or even exceed the efficiency of noble metals.

In this chapter, we will elucidate the operation of aerated cathodes, their advantages over non-aerated counterparts, the significance of materials in their construction, and explore potential substitutes for noble metals that could offer cost reduction while ensuring robust efficiency.

3.2 Operation and Structure of the Aerated Cathode

Biofuel cell systems with aerated cathode are characterized by keeping most or all of their surface exposed to the air, primarily it is required that the cathode surface be free of moisture that would prevent the flow of oxygen to the surface; however, this humidity is not usually removed, since the same airflow allows the evaporation of the humidity accumulated on the surface. Other than this, aerated cathodes could be considered maintenance free.

Aerated electrodes can be made up of multiple layers that fulfill various functions (Figure 3.2), the most common usually have 3 layers (Zhang et al., 2014):

A layer of diffusive material is responsible for both gathering and diffusing the oxygen present in the air, as well as assisting in moisture elimination, as it is typically composed of a hydrophobic material (Wei et al., 2011).

The next is a layer of catalytic material, which is responsible for accelerating the speed of the oxygen reduction reaction since it is usually very slow. Also, it can lower the activation energy of the reaction (Duteanu et al., 2010), in this layer, the surface concentration of the catalytic material must be taken care of, since it must have a balance between the active sites and the diffusing material.

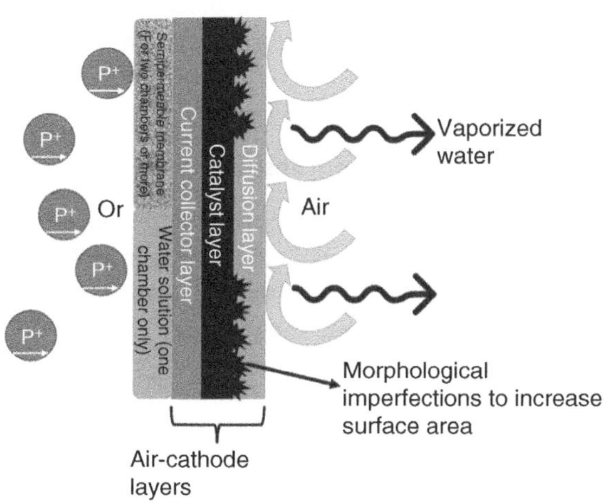

Figure 3.2 Air-cathode layers.

Finally, the layer of current collector material, in which the electrodes and protons coming from the anode are accepted, these materials must be highly conductive, since this would represent a significant ohmic load of the total ohmic load, which is detrimental to the current density generated by the biocell (Rozendal et al., 2008).

Certain authors make reference to additional component layers within the cathode, such as those composed of binding materials, which fulfill the function of a sealant, preventing the passage of electrolyte to the outermost layer of the cathode in single-chamber systems (Shi et al., 2012), in addition, said layer must be permeable to the protons that need to pass to the layer where the oxygen reduction reaction is taking place. This layer is not necessary in 2- or 3-chamber systems.

3.2.1 Advantages of Aerated Versus Non-aerated Cathodes

One of the main advantages of using an aerated cathode against a non-aerated one is the availability of oxygen that results in a greater flow of this, since flows up to 3.7 times greater have been reported in an aerated cathode system than in an aqueous one (Cheng et al., 2006), in this sense, when the available oxygen increases, it affects high power densities generated (Chatterjee et al., 2018). Added to this is the fact that the regeneration needs of these are less than those of non-aerated ones, because they are not in intimate contact with electrolyte capable of producing scale. Among its drawbacks, we encounter the necessity of utilizing a catalyst material to enhance the efficiency of oxygen reduction reactions (Freguia et al., 2007), thereby elevating construction costs. Additionally, in single-chamber cells, occasional electrolyte leakage to the aerated external environment and/or excessive oxygen diffusion toward the liquid phase of the biocell may occur, potentially impacting oxygen chemical demand or interfering with the metabolism of the anaerobic anode chamber.

3.2.2 Oxygen Reduction Reactions and Electron Transport

Oxygen reduction reactions can occur in two ways, one directly with 4 electrons and one via peroxide with 2 electrons (Neburchilov et al., 2010; Yeager, 1986).

The 4-electron process reactions are further divided into alkaline and acidic systems, the alkaline reduction reaction being the one shown in Eq. (3.1):

$$O_2 + 2H_2O + 4e^- \rightarrow 4OH^- \tag{3.1}$$

Oxygen reduction can also occur in an acid environment, although it is less likely to happen since its required activation energy is three times higher than its alkaline counterpart, this is shown in Eq. (3.2):

$$O_2 + 4H^+ + 4e^- \rightarrow 2H_2O \qquad (3.2)$$

On the other hand, when the reduction process occurs through the peroxide process, that is, with a pair of electrons, it can also be carried out in alkaline and acid media. The reaction mechanism in two steps of 2 electrons in alkaline media is as follows (Eqs. (3.3) and (3.4)):

$$O_2 + H_2O + 2e^- \rightarrow HO_2^- + OH^- \qquad (3.3)$$

This reaction is followed by the reduction reaction as such:

$$HO_2^- + H_2O + 2e^- \rightarrow 3OH^- \qquad (3.4)$$

Or the decomposition reaction (Eq. 3.5):

$$2HO_2^- \rightarrow 2OH^- + O_2 \qquad (3.5)$$

In alkaline media, the 4-electron process is again the one that requires a higher activation energy, the two-step mechanism is shown below (Eqs. (3.6) and (3.7)):

$$O_2 + 2H^+ + 2e^- \rightarrow H_2O_2 \qquad (3.6)$$

Followed by:

$$H_2O_2 + 2H^+ + 2e^- \rightarrow 2H_2O \qquad (3.7)$$

Or by decomposition (Eq. 3.8):

$$2H_2O_2 \rightarrow 2H_2O + O_2 \qquad (3.8)$$

Additionally, both the 4- and 2-electron processes can be carried out in parallel or, alternatively, they can be carried out through reactions in series.

These reactions depend directly on the pH of the environment in which the electrodes are located, but primarily, they depend on the catalyst materials that are present, it is considered that carbon-based platinum materials are the ones that present the best performance to carry out the reaction. oxygen depletion (Vij et al., 2017).

On the other hand, the transport of electrons is a function of the current collector material and the base conductive material, carbon-based materials such as felts and paper (Martin et al., 2011); however, these tend to have low mechanical strength and are generally expensive. In this sense, metal-based electrodes, such as stainless steel, can also be used, although these have good conductivity and are not expensive, their chemical stability is not very good, especially since they are susceptible to the presence of acids, of the use of electrolytes with wastewater (Wang et al., 2015). As can be seen, there is no material that is optimal for all types of processes within biocells, since many characteristics of each system influence such as pH, analyte, catalyst load, temperature, and mode of operation, for this reason. The materials must be evaluated for each condition to which they are exposed, since the reduction of the cell's ohmic load depends on this to a great extent, which directly affects the efficiency of the biocell (Zhuang et al., 2010).

3.3 Importance of Materials in the Construction of Catalytic Electrodes for Hydrogen Reduction

As has been seen, the reduction of oxygen is crucial to achieving optimal performance in fuel biocells, and this is a function of the presence of a good catalyst that accelerates the reduction reactions. It is understood that the characteristics of chemical resistance, mechanical resistance, cost, and efficiency to carry out the reaction are the pillars on which the choice of this material should be based. In this sense, multiple materials have been studied, ranging from noble metals and metal alloys, carbons, and transition metal oxides/chalcogenides to macrocyclic metallic compounds (Cheng & Chen, 2012).

3.3.1 Use of Noble Metals as Catalytic Materials and Their Performance

Various investigations have shown that the so-called noble metals have unique chemical and corrosion resistance characteristics, in this select group of elements are silver (Ag), gold (Au), and metals of the platinum subgroup, including iridium (Ir), osmium (Os), palladium (Pd), platinum (Pt), rhodium (Rh), and ruthenium (Ru) (Dinh et al., 2022). In addition to this, it has been proven that platinum, rhodium, palladium, and ruthenium themselves have good characteristics as catalysts to promote or accelerate reactions.

Undoubtedly, the noble metal that has been used the most is platinum, since it has high chemical stability, much higher than other materials. By itself, platinum has proven to be a good catalyst, although in biocell systems for oxygen reduction, it is not so good (Chatterjee et al., 2018). In addition to this, if biocell technology becomes popular, and platinum is mainly used as a catalyst material in oxygen reduction, its demand would increase and therefore its value, which is already high (Nie et al., 2015), making the use of this material unfeasible. This would mean that the substitution of this material for others that exceed its catalytic characteristics and lower costs should prevail, but there is definitely not a wide variety of materials that have the characteristics of chemical stability that platinum has. In this sense, it is possible to think of gradually substituting platinum, using

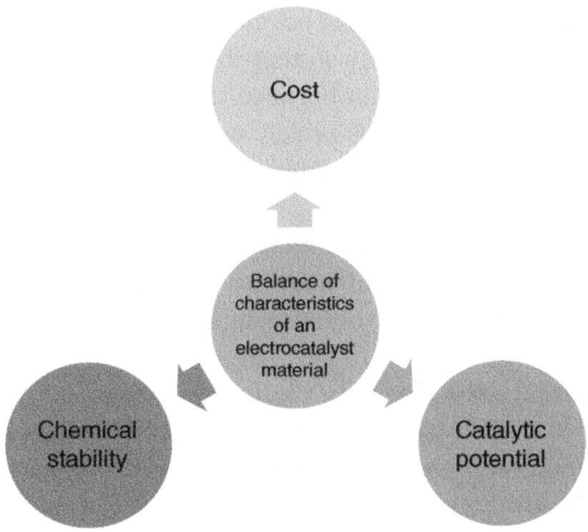

Figure 3.3 Balance between the three main characteristics of an electrocatalyst.

mixtures with other compounds that achieve a balance between cost, catalyzing efficiency, and chemical stability (Figure 3.3). It is believed that if the surface electronic properties of platinum-based materials are altered to a certain extent, their catalytic potential could be optimized and thus use less and less mass of platinum, which would help to minimize costs (Strasser et al., 2010). One way to achieve these electronic surface changes could be to use mixtures of platinum and metals such as transition metals (Greeley et al., 2009) or alternatively, it can be with a restructuring of platinum and its reduction to nanometric particles (Liu et al., 2019; Mou et al., 2013).

3.4 Disadvantages of Noble Metal Electrocatalysts

At present, various processes are being developed in favor of the care of the environment, among them, the generation of clean energies using fuel cells. Within this process, catalysts are used, which can be synthesized from noble metals or materials free of noble metals. Noble metals such as Pt and Pd have been widely used for the synthesis of catalysts for fuel cell reactions. However, the use of noble metals within the synthesis of catalysts has disadvantages, among which the high cost stands out (Rodriguez et al., 2013; Shit et al., 2021).

3.4.1 Synthesis

To begin with, it is imperative to note that the quality of an electrocatalyst hinges on characteristics such as the following: high stability, corrosion resistance, good conductor of electrons and protons, high surface area (to maximize the surface area of the catalysts), the size of these catalysts needs to be reduced to increase the area/volume ratio (Phan et al., 2021), high porosity so that the mass transfer is carried out in an adequate way, a strong interaction between the (metal or non-metal) and the support and, in addition, that it is widely available in nature. Mainly the shortages, high price, and low stability of electrocatalysts based on precious metals restrict their large-scale applications.

Undoubtedly, the main disadvantage of using noble metals as catalysts is their limited availability and exorbitant cost, these are the main reasons why for years they have been working with materials that replace noble metals and whose yields are comparable.

Among the synthesis methods that exist to obtain electrocatalysts based on noble metals are the top–down approaches, which include conventional methods, melt quenching, dealloying method, and thermal plastic forming techniques. Conventional methods have the disadvantage that their process time is extremely large (around 40 hours) and extremely energy intensive. Moreover, it is wasteful. Melt quenching is a traditional method used to produce amorphous metal alloys. This method, although faster than mechanical alloying, is still energy intensive. Dealloying method is a top–down process used for the manufacture of amorphous catalysts at the nanoscale level. The catalysts synthesized by this method are generally based on Pt. This method has the disadvantage that it does not induce an amorphous structure. Its main purpose is to increase the surface area of the catalyst while maintaining its amorphous characteristics (Phan et al., 2021).

The other large group that includes synthesis methods based on noble metals is the bottom–up approaches in which the methods of wet chemical reduction and thermal annealing are included, although the latter are the most used methods that present various problems due to the use of various chemical reagents, including capping agents (polyvinylpyrrolidone, cetyltrimethylammonium chloride, oleic acid), recutor agents (borohydride, superhydride), and solvents (oleylamine, ethylene glycol). Capping agents can block active sites and generate intermetallic compounds. Reducing agents can induce rapid nucleation and overgrowth of crystals presenting problems in the control of morphology. In addition, reaction temperatures in wet chemistry synthesis are limited by the boiling point of solvents, which are often sufficient in required synthesis reactions, resulting in compounds with a partial order. For thermal annealing method, the annealing temperature exceeding 500 °C causes a series of agglomeration of metal particles, practically avoiding the control of particle size (Phan et al., 2021; Zhao et al., 2021).

Many of the metals used in conventional fuel cells and water electrolyzers are unstable under certain catalytic reaction conditions used in these types of fuel cells. In this sense, Pt (the most widely used noble metal) is susceptible to degradation after long exposure to chemical processes that occur in fuel cells; for example, the oxidation of methanol in direct methanol bellows cells, which can form

CO, known as "poison" for Pt. The catalytic activity and selectivity of metals such as Pt can be compromised by the fact that the metal can be oxidized, dissolved, and subjected to some other degradation processes (Asefa, 2016).

3.4.2 Economy

The main noble metals used in the synthesis of electrocatalysts are Pt, Pd, and Ir; these metals have a high cost, little availability in nature, and susceptibility to generate toxic gases that cause negative effects on the environment (Phan et al., 2021; Zhao et al., 2021; Asefa, 2016; Zhu et al., 2017; Abdelkader-Fernández et al., 2019; Yu et al., 2010; Zhang et al., 2021; Yu et al., 2019).

All authors concur that noble metals exhibit excellent performance; however, their prohibitively high costs render them unsuitable materials for large-scale processes.

3.4.3 Performance

A disadvantage of these materials is that the high temperatures required in their processes could cause severe aggregation which would cause obtaining a low catalytic active surface area (Zhao et al., 2021).

Finally, another aspect that represents a disadvantage in terms of the performance of these materials is that they are unstable materials (Zhang et al., 2021) and have insufficient durability, which limits the applications that could be granted to them (Yu et al., 2019).

3.5 Synthesis of Non-noble Electrocatalysts and Their Performance

3.5.1 Metals

Noble metals such as Pt, Rh, and Ru show superior electrochemical activity (Rodriguez et al., 2013). However, bimetallic electrodes composed of a noble metal and a promoter metal such as Cu and Sn have been investigated in recent years and the results obtained are really promising (Zhang et al., 2021).

On the other hand, the combination of two metals in the synthesis of electrocatalysts have shown an improvement in the performance of the catalyst since there is a synergistic effect between both metals.

It is important to mention that most of the reactions where this type of catalysts is used are carried out in an acidic medium, so it makes it to some extent another restriction as to the nature of the materials used as catalysts. Among them is the

Mo, which is low cost and has excellent electrocatalytic properties. However, Mo-based materials have a lower availability of active sites and electrical conductivity compared to those presented by noble metals (Li et al., 2015).

If the metals used in the synthesis of catalysts are non-noble metals, it is best to use metals in transition, since they present many active sites, promote a rapid transfer of electrons, and are low cost; among the most commonly used transition metals are Fe, Ni, Co, and Mn (Tang et al., 2022; Yu et al., 2021).

In this sense and finally, it is important to mention that generally non-noble metals are used within complexes, which in this case would serve as composites or amalgams, one of the types of materials used to synthesize catalysts that do not use noble metals.

Within these complex materials in which non-noble metals can be used are also nitrides, carbides, phosphides, and chalcogenides, among others, complexes that will mostly be treated in this chapter (Zhang et al., 2021; Yu et al., 2021).

3.5.2 Carbonaceous

The pursuit of efficient electrocatalysts utilizing non-noble metals, characterized by their cost-effective production and potential contribution to environmentally friendly hydrogen production as a renewable energy source, stands as a paramount objective for numerous contemporary researchers (Raut et al., 2021).

A carbonaceous material to be used as a catalyst must have the following characteristics: high specific area, this to ensure the availability of catalytic active sites so that reactions occur efficiently, and high electrical conductivity so that the charge can be transferred faster and an adequate porous structure to favor mass transport (Yuana et al., 2019).

Electrochemical water splitting is the most preferred way to produce hydrogen. In addition to Pt, whose use has several disadvantages, so research has been directed toward the use of compounds based on transition metals, such as the study in which they encapsulated Mo particles in supported carbon, demonstrating that the incorporation of both metal and nonmetal elements in the carbon coating along with Mo_2C improves reaction performance for hydrogen production (Zhang et al., 2021). In this same study, it was also concluded that the graphitization of carbon improves with temperature which in turn favors charge transfer kinetics.

The carbonaceous materials generally used as a support during the process of synthesis of catalysts have presented good results, since it has been shown that the electronic conductivities of these materials are larger than those of other materials used as supports, as is the case of TiO_2. As is already known, one of the main characteristics that a catalyst must have is an amorphous surface, which generates a large surface area and in this sense, activated carbon has high surface areas, in addition to high conductivity and considerable resistance (Borbath et al., 2021).

In addition, activated carbon is a material that, when functionalized, has certain advantages over other materials during the deposition process, whether of metals, metal oxides, or other elements that favor its performance in the reaction; one of these advantages is good performance in both acidic and basic media (Yuana et al., 2019). Not only activated carbon has been used as a precursor to synthesize catalysts but also other types of carbonaceous materials such as carbon nanotubes, graphene, and reduced graphene oxide, among others, due to its special electronic structure (Borbath et al., 2021; Yu et al., 2020).

Carbonaceous materials can also be prepared from biomass, as they are environmentally friendly, have low cost, and are found in an abundant way in nature, so their availability would not be a problem.

In the literature, there is a great diversity of research in which biochars are synthesized through various techniques such as hydrothermal, chemical electrodeposition, and adsorption, among others, demonstrating these materials high electrical conductivities chemical stabilities, and excellent reaction kinetics. A work reported by Raout et al. in which they used coconut shell as a precursor to synthesize a material that was used as a catalyst obtaining excellent yields which was attributed to a high surface area and therefore, the availability of many active sites (Raut et al., 2021).

Finally, carbonaceous materials can also be synthesized using waste, such as the case where they use plastics as a source of carbon (Mir et al., 2020).

3.5.3 Amalgams

Another type of materials used for the synthesis of electrocatalysts free of noble metals are materials called amalgam or composite materials. These materials encompass compounds derived from zeolites, both organic and inorganic compounds, and nanofibers primarily doped with transition metals like Au, Ni, Fe, La, Mn, and Co. These metals as well as their compounds, such as phosphides, oxides, and among others, have been widely used due to their low cost and high efficiency. Nonetheless, they exhibit a notable propensity for agglomeration, impeding their efficacy in chemical reactions. Consequently, the utilization of composite structures, either bimetallic or involving a metal in conjunction with an organic or inorganic compound, is primarily warranted. In this sense, it is important to mention that when it comes to bimetallic structures, they have physical and chemical properties that make them materials with better performance, even the presence of one metal can improve the performance of the other metal, which is equivalent to a synergistic effect between them (Amiripour et al., 2019).

Zeolites are inorganic materials that generally have resistant structures. Their pore distribution makes them an excellent choice to play the role of support, since they generally promote ion exchange between surface groups and cations to which they are exposed. In the study reported by Amiripour et al., they used nanozeolites that proved to be an excellent support for a bimetallic compound containing Au and Ni (Amiripour et al., 2019).

Other compounds, such as chalcogenide metal (Ni_3S_2), constitute minerals whose behavior can be attributed to the existence of a continuous network of linkages within their structural framework, in addition to considering that it is a low-cost material compared to its performance as an electrocatalyst and despite the fact that it has presented a lower catalytic activity than other materials, its catalytic active sites promote both hydrogen production and oxidation–reduction reaction at the same time, what does not happen with other materials that are used as electrocatalysts (Feng et al., 2015).

Nanofibers doped with La, Mn, and Co with an oxygen deficiency have also been used and their results show performances that can be compared with commercial electrocatalysts. This material presented good stability due to its novel structure, a large number of metal ions with a high oxidation state, and a high oxygen deficiency (Bian et al., 2019).

Finally, organic–inorganic compounds have also been used, such as polyoxometalate and anisidine and the metals Ni and Fe, materials that showed stable characteristics in basic media (Imani et al., 2018).

Undoubtedly, it is necessary to develop new materials and techniques that promote the generation of clean energy and composites, or composite structures are a group of materials that meet these expectations, since they also are found abundantly in nature.

3.5.4 Carbides

In order to replace noble metals within the synthesis of electrocatalysts, it is essential to develop effective materials at low cost and with high performance. Transition metal carbides have attracted special interest due to their high conductivity and thermochemical stability. Transition metal carbides have variable oxidation states, which is why they have the ability to exchange between oxidation states and an electron source, so transition metals demonstrate good catalytic activity.

Specifically, titanium carbide catalysts show great influence on cost, performance, and stability. These materials have been widely used as electrodes, since they have high conductivity, electrochemical stability, high resistance, and good interaction with metals and although it has a lower catalytic activity with respect to that presented by the Pt, if it has presented a great performance as an electrocatalyst support (Saha et al., 2021).

Another study shows that when working with transition metal carbides, the presence of an ionic liquid is necessary since the latter can act as an inhibitor or as a promoter of oxygen growth on the surface of the carbide depending on its nature; considering the above, there will be an increase or decrease in the main particle size of the material and therefore, the catalytic performance will be modified. Taking into account the above, it can be concluded that the nature of the carbide will depend on the interaction with the ionic liquid and therefore, the catalytic properties of the material (Díaz-Coello et al., 2021). This behavior can also be observed in tungsten carbides (Su et al., 2020; Chen et al., 2020; Wang et al., 2019).

Another transition metal that has been widely used with carbides to act as a catalyst is Mo, which has aroused great interest due to its abundance, availability, and tunable electronic structures. In fact, its unique electronic structure and high conductivity make it not only a good support for other active centers but also capable of modifying the electronic structure of other components so that it synergistically improves the catalytic activity in the oxygen evolution reaction (Li et al., 2021; Wang et al., 2019; Tang et al., 2017).

Finally, boron carbide has also been used as a support presenting advantages such as adequate electrical conductivity, good electrochemical, and thermal stability. It even retains its good stability in acidic media and is far superior to conventional carbonaceous materials (Islam et al., 2021).

3.5.5 Nitrides

In our pursuit of developing cost-effective electrocatalysts with superior activity, stability, and efficacy, we have conducted research involving transition metal nitrides. These materials exhibit excellent conductivity, remarkable resistance to corrosion, and a bifunctional catalytic performance.

Transition metal nitrides retain the physical properties of metals and chemical properties with unique electronic structures. The introduction of nitrogen into the transition metals can increase the electron density of the d-bands and contract the d-band of the transition metals, resulting in an electronic structure similar to Pt. In addition, transition metal nitrides can be used as efficient electrocatalysts in energy-related processes (Du et al., 2022; Theerthagiria et al., 2020; Chen et al., 2018).

Within the transition, metal nitrides are the nitride of Mo complexes, which have low electrical resistance and high resistance to corrosion. It could be verified that due to the layered nature of this type of materials, 3d metal transitions are presented in its structure and therefore change the electronic states of the Mo on the catalytic surface without harming the catalytic activity of the material (Cao et al., 2013).

Nitrides of bimetallic compounds have been shown to have not only a large surface area and provide a large number of active sites but also to have good electrical conductivity, contributing to reaction kinetics. In addition, the synergistic effect between Co and Mo promotes charge transfer and improves the catalytic activity of the material (Liu et al., 2016; Lu et al., 2021).

3.5.6 Oxides

Oxides as possible materials to be used as electrocatalysts have had a great boom in recent years. In this type of materials, it is important to mention that it is possible to have an oxide as a support (reduced graphene oxide) or as a transition metal (oxides of Fe, Co, and Ni). In the first case, a compound based on reduced graphene oxide and polyoxomolybdato was synthesized, which presented a good performance in the kinetics of the oxidation–reduction reaction. These materials were prepared with low-cost methodologies and effectively presenting a good catalytic activity (Ingavale et al., 2019). In the second case, this group of transition metals has shown positive results since the electrocatalysts synthesized from these metals have high electrocatalytic activity, are low cost, abundant in nature, which makes the oxides of these transition metals an excellent option (Yi et al., 2019).

3.6 Conclusions and Perspectives

In short, the transition to the disuse of noble metals as catalysts in aerated cathodes is slow, as a first step it has tried to use them in a smaller proportion, being only the basis to create new composite materials trying to overcome the intrinsic catalytic efficiency of noble metals by itself alone; however, this has brought with it new problems such as the release of decorative components other than noble metals, such as heavy metals, rare earths, or oxides, which could impair the operation of biocells.

In summary, the transition away from the use of noble metals as catalysts in aerated cathodes is proceeding gradually. As an initial step, efforts have been made to employ them in reduced quantities, being only the basis to create new composite materials trying to overcome the intrinsic catalytic efficiency of noble metals by itself alone; however, this has brought with it new problems such as the release of decorative components other than noble metals, such as heavy metals, rare earths, or oxides, which could impair the operation of biocells.

References

Abdelkader-Fernández, V.K., Fernandes, D.M., Balula, S.S. et al. (2019). Noble-metal-free MOF-74-derived nanocarbons: insights on metal composition and doping effects on the electrocatalytic activity toward oxygen reactions. *ACS Applied Energy Materials* 2: 1854–1867.

Amiripour, F., Ghasemi, S., and Azizi, S.N. (2019). Novel composite based on bimetallic AuNi-embedded nano X zeolite/MWCNT as a superior electrocatalyst for oxygen evolution reaction. *ACS Sustainable Chemistry & Engineering* 7: 19384–19395.

Asefa, T. (2016). Metal-free and noble metal-free heteroatom-doped nanostructured carbons as prospective sustainable electrocatalysts. *Accounts of Chemical Research* 49: 1873–1883.

Bian, J., Li, Z., Li, N., and Sun, C. (2019). Oxygen deficient $LaMn_{0.75}Co_{0.25}O_{3-\delta}$ nanofibers as an efficient electrocatalyst for oxygen evolution reaction and zinc–air batteries. *Inorganic Chemistry* 58: 8208–8214.

Borbath, I., Talas, E., Paszti, Z. et al. (2021). Investigation of Ti-Mo mixed oxide-carbon composite supported Pt electrocatalysts: effect of the type of carbonaceous materials. *Applied Catalysis A: General* 620: 118155.

Cao, B., Veith, G.M., Neuefeind, J.C. et al. (2013). Mixed close-packed cobalt molybdenum nitrides as non-noble metal electrocatalysts for the hydrogen evolution reaction. *Journal of the American Chemical Society* 135: 19186–19192.

Chatterjee, P., Ghangrekar, M.M., and Leech, D. (2018). A brief review on recent advances in air-cathode microbial fuel cells. *Environmental Engineering and Management Journal* 17: 1531–1544.

Chen, Z., Ha, Y., Liu, Y. et al. (2018). In-situ formation of cobalt nitrides/graphitic carbon composites as efficient bifunctional electrocatalyst for overall water splitting. *ACS Applied Materials & Interfaces* 10: 7134–7144.

Chen, Y., Zheng, Y., Yue, X., and Huang, S. (2020). Hydrogen evolution reaction in full pH range on nickel doped tungsten carbide nanocubes as efficient and durable non-precious metal electrocatalysts. *International Journal of Hydrogen Energy* 45: 8695–8702.

Cheng, F. and Chen, J. (2012). Metal–air batteries: from oxygen reduction electrochemistry to cathode catalysts. *Chemical Society Reviews* 41: 2172.

Cheng, S., Liu, H., and Logan, B.E. (2006). Increased performance of single-chamber microbial fuel cells using an improved cathode structure. *Electrochemistry Communications* 8: 489–494.

Díaz-Coello, S., Afonso, M.M., Palenzuela, J.A. et al. (2021). Composite materials from transition metal carbides and ionic liquids as electrocatalyst for hydrogen evolution in alkaline media. *Journal of Electroanalytical Chemistry* 898: 115620.

Dinh, T., Dobo, Z., and Kovacs, H. (2022). Phytomining of noble metals – a review. *Chemosphere* 286: 131805. https://doi.org/10.1016/j.chemosphere.2021.131805.

Du, Q., Gong, Y., Khan, M.A. et al. (2022). Regulating non-precious transition metal nitrides bifunctional electrocatalysts through surface/interface nanoengineering for air-cathodes of Zn–air batteries. *Green Energy & Environment* 7: 16–34.

Duteanu, N., Erable, B., Senthil Kumar, S.M. et al. (2010). Effect of chemically modified Vulcan XC-72R on the performance of air-breathing cathode in a single-chamber microbial fuel cell. *Bioresource Technology* 101: 5250–5255.

Feng, L., Yu, G., Wu, Y. et al. (2015). High-index faceted Ni3S2 nanosheet arrays as highly active and ultrastable electrocatalysts for water splitting. *Journal of the American Chemical Society* 137: 14023–14026.

Freguia, S., Rabaey, K., Yuan, Z., and Keller, J. (2007). Non-catalyzed cathodic oxygen reduction at graphite granules in microbial fuel cells. *Electrochimica Acta* 53: 598–603.

Gao, F., Zhang, Y., Wu, Z. et al. (2021). Universal strategies to multi-dimensional noble-metal-based catalysts for electrocatalysis. *Coordination Chemistry Reviews* 436: 213825.

González Vázquez, O.F., Frausto Reyes, C., Ortiz Morales, M. et al. (2022). Facile scalable manufacture of improved electrodes using structured surface coatings of nickel oxide as cathode and reduced graphene oxide as anode for evaluation in a prototype development on microbial fuel cells. *International Journal of Hydrogen Energy* 47: 30248–30261.

Greeley, J., Stephens, I.E.L., Bondarenko, A.S. et al. (2009). Alloys of platinum and early transition metals as oxygen reduction electrocatalysts. *Nature Chemistry* 1: 552–556.

HaoYu, E., Cheng, S., Scott, K., and Logan, B. (2007). Microbial fuel cell performance with non-Pt cathode catalysts. *Journal of Power Sources* 171: 275–281.

Imani, A.H., Ojani, R., and Raoof, J.B. (2018). In situ synthesis of a novel organic–inorganic composite as a non-noble metal electrocatalyst for the oxygen evolution reaction. *International Journal of Hydrogen Energy* 43: 8267–8277.

Ingavale, S.B., Patil, I., Parse, H. et al. (2019). Polyoxomolybdate anchored graphite oxide: noble metal-free electrocatalyst for oxygen reduction reaction. *International Journal of Hydrogen Energy* 44: 24922–24933.

Islam, J., Kim, S.K., Thien, P.T. et al. (2021). Enhancing the activity and durability of iridium electrocatalyst supported on boron carbide by tuning the chemical state of iridium for oxygen evolution reaction. *Journal of Power Sources* 512: 230506.

Kaur, R., Marwaha, A., Chhabra, V.A. et al. (2020). Recent developments on functional nanomaterial-based electrodes for microbial fuel cells. *Renewable and Sustainable Energy Reviews* 119: 109551. https://doi.org/10.1016/j.rser.2019.109551.

Li, F., Zhang, L., Li, J. et al. (2015). Synthesis of $CueMoS_2$/rGO hybrid as non-noble metal electrocatalysts for the hydrogen evolution reaction. *Journal of Power Sources* 292: 15–22.

Li, Y., Yin, Z., Liu, X. et al. (2021). Current progress of molybdenum carbide-based materials for electrocatalysis: potential electrocatalysts with diverse applications. *Materials Today Chemistry* 19: 100411.

Liu, J., Huang, K., Tang, H.L., and Lei, M. (2016). Porous and single-crystalline-like molybdenum nitride nanobelts as a non-noble electrocatalyst for alkaline fuel cells and electrode materials for supercapacitors. *International Journal of Hydrogen Energy* 41: 996–1001.

Liu, M., Zhao, Z., Duan, X., and Huang, Y. (2019). Nanoscale structure design for high-performance Pt-based ORR catalysts. *Advanced Materials* 31: 1802234. https://doi.org/10.1002/adma.201802234.

Lu, Y., Li, Z., Xu, Y. et al. (2021). Bimetallic Co–Mo nitride nanosheet arrays as high-performance bifunctional electrocatalysts for overall water splitting. *Chemical Engineering Journal* 411: 128433.

Martin, E., Tartakovsky, B., and Savadogo, O. (2011). Cathode materials evaluation in microbial fuel cells: a comparison of carbon, Mn_2O_3, Fe_2O_3 and platinum materials. *Electrochimica Acta* 58: 58–66.

Mir, R.A., Kaur, G., and Pandey, O.P. (2020). Facile process to utilize carbonaceous waste as a carbon source for the synthesis of low cost electrocatalyst for hydrogen production. *International Journal of Hydrogen Energy* 45: 23908–23919.

Mou, Z., Han, M., Li, G. et al. (2013). Pt nanoparticles modified by rare earth oxides: electronic effect and influence to catalytic hydrogenation of 3-phenoxybenzaldehyde. *Materials Research Bulletin* 48: 4780–4784.

Neburchilov, V., Wang, H., Martin, J.J., and Qu, W. (2010). A review on air cathodes for zinc–air fuel cells. *Journal of Power Sources* 195: 1271–1291.

Nie, Y., Li, L., and Wei, Z. (2015). Recent advancements in Pt and Pt-free catalysts for oxygen reduction reaction. *Chemical Society Reviews* 44: 2168–2201.

Phan, Q.T., Poon, K.C., and Sato, H. (2021). A review on amorphous noble-metal-based electrocatalysts for fuel cells: synthesis, characterization, performance, and future perspective. *International Journal of Hydrogen Energy* 46: 14190–14211.

Raut, S.D., Shinde, N.M., Nakate, Y.T. et al. (2021). Coconut-water-mediated carbonaceous electrode: a promising eco-friendly material for bifunctional water splitting application. *ACS Omega* 6: 12623–12630.

Rodriguez, J.R., Félix, R.M., Reynoso, E.A. et al. (2013). Coordination complex synthesis of noble metals in the preparation of nanoparticles supported on MWCNTs used as electrocatalysts. *Inorganica Chimica Acta* 406: 138–145.

Rozendal, R.A., Hamelers, H.V.M., Rabaey, K. et al. (2008). Towards practical implementation of bioelectrochemical wastewater treatment. *Trends in Biotechnology* 26: 450–459.

Saha, S., Rajbongshi, B.M., Ramani, V., and Verma, A. (2021). Titanium carbide: an emerging electrocatalyst for fuel cell and electrolyser. *International Journal of Hydrogen Energy* 46: 12801–12821.

Shi, X., Feng, Y., Wang, X. et al. (2012). Application of nitrogen-doped carbon powders as low-cost and durable cathodic catalyst to air–cathode microbial fuel cells. *Bioresource Technology* 108: 89–93.

Shit, S., Bolar, S., Chandra Murmu, N., and Kuila, T. (2021). An account of the strategies to enhance the water splitting efficiency of noble-metal-free electrocatalysts. *Journal of Energy Chemistry* 59: 160–190.

Strasser, P., Koh, S., Anniyev, T. et al. (2010). Lattice-strain control of the activity in dealloyed core–shell fuel cell catalysts. *Nature Chemistry* 2: 454–460.

Su, W., Yan, P., Wei, X. et al. (2020). Facile one-step synthesis of nitrogen-doped carbon sheets supported tungsten carbide nanoparticles electrocatalyst for hydrogen evolution reaction. *International Journal of Hydrogen Energy* 45: 33430–33439.

Tang, Y.J., Liu, C.H., Huang, W. et al. (2017). A bimetallic carbides-based nanocomposite as superior electrocatalyst for oxygen evolution reaction. *Applied Materials & Interfaces* 9: 16977–16985.

Tang, W., Li, B., Teng, K. et al. (2022). Advanced noble-metal-free bifunctional electrocatalysts for metal–air batteries. *Journal of Materiomics* 8: 454–474.

Theerthagiria, J., Leea, S.J., Murthy, A.P. et al. (2020). Fundamental aspects and recent advances in transition metal nitrides as electrocatalysts for hydrogen evolution reaction: a review. *Current Opinion in Solid State & Materials Science* 24: 100805.

Vij, V., Sultan, S., Harzandi, A.M. et al. (2017). Nickel-based electrocatalysts for energy-related applications: oxygen reduction, oxygen evolution, and hydrogen evolution reactions. *ACS Catalysis* 7: 7196–7225.

Wang, K., Wang, J., and Hu, W. (2015). Evaluation of temperature effect on the corrosion process of 304 stainless steel in high temperature water with electrochemical noise. *Materials and Design* 82: 155–163.

Wang, Y., Zhang, L., Meng, X. et al. (2019). Scalable processing hollow tungsten carbide spherical superstructure as an enhanced electrocatalyst for hydrogen evolution reaction over a wide pH range. *Electrochimica Acta* 319: 775–782.

Wang, W., Liu, C., Zhou, D. et al. (2019). In-situ synthesis of coupled molybdenum carbide and molybdenum nitride as electrocatalyst for hydrogen evolution reaction. *Journal of Alloys and Compounds* 792: 230–239.

Wang, H., Long, X., Zhang, J. et al. (2020). Relationship between bioelectrochemical copper migration, reduction and electricity in a three-chamber microbial fuel cell. *Chemosphere* 241: 125097.

Wei, J., Liang, P., and Huang, X. (2011). Recent progress in electrodes for microbial fuel cells. *Bioresource Technology* 102: 9335–9344.

Yeager, E. (1986). Dioxygen electrocatalysis: mechanisms in relation to catalyst structure. *Journal of Molecular Catalysis* 38: 5–25.

Yi, J., Liu, X., Liang, P. et al. (2019). Non-noble iron group (Fe, Co, Ni)-based oxide electrocatalysts for aqueous zinc–air batteries: recent progress, challenges, and perspectives. *Organometallics* 38: 1186–1199.

Yu, D., Nagelli, E., Du, F., and Dai, L. (2010). Metal-free carbon nanomaterials become more active than metal catalysts and last longer. *The Journal of Physical Chemistry Letters* 1: 2165–2173.

Yu, J., Wu, X., Zhang, H. et al. (2019). Core effect on the performance on N/P codoped carbon encapsulating noble-metal phosphide nanostructures for hydrogen evolution reaction. *ACS Applied Energy Materials* 2: 2645–2653.

Yu, J., Zhang, T., Sun, Y. et al. (2020). Hollow FeP/Fe_3O_4 hybrid nanoparticles on carbon nanotubes as efficient electrocatalysts for the oxygen evolution reaction. *ACS Applied Materials & Interfaces* 12: 12783–12792.

Yu, J., Dai, Y., He, Q. et al. (2021). A mini-review of noble-metal-free electrocatalysts for overall water splitting in non-alkaline electrolytes. *Materials Reports: Energy* 1: 100024.

Yuana, K., Luc, C., Sfaeloud, S. et al. (2019). In situ nanoarchitecturing and active-site engineering toward highly efficient carbonaceous electrocatalysts. *Nano Energy* 59: 207–215.

Zhang, X., Pant, D., Zhang, F. et al. (2014). Long-term performance of chemically and physically modified activated carbons in air cathodes of microbial fuel cells. *ChemElectroChem* 1: 1859–1866.

Zhang, Q., Hu, J., and Lee, D.-J. (2016). Microbial fuel cells as pollutant treatment units: research updates. *Bioresource Technology* 217: 121–128.

Zhang, S., Yang, W., Liang, Y. et al. (2021). Template-free synthesis of non-noble metal single-atom electrocatalyst with N-doped holey carbon matrix for highly efficient oxygen reduction reaction in zinc–air batteries. *Applied Catalysis. B, Environmental* 285: 119780.

Zhang, X., Wang, Y., Liu, C. et al. (2021). Recent advances in non-noble metal electrocatalysts for nitrate reduction. *Chemical Engineering Journal* 403: 126269.

Zhao, L., Wu, R., Wang, J. et al. (2021). Synthesis of noble metal-based intermetallic electrocatalysts by space-confined pyrolysis: recent progress and future perspective. *Journal of Energy Chemistry* 60: 61–74.

Zhu, Y.P., Guo, C., Zheng, Y., and Qiao, S.Z. (2017). Surface and interface engineering of noble-metal-free electrocatalysts for efficient energy conversion processes. *Accounts of Chemical Research* 50: 915–923.

Zhuang, L., Feng, C., Zhou, S. et al. (2010). Comparison of membrane- and cloth-cathode assembly for scalable microbial fuel cells: construction, performance and cost. *Process Biochemistry* 45: 929–934.

4

Configurations of Plant-Based Microbial Fuel Cell System and Its Impact on Power Density

Mohnish M. Borker

Department of Mechanical Engineering, Padre Conceicao College of Engineering, Affiliated to Goa University, Verna, Goa, India

4.1 Introduction

In the current world scenario, we see that the world relies basically on two forms of energy, which are (i) nonrenewable fossil fuels with increasing emissions of greenhouse gasses, (ii) biofuels such as biodiesel which influences the current agricultural practices competing with food production and requiring additional energy input. Bioenergy is a renewable source of energy that has been exploited over the years, but its sustainability is one big question.

The world today needs a source of energy that is easily available, cheaper, and possesses minimum social and environmental damage. An idea was put forward by a group of researchers at Wageningen University, The Netherlands, in the year 2007. This approach focuses on generating in-situ bioelectricity using living plants without affecting the plant's ecology in a plant microbial fuel cell (PMFC). Microbe-based fuel cells operate as biocatalysts and generate bioelectricity using microbes in the lower soil. Thus, biochemical energy is the plant exudates is converted into a useful form of energy like electricity. Many microbes have the ability to breakdown the organic waste matter into electrons and protons. These microbes are called electrochemical bacteria (ECBs) due to their tendency to breakdown the organic waste matter into electrons and protons. These bacteria find themselves in

Photosynthesis-Assisted Energy Generation: From Fundamentals to Lab Scale and In-Field Applications, First Edition. Edited by Sathish-Kumar Kamaraj and Iryna Rusyn.

abundance in soil, wastewater, and also in activated sludge or waste slurry (Strik et al., 2008). Plant MFCs are the enhanced version of a MFC which feeds these ECBs with root exudates. Thus, the PMFCs convert solar energy into electricity by the anaerobic degradation of organic matter produced by photosynthesis.

Plants produce organic matter in the presence of sunlight and CO_2 by photosynthesis. Of this organic matter 60% is lost by the plant as exudates in the lower soil on which the ECBs feed. Nearly 20–30% of the organic matter ($C_6H_{12}O_6$) generated during the process of photosynthesis is converted into exudates known as rhizodeposites (Rabaey & Verstraete, 2005). Anaerobic degradation of these rhizodeposits and other biomass in the soil is carried out by the ECBs. They break down the carbohydrates and sugars to release electrons essential for the bioelectricity generation (De Schamphelaire et al., 2008). ECBs like *cyanophyta* and *chlorophyaetic* form a layer around the anode surface called as the biofilm (Elmy, 2012). PMFCs are broadly classified into two different configurations based on the anode and cathode chamber: (i) single chamber PMFC – it contains the anode and cathode maintained in a single chamber, (ii) dual chamber PMFC – it contains separate chambers for the anode and cathode chamber divided by a Proton Exchange Membrane (Nafion) (Park & Zeikus, 2003).

4.2 Operating Principle

Plants generate a decent quantity of organic matter like carbohydrates ($C_6H_{12}O_6$) during photosynthesis. About 50–60% of the fixed carbon from the leaves of the plant is directed to the roots, depending on the type of plant selected, the potting media, and the ambient conditions (Lynch & Whipps, 1990). Electrochemically active microbes located around the roots in the lower soil separate this organic waste, releasing protons, electrons, and small quantities of CO_2.

Electricity is generated when the electrons are channeled through an external circuit by means of an electron acceptor like that of an electrode which is placed close to these bacteria in the anode chamber (Timberlake, 2009). The anode region is made suitable for the growth of the plant while the cathode region consists of water. Figure 4.1 depicts the operation of a simple PMFC.

Following are the chemical reactions taking place in both, the anode and the cathode chamber.

Anode:

$$2C_6H_{12}O_6 \rightarrow 2C_6H_{10}O_6 + 4H^+ + 4e^- \tag{4.1}$$

These electrons integrate with the protons that have permeated across the membrane in the cathode chamber.

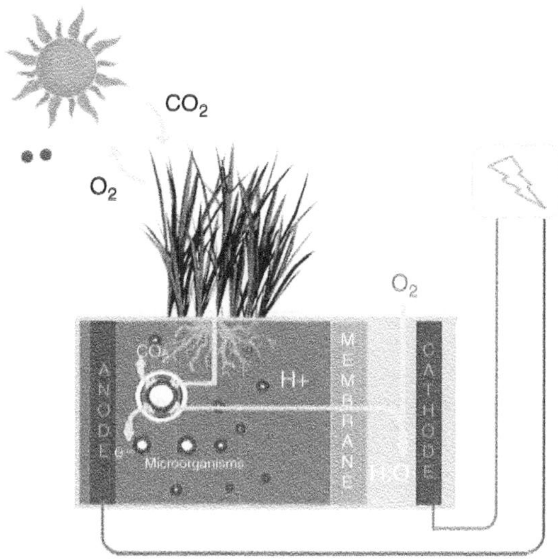

Figure 4.1 Basic operation of a plant microbial fuel cell (Borker et al., 2017).

Cathode:

$$O_2 + 4H^+ + 4e^- \rightarrow 2H_2O \tag{4.2}$$

Net Reaction:

$$2C_6H_{12}O_6 + O_2 \rightarrow 2C_6H_{10}O_6 + 2H_2O \tag{4.3}$$

The power density for a PMFC is less ($\simeq$10 W/m^2) as compared to wind ($\simeq$8 MW/m^2) and solar ($\simeq$6 MW/m^2); however, it is inexpensive when compared to solar and wind. Also, the environmental impact of both wind energy converters and solar power plants is considerable and is a source of social debate (McGowan & Connors, 2000).

4.3 PMFC Configurations

As discussed earlier, PMFCs are broadly classified into single-chamber and dual-chamber systems. The classification is solely done based on the separation of the anode and cathode chamber as can be seen in Figure 4.2.

Besides the general classification, there are various other configurations of the dual chamber PMFCs that have been tested with different electrode materials and

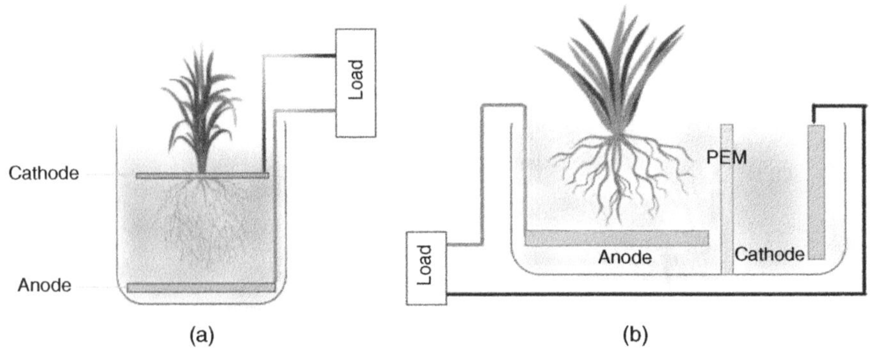

Figure 4.2 PMFC classification (a) sediment type single chamber PMFC, (b) dual chamber PMFC.

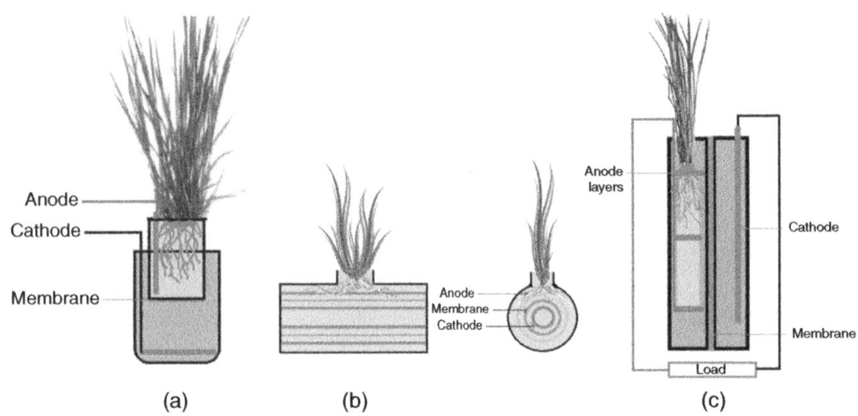

Figure 4.3 Plant microbial fuel cell configurations. (a) Cylindrical PMFC, (b) tubular PMFC, and (c) flat plate PMFC.

plant species as seen in Figure 4.3 and the power densities shown in Table 4.1 (Borker et al., 2017).

The system's performance can be determined by measuring the open circuit voltage and the short circuit current. Polarization curves are obtained to determine the maximum power density using various resistance values. In plant-MFCs, the cell potential is a linear function of the maximum theoretical cell potential, current density, and the internal resistance (Helder et al., 2012).

The three systems of sediment-type plant MFCs can be tested for various performance parameters. The basic performance parameters required to evaluate any MFC are the open circuit voltage, the short circuit current, the current density,

Table 4.1 PMFC configurations.

Configuration	Anode	Cathode	Power density (W/m^2)
Cylindrical PMFC	Graphite rod and granules	Graphite felt	$22\,W/m^2$ (Logan et al., 2006)
Tubular PMFC	Carbon cloth	Carbon cloth	$60\,W/m^2$ (Timmers et al., 2013)
Flat plate PMFC	Graphite sheet	Graphite sheet	$679\,W/m^2$ (Helder et al., 2012)

Figure 4.4 Sediment-type PMFC system with (a) *Setaria faberi* (fox tail), (b) *Cymbopogo citratus* (lemon grass), and (c) *Eleusine indica* (goose grass) species.

internal resistance, and the power density for various values of external resistances. All three systems have to be evaluated for the same time period.

4.3.1 Open Circuit Voltage (V_{oc})

Figure 4.4 depicts sediment type PMFC systems with different plant species. The open circuit voltage gives the maximum potential developed by the system with no load connected to the system. The three systems are measured for their average open circuit voltage (V_{oc}) values for 100 days. Figure 4.5 gives the comparison in the V_{oc} values of the three systems.

4.3.2 Polarization Curves

Polarization curves are obtained for individual sediment PMFCs by connecting external resistance values of 100, 500, 1000, 1500, and 2000 Ω. Polarization curves are plotted to understand the I–V relationship of the fuel cell. It also enables to

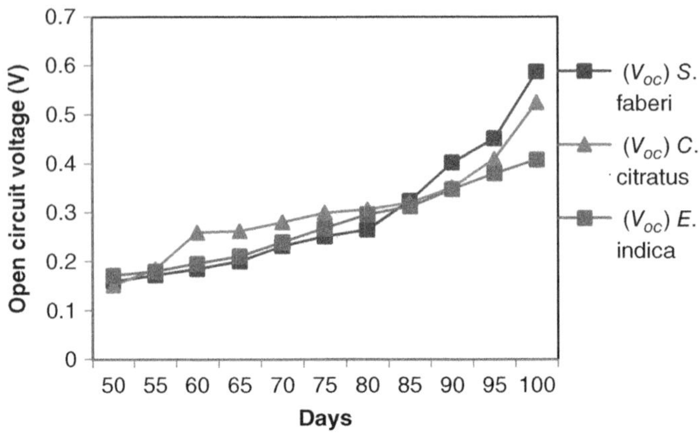

Figure 4.5 Open circuit voltage variation for *S. faberi*, *C. citratus*, and *E. indica* (upto 100 days).

enhance the maximum power density of the system by evaluating the internal resistance of the system. The internal resistance is equivalent to the external resistance corresponding to the maximum power density value. Figure 4.6a–c give the polarization curves of the three systems.

Sediment-type plant MFCs are tested with three different plant species. The plants are tested for their compatibility in the system and their ability to adapt to the PMFC system. The systems are evaluated for their power density values and their internal resistances are compared. It was observed that *Elusine indica* (goose grass) species provided the highest maximum power density of 53.64 mW/m^2 for an internal resistance of 500 Ω, compared to 4.6 mW/m^2 maximum power density at 1500 Ω of *Setaria faberi* (fox tail) and 3.274 mW/m^2 maximum power density at 2000 Ω of *Cymbopogo citratus* (lemon grass). The performance of a sediment PMFC is always hindered by the absence of separate cathode and anode chambers using a proton exchange membrane. This results in vast contamination of the electrons and protons and no proper channeling of the electrons in the external circuitry. Hence, it is recommended to test the performance parameters in a dual-chambered system for better results.

4.4 Cylindrical PMFC

A cylindrical PMFC set up was tested with *Arundinella anomala* grass species having graphite rod + granules as the anode in the anode section and graphite felt

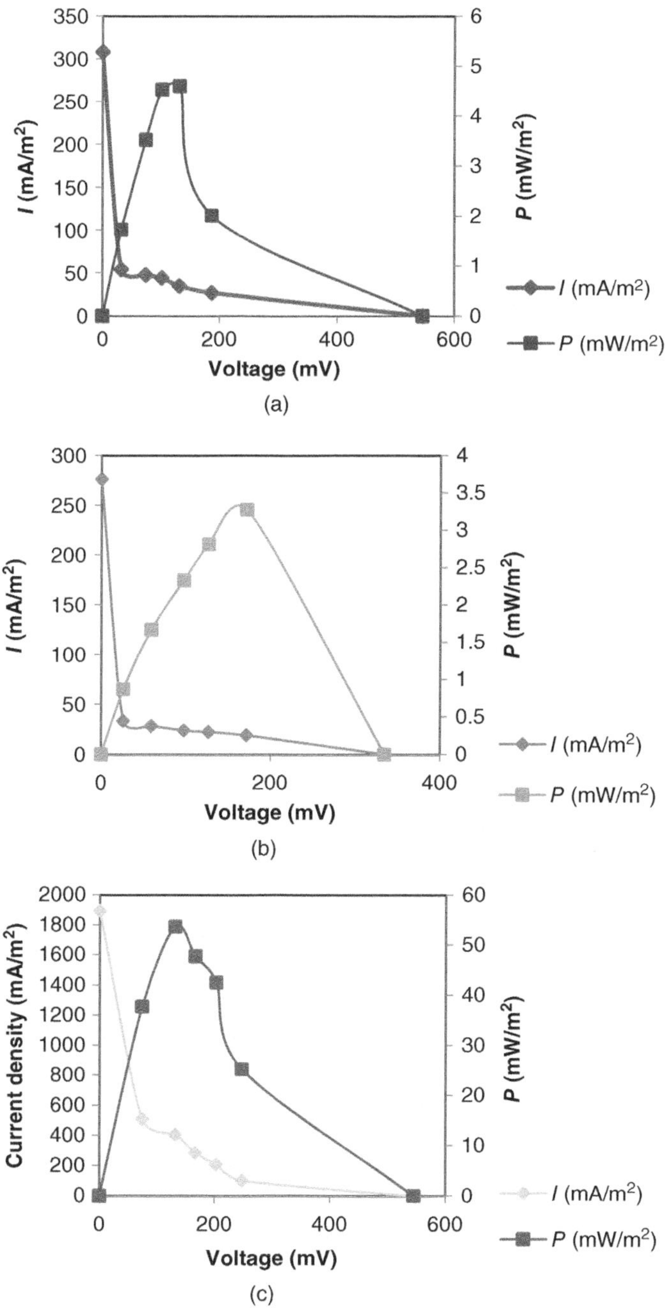

Figure 4.6 (a) The polarization curve for *S. faberi* (Fox tail) grass species. (b) The polarization curve for *C. citratus* (Lemon grass) grass species. (c) The polarization curve for *E. indica* (goose grass) grass species.

Figure 4.7 Cylindrical PMFC.

as the cathode in the cathode section. The system consisted of a cylindrical anode chamber placed inside a cylindrical cathode chamber, as can be seen in Figure 4.7.

The anode chamber is placed at a height of 70 mm from the base of the cathode chamber. The two chambers are separated by sealing the open anode base with a proton exchange membrane (PEM) (Nafion 117A) membrane. Coir pith or Cocopeat is used as the potting media for the plant in the anode chamber. The cathode chamber is filled with water and a graphite sheet is placed at the base as cathode.

The system was evaluated for its performance in terms of voltage, current density, and power density. Polarization curve gives the *I–V* relation for the system

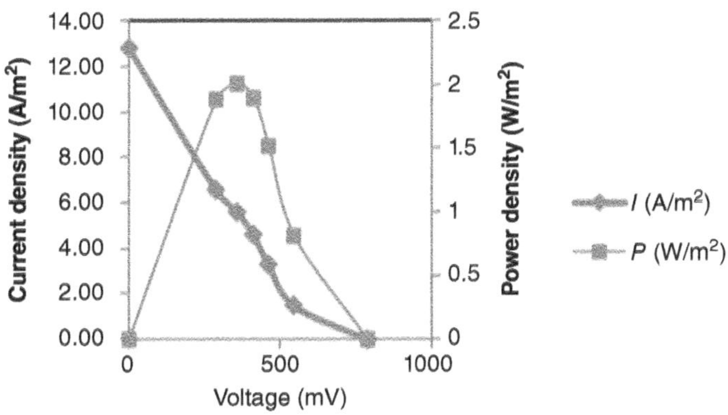

Figure 4.8 Polarization curve for cylindrical PMFC.

and indicates the current density per unit surface area of the anode versus the voltage. From the polarization curve, one can determine the internal resistance of the system. The internal resistance (which is to be reduced) is identified as the external resistance corresponding to the maximum power density Figure 4.8 depicts the polarization curves of the cylindrical PMFC which attained a power density of 2.01 A/m^2 anode GA at a low resistance of 500 Ω, which is the internal resistance of the system.

4.5 Conclusion

This chapter discusses the PMFC systems with reference to their working principles and the upcoming research field. It was identified as a sustainable source of bioenergy that does not compete with arable land for the purpose of power. Also, since there is no damage to the ecosystem in the power generation, the PMFC system is called as an in-situ power generation system. The different configurations are depicted. Most of the research work is done on dual-chamber PMFCs and one such example studied here is a cylindrical PMFC system. Its performance parameters are evaluated with the help of polarization curves which represent a power density of 2.01 A/m^2 anode GA for 500 Ω load resistance.

PMFC system is currently in its research phase and needs intensive study to compete with other renewable sources. The applications are not limited as they already have good advantages in comparison to solar and wind energy. Rooftop PMFC systems are also being studied for an additional benefit of maintaining the temperature of the buildings while providing sustainable green electricity.

References

Borker, M., Suchithra, T.V., Srinivas, M., and Jayaraj, S. (2017). Sustainable bioelectricity generation from living plants. In: *Microbial Biotechnology*. Springer.

De Schamphelaire, L., Bossche, L.V.D., Dang, H.S. et al. (2008). Microbial fuel cells generating electricity from Rhizodeposits of rice plants. *Environmental Science & Technology* 42: 3053.

Elmy, E. (2012). Generating electricity by nature's way. *Salt 'B' Online Magazine* 1.

Helder, M., Strik, D.P., Hamelers, H.V. et al. (2012). Year-round performance of the flat-plate plant-microbial fuel cell. *Bioresource Technology* 104: 417–423.

Logan, B.E., Hamelers, H.W.M., Rozendal, R., and Schröder, U. (2006). Microbial fuel cells: methodology and technology. *Environmental Science and Technology* 40: 181–5192.

Lynch, J.M. and Whipps, J.M. (1990). Substrate flow in the rhizosphere. *Plant and Soil* 129: 1–10.

McGowan, J.G. and Connors, S.R. (2000). Wind power: a turn of the century review. *Annual Review of Energy and the Environment* 25: 147–197.

Park, D.H. and Zeikus, J.G. (2003). Improved fuel cell and electrode designs for producing electricity from microbial degradation. *Biotechnology and Bioengineering* 81: 348.

Rabaey, K. and Verstraete, W. (2005). Microbial fuel cells: novel biotechnology for energy generation. *Trends in Biotechnology* 23: 291.

Strik, D.P.B.T.B., Hamelers, H.V.M., Snel, J.F.H., and Buisman, C.J.N. (2008). Green electricity production with living plants and bacteria in a fuel cell. *International Journal of Energy Research* 32: 870.

Timberlake, K.C. (2009). *Basic Chemistry*, 3e. Prentice Hall.

Timmers, R., Strik, D., and Hamelers, H. (2013). Electricity generation by a novel design tubular plant microbial fuel cell. *Journal of Biomass and Bioenergy* 51: 60–67.

5

The Critical Impact of Photosynthetic Pathway of Plants on the Performance of PMFC

Julio C. Gómora-Hernández[1], Nicolas Flores-Álamo[2], L.A. Díaz-Colín[1], S. Ventura-Cruz[1], and Miriam J. Jiménez-Cedillo[1]

[1] *Division of Environmental Engineering, Technological of Higher Studies of Tianguistenco, Tianguistenco, México*
[2] *Department of Chemical and Biochemical Engineering, Technological Institute of Toluca, National Technological of México, State of México, México*

To our families and friends.
To all people interested in renewable energy, plant microbial fuel cells and science.

5.1 Introduction

Environmental pollution affects soil, water, and air, this is a major problem faced by society, since it diminishes human comfort and promotes several diseases and even death. Some of the most important contributors to environmental damage are greenhouse gasses derived from fossil fuels such as NOx, SOx, and CO_2, being this last one the most abundant atmospheric pollutant which acts as either a precursor for cardiovascular and respiratory diseases or a major contributor to global warming, urban smog, acid rain, and among others (Li et al., 2013). In addition, the cost of fossil fuels has increasing in the last years due to the depletion of petroleum reserves worldwide, which contributes to the negative affectation of the global energy security (Martins et al., 2019). The aforementioned highlights the

Photosynthesis-Assisted Energy Generation: From Fundamentals to Lab Scale and In-Field Applications, First Edition. Edited by Sathish-Kumar Kamaraj and Iryna Rusyn.

need of new, efficient, environmental friendly and sustainable energy sources that can be operated continuously in a long term and everywhere, and allow us to power up common electronic devices. Bioelectrochemical systems (BES) are electrochemical complex devices that use microorganism's activity to produce bioenergy and chemicals from several organic sources by using either galvanic or electrolysis cells. BES are novel technology capable to transform wastes into added-value products as a part of a circular economy and zero waste generation processes (Jung et al., 2020). There are different kinds of BES such as microbial fuel cell (MFC) (He et al., 2017), sediment microbial fuel cell (SMFC) (Yang & Chen, 2021), algal microbial fuel cell (AMFC) (Arun et al., 2020), microbial electrolysis cell (MEC) (Katuri et al., 2019), microbial desalination cell (MDC) (Al-Mamun et al., 2018), plant microbial fuel cell (PMFC) (Prasad & Kalla, 2021), constructed wetland microbial fuel cell (CW-MFC) (Saeed et al., 2022), photo-bioelectrochemical fuel cell (PBFC) (Sun et al., 2019), and bio-electro-Fenton microbial fuel cell (BEF-MFC) (Sathe et al., 2022), these systems serve as bifunctional technologies to produce simultaneously green energy and to treat organic wastes.

PMFC is a type of BES able to produce sustainable energy from complex carbohydrates and exudates present in plant roots which were naturally produced during photosynthetic process. In rhizosphere, a symbiosis between plant and electrochemically active microorganisms is carried out transforming these complex carbohydrates into CO_2, protons, and electrons, these last ones are delivered into a final electron acceptor commonly known as anode and transferred to the cathode by an external electrical circuit (Kabutey et al., 2019; Regmi et al., 2018). Carbon-based materials are usually employed as electrodes in PMFCs, since they are biodegradable, biocompatible, and non-contaminant, do not affect the vitality, performance, and life cycle of the plants, and promote interfacial colonization of the microbes and the subsequent extracellular transfer of electrons, however, the main disadvantage of these materials is the low energetic yield attributed to the high electric resistance and low potential (Li et al., 2017).

Currently, several plants have been evaluated for energy production by PMFC including terrestrial, aquatic, wetland, and even drought-resistant plants. The power generation yield is affected by different factors that can be categorized into two main groups: biotic or abiotic elements. For abiotic some factors such as electrode materials, radiation intensity, soil or water properties, and conformation of the fuel cell are usually explored. Biotic elements are related to microorganisms, plants, and the natural symbiosis between them. It is well known that different plants produce different amount of voltage and electric current, and the most efficient plants are related to high rhizomes density on rhizosphere zone (Nitisoravut & Regmi, 2017; Strik et al., 2008). Rhizodeposition process is directly related to photosynthetic pathway.

PMFC technology has become an interesting proposal to produce green and renewable energy employing an abundant natural source. From 2008; time at which the first PMFC was constructed to nowadays, several works have been made focusing on design, novel configurations, materials technology, bifunctional process to treat organic streams, optimal operational conditions, long-term performing, and real applications, however, despite these efforts more research studies are still needed to better understand the phenomena and to use plants energy to power up common electronic devices.

This chapter comprises a brief history of PMFC development, its conventional configurations, the materials commonly employed as electrodes, and the basic elements. As biotic elements are important factors, microorganism's community and plant's metabolisms are also described and related to power performance. The enormous diversity of plants that have been used to produce energy by PMFC is categorized according to their photosynthetic pathway in Carbon 3 (C3), Carbon 4 (C4), and Crassulacean Acid Metabolism (CAM) and is described in depth. Extreme weather plants like those that grow in wetlands and in water scarcity environments are briefly summarized. Finally, the current challenges and future trends of PMFC technology are also discussed.

5.2 Brief History of PMFC

The earliest researches on the electrical effects observed in organic compounds decomposition and during plant metabolisms process (respiration and carbon assimilation) date back to the work of Potter in 1911 (Potter, 1911), who refer to "the signs of life" of Waller, in whose work he mentions the realization of the mechanical excitability of vegetables protoplasm are accompanied by electrical effects (Waller, 1903).

It is all based on the exciting ability of plants to capture and convert CO_2 from the environment into energy, a process that also produces breathable air through photosynthesis. Hamelers from Wageningen University in the Netherlands explains that electricity produced by living plants occurs by using the natural process in which the plants produce organic matter via photosynthesis, part of this matter is used for the plant growth and the rest is secreted through the roots where the microorganism feed from it, and this process releases electrons that can be captured by an electrode using an electric circuit (Hamelers et al., 2010).

BES comprise MFC, a novel technology that uses microorganisms attached to one electrode known as an anode and supports the functioning of PMFC, which is a new technology that uses living plants and microorganisms to generate electricity in a sustainable way (Rabaey et al., 2006).

As described before, bioelectricity was discovered by Potter in 1911 (Potter, 1911). However, it was not until 1960s when NASA showed great interest in transforming organic wastes into electricity on space missions. Years after, in 1980s electron transfer mechanisms were the focus of several research studies. Interest in bioelectricity and MFC grew exponentially in the early years of the 21st century, and since then thousands of papers have been reported including mechanisms, systems optimization, bacteria involved in electron transfer, organic waste transformation, and secondary BES derived from conventional MFC (Santoro et al., 2017).

PMFC is a BES derived from MFC. The term PMFC was proposed for the first time by Strik et al. (Strik et al., 2008), who adapted reed sweet-grass into MFC. Plant roots serving as a continuous carbon source for microbe's activity is the main idea of this fuel cell. After that, aquatic plants were considered in simulated wetlands in which not only green energy production at high yields was achieved but also wastewater treatment was efficiently proven. So, constructed wetland PMFC appeared for the first time in 2010 (Yaday, 2010). The use of desert plants on PMFC began to be studied in recent years. Tapia et al. (Tapia et al., 2017) was pioneer in using CAM plants in BES to recover energy from arid regions. Since PMFC has evolved exponentially since its appearance, enormous advances have been reached on topics such as electrode materials, microbial community, microbes–plant interaction, electronic devices, BES configurations, mechanisms, and application at large scale; however, until now this technology needs optimized to power up useful electronics.

5.3 Conformation of Conventional PMFC, Electrode Materials, and Basic Elements

The PMFC consists of a living plant with the capacity to carry out the process of photosynthesis through which the plant produces organic matter from CO_2 and solar energy, captured from the environment. The plant is placed in a test cell made of acrylic, through which the growth of the roots can be observed, where an exchange of electrons takes place due to the oxidation of organic matter by the interaction of the bacteria that live around or in the rhizosphere. As long as the plant keeps growing with its roots in the PMFC, electricity is continuously generated, to capture the energy, a cathode and an anode must be installed. A conventional PMFC must contain the following elements: an anode installed at the roots of the plants, air-cathode, soil-coupled to organic substrates (water for wetland PMFC), living plant, microbial consortia, metal wires to collect electrons, and electrical circuit. Sometimes, PMFC operates at open circuit voltage (Figure 5.1, above), where data acquisition system is directly intertwined with electrodes. When electrodes are intertwined between them and resistors are applied, energy as either current or power is obtained (Figure 5.1, below).

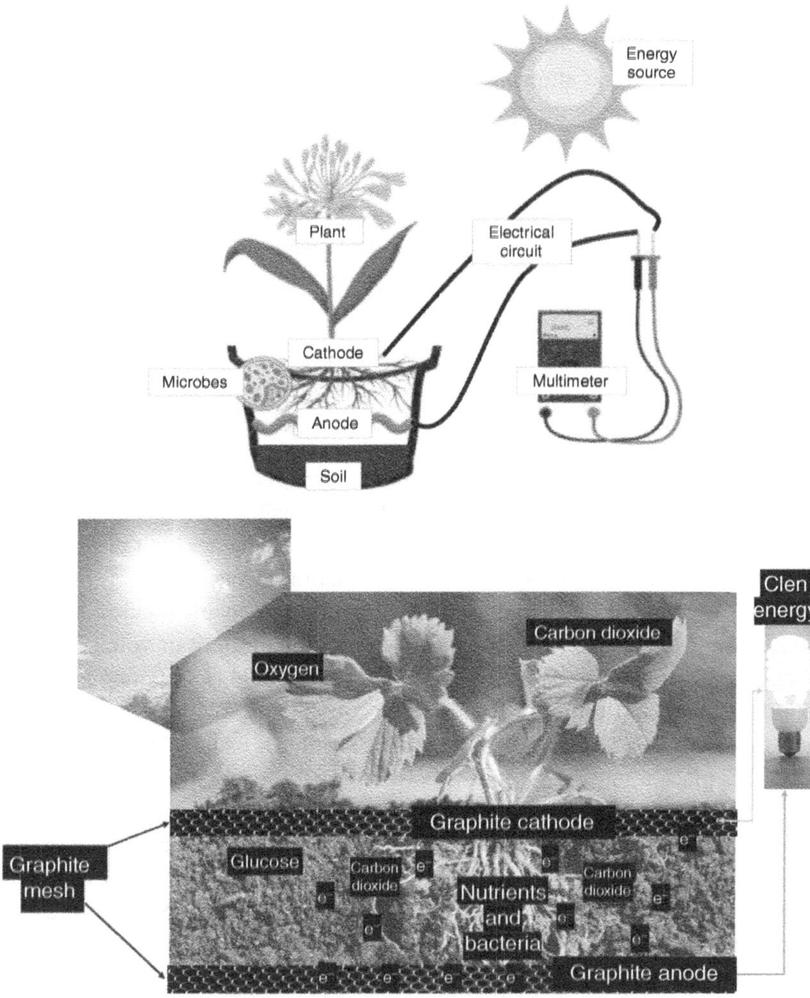

Figure 5.1 Schematic representation of conventional PMFC.

To avoid energy production losses, it is recommended that PMFC be flooded (sediments or sludge) to reduce oxygen intrusion, which is present during the electron exchange (Helder et al., 2010). Soil water logging reduces the exchange of gases from soil and the atmosphere, a good example of a plant growing in this condition is Brachiaria spp. used as forage in tropical America (Cardoso et al., 2013).

The anode commonly used is made of graphite mesh, which has shown good results in energy production, generating power densities in the range of 0.02 and 9.1 mW/m^2. Several different materials have been used as anodes, including graphite blocks, graphite fibers, carbon cloth, carbon paper, and carbon foams (Pant et al., 2011; Apollon et al., 2021). The use of carbon based electrodes containing platinum as catalyst has shown high power output potential in PMFC systems.

In addition to the anode and cathode, it is necessary to apply substrates to improve the growth of the plant, commonly used substrates are organic compounds such as amino acids, organic acids, and sugars (Hamelers et al., 2010). Biotic elements like bacteria and electrochemically active bacteria (EAB's) and abiotic factors such as plant nutrients and gases (carbon dioxide and oxygen) in the surrounding air can act as electron acceptors and diminishing the current generation in PMFC's (Timmers et al., 2012).

5.4 Bacterial Community

In nature, there are wide range of native microorganisms reaching around 10^9 to 10^{12} bacteria per gram of soil; some of them are EAB responsible for transforming root exudates into free electrons (Narasimhan et al., 2003). The generation of free electrons during bacteria metabolisms depends on rhizome density, composition and concentration of root exudates, environmental conditions, plant species, bacteria community and exoelectrogenic mechanism, hence, the amount of voltage generated vary from plant to plant, i.e. different power yields could be obtained from the same kind of plant and different bacteria consortia (Nitisoravut & Regmi, 2017; Maddalwar et al., 2021). The plant root's surrounding area is usually known as rhizosphere; in this place a natural symbiosis between microorganisms and plant takes place. Naturally, microorganisms are adhered to plant rhizomes and use root exudates as raw material for their metabolic activities producing free electrons and other by-products such as humic acids, carbon dioxide, and protons and reducing the complex nutrients into ionic compounds that are amenable to the plants (Strik et al., 2008; Chiranjeevi et al., 2019). Bacteria able to produce free electrons during their natural metabolisms are known as EAB, which are classified as electricigens, anodophiles, and exoelectrogens based on their electron transfer mechanisms. Review articles focused on the diversity of EABs acting on either PMFC or conventional MFC systems have been previously reported (Logan & Regan, 2006; Rusyn, 2021).

Some EAB isolated from natural environments involves the following: α, β, γ, and δ-*Proteobacteria*, *Firmicutes*, *Bacteroidetes*, *acidobacteria*, and *actinobacteria* (Nitisoravut & Regmi, 2017; Kadier et al., 2016; Shaikh et al., 2021). In agreement with Zhi et al. (Zhi et al., 2014), around 50 bacteria species belonging *Proteobacteria*, *Firmicutes*, and *Acidobacteria* have been isolated and identified as exoelectrogenic microorganisms. These bacteria were isolated from wastewater, sediments of water bodies (rivers and lakes), fermented sludge, and even soil. The most important microorganisms able to produce electricity by PMFC are the following: *Geobacter spp. Bacillus, Geothrix, Shewanella, Desulfobulbus spp. Pseudomonas, Acidoba*,

Ruminococcaceae, *Natronocella acetinitrilica*, and *Rhizobiales* (Apollon et al., 2021; Maddalwar et al., 2021), most of them have the capability to produce microbial biofilms around electrode and promote direct electron transfer to the anode. The formation of bacterial biofilms around electrodes depends on several factors such as soil properties, temperature, humidity, pH, dissolved oxygen, amount and composition of root exudates, as well as microbial consortia, addition of mineral salts, and organic matter supplements (Nitisoravut & Regmi, 2017; El Gamal et al., 2018).

Microbial communities differ from PMFC to PMFC which is caused by the different composition and type of root exudates, thus different carbohydrates promote different microbial consortia around the roots. During PMFC operation in rice paddy field supplemented with acetate, EAB were isolated and identified as *Natronocella acetinitrilica*, *Beijerinckiaceae*, *Rhizobiales*, and *Rhodobacter gluconicum*. These bacteria enhanced the electricity production in the acetate environment (Kaku et al., 2008). A subsequent work made by Cabezas et al. (Cabezas et al., 2015) on rice paddy field showed the influence of electrical circuit arrangement on bacterial community, since during power generation in a PMFC the most common bacteria were related to clones of *δ-Proteobacteria*, *Geobacter*, *Anaerolineae*, *Anaeromyxobacter*, and *Chloroflexi*. However, microbial community was different when PMFC operated in an open circuit voltage being α, β, and δ-*Proteobacteria* and *Firmicutes* the most abundant microorganisms. On the other hand, two different plants were evaluated in a wetland PMFC (*Pennisetum alopecuroides* and *Phragmites communis*) to produce green energy and simultaneous removal of hexavalent Chromium (Cr VI). The bacterial community predominant in anode surface was composed of α, δ and *ϒ-Proteobacteria*, *Firmicutes*, *Bacteroidetes*, *Patescibacteria*, *Acidobacteria*, *Euryarchaeota*, and *Cyanobacteria* (Guan et al., 2019). The bacterial diversity was studied by 16S rDNA technique during PMFC operation with two essential food plants: *Amaranthus viridis* and *Triticum aestivum*, in a long-term process. In this work, there are two main representative genera; phylum *Firmicutes* and phylum *Proteobacteria*, from which *Firmicutes* was the dominant for both plant species. The most abundant bacteria in anodic regions were *Bacillus* and *Pseudomonas*, while the predominant in the plant-mediated PMFC were *Bacillus safensis*, *Bacillus licheniformis*, *Pseudomonas aeruginosa*, and *Bacillus pumilus* (Arulmani et al., 2021). Other studies have been performed evaluating the addition of pure cultures on plant soil. *Bacillus thuringiensis* bacteria were inoculated on *Echinodorus cordifolius* roots. The symbiosis between bacteria and plant coupled to a serial electrical circuit increased power density during 160 days operation time (Treesubsuntorn & Thiravetyan, 2021).

In the literature, there are several papers focusing on EAB metabolisms, diversity, identification, mechanisms, and the optimal conditions at which they improve electrons transfer for further applications in BES (Logan et al., 2019; Xiao & Yu, 2020; Hirose et al., 2019; Hu et al., 2022; Cao et al., 2019).

5.5 Rhizodeposition Process and Photosynthetic Pathways

According to their carbon fixation mechanisms, higher plants are classified as C3, C4, and CAM. The first metabolic route acquired by plants corresponds to C3 path, which is a catalytic process that begins by incorporating atmospheric carbon dioxide inside plant cells to produce glyceraldehyde 3-phosphate (G3P), which is subsequently used to form saccharide molecules such as glucose and sucrose (Cardemil & Squeo, 2007).

C3 metabolism involves the dark phase of photosynthesis, also called Calvin–Benson cycle, it depends on the products obtained during light phase. In diurnal phase, solar energy is absorbed, fixed, and then converted into chemical potential (Bidwell, 1979). Thylakoid membrane performs this process by a series of electron transfer reactions in which organized chlorophyll participates in photosystem I and photosystem II, naturally designed to capture electromagnetic spectra at a certain wavelength (Reece et al., 2014).

Photolysis of water occurs in photosystems, this process delivers oxygen to the atmosphere and generates protons that are used in photosynthetic phosphorylation to form molecules of NADPH coenzyme from $NADP^+$ and ATP molecules from ADP. At the end of the light stage, both ATP and NADPH are found in chloroplast stromas to be used either on dark- or light-independent phase (Figure 5.2). According to Reece et al. (Reece et al., 2014), the Calvin–Benson cycle comprises three stages:

Figure 5.2 C3 photosynthetic pathway and common plants (Yamori et al., 2014). Source: Figure modified.

1. Carboxylation of six units of carbon dioxide with six units of ribulose 1,5-diphosphate (RuBP) to form twelve molecules of 3-phosphoglycerate (3-PGA), this reaction is catalyzed by the enzyme ribulose 1,5-diphosphate carboxylase-oxygenase also known as RuBisCO.
2. Reduction of 3-PGA molecules to twelve molecules of glyceraldehyde 3-phosphate (G3P) by using 3-phosphoglycerate kinase and glyceraldehyde 3-phosphate dehydrogenase enzymes.
3. Regeneration of six RuBP units by using ten of the twelve G3P molecules. The remaining two molecules are stored as starch or exported as sucrose to the rest of the plant.

Due to their structural affinity, RuBisCO enzyme can catalyze oxygen and carbon dioxide. The oxygenase pathway occurs as a function of the relative concentration of both gases. There is evidence that this mechanism operates on water-stress situations and when the organism is exposed to light. However, this mechanism also called photorespiration is between 20% and 30% less efficient for carbon fixation than the pentose cycle (Cardemil & Squeo, 2007; Lara et al., 2010).

Photosynthetic organisms increased oxygen concentration on primitive atmosphere; moreover, adaptive efforts were made by different plants particularly drought resistant plants to limit the negative effects of photorespiration improving photosynthetic efficiency. Derived from these adaptations, some groups of plants successfully developed the photosynthetic metabolisms C4 and CAM (Hohmann-Marriorr & Blankenship, 2011).

For C4, stomata must be open to capture CO_2, this fact represents a risk that the plant loses water during this process, since diffusion of water is faster than carbon dioxide. The C4 pathway increases photosynthetic efficiency and minimizes enzymatic evapotranspiration, consequently, C4 plants lack photorespiration (Lara et al., 2010; Salisbury & Ross, 1991).

The acquired adaptation of C4 plants consists of the anatomical separation of photosynthetic processes in response to biochemical specialization. Mesophyll cells (MC) lack the RuBisCO enzyme, so carbon fixation begins with the cytoplasmic enzyme phosphoenolpyruvate-carboxylase, which fixes carbon dioxide to a molecule of phosphoenolpyruvate (3C) causing the synthesis of a molecule of oxaloacetate (4C), this molecule gives its name to C4 metabolism (Figure 5.3) (Raya-Pérez & Aguirre-Mancilla, 2008).

In chloroplast stroma of MC, oxaloacetate is reduced by the enzymatic action of $NADPH^+$-dependent malate dehydrogenase to malate or malic acid. Malate passes from the MC to the vascular sheath cells (VC) through plasmodesmata formed

Figure 5.3 C4 photosynthetic pathway and common plants (Yamori et al., 2014). Source: Figure modified.

between the cell wall of both cell types. Within the VC, malate will be converted by malate dehydrogenase to carbon dioxide and pyruvate. Then, pyruvate is catalyzed to phosphoenolpyruvate by pyruvate orthophosphate dikinase and returned to MC.

The synthesized carbon dioxide is used within the VC by the original Calvin–Benson cycle for carbohydrate production. This metabolic process ensures that the concentration of carbon dioxide within the sheath cells be three to four times higher than in cells with C3 cycle (Reece et al., 2014; Raya-Pérez & Aguirre-Mancilla, 2008).

Finally, CAM plants adapted to a temporal separation of activities. Due to extreme climatic conditions, CAM plants keep their stomata open at night avoiding water loss by evapotranspiration caused by the high temperatures during the day (Solomon et al., 2013).

In the first stage, carbon dioxide is captured overnight and fixed by phosphoenolpyruvate carboxylase to phosphoenolpyruvate to produce oxaloacetate molecules, then it is catalyzed by $NADH^+$-dependent malate dehydrogenase to form malic acid which is stored in the vacuole cell (Reece et al., 2014).

During the day, cytosolic alkalinization inactivates the proton pump promoting that malate begins to be released passively and travels to the chloroplasts stroma. In the presence of decarboxylating enzymes, malate is fragmented into CO_2 and pyruvate, this last one is recycled to form phosphoenolpyruvate and then returned to the previous stage, while carbon dioxide is fixed by Calvin–Benson cycle. A resume of photosynthetic CAM mechanism as well as some common plants is shown in Figure 5.4 (Cardemil & Squeo, 2007; Lara et al., 2010).

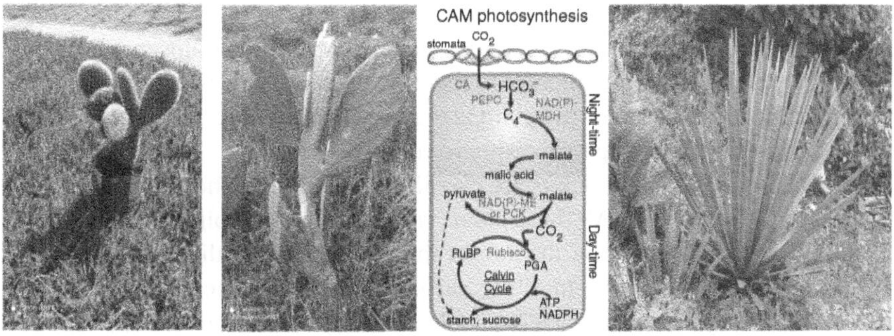

Figure 5.4 CAM photosynthetic pathway and common plants (Yamori et al., 2014). Source: Figure modified.

5.6 The Role of C3, C4, and CAM Plants in PMFC

Strik et al. (Strik et al., 2008) were the first to propose and construct a PMFC with *Glyceria maxima* plant and carbon-based electrodes. Since then, hundreds and even thousands of papers have reported the capability of plants to produce energy in a green way employing plant–bacteria symbiosis as a driving force. It is important to note that one special feature of PMFC technology is in fact the carbon adsorption and glucose production process during photosynthetic activity, which promotes food self-sufficiency for the plant and soil bacteria. Thus, in PMFC the substrate for microbes is constantly generated in natural way. Approximately 70% of the carbon dioxide absorbed during carbon fixation phenomena is transferred to the plant roots generating rhizodeposits including gases such as carbon dioxide, secretions (oligomeric sugars, enzymes), exudates (carbohydrates, monomeric sugars, and low molecular weight organic acids), and dead organic matter (dead plants and microbes) serving as substrate for soil bacteria (Chicas et al., 2018; Bais et al., 2006).

The performance of PMFC to generate energy depends on abiotic and biotic factors. Abiotic factors are related to environmental conditions such as solar irradiation, ambient temperature, humidity, soil properties, dissolved oxygen as well as the arrangement of the electrical circuit, electrode materials, and the configuration and design of the bioelectrochemical cell. Biotic factors are primarily associated with microbial community and its electron transfer mechanisms, and the type of plant and their photosynthetic route. Moreover, the site of installation, constructed wetland, controlled environment, greenhouse, laboratory, or gardens affects the vitality of the plant and changes the energy performance since the amount of carbohydrates in root plants differs from environments to environments (Nitisoravut & Regmi, 2017; Strik et al., 2008; Doherty et al., 2015). The energy produced in PMFCs is directly dependent on the photosynthetic pathway of the plant. Therefore, in order to increase the energy recovery from the fuel cell, an adequate selection of the plant is required and is expected to obtain major energy from plants with high rhizosphere volume.

As was described before, plants are classified according to their photosynthetic pathway in C3, C4, and CAM; these plants vary in their metabolic route and times. Both C3 and C4 fixed carbon dioxide in the presence of sunlight; however, the main difference between them is the carbon molecule formed during the first stage of Calvin cycle. For C3, 3-PGA molecule is formed in bundle sheath cells. On the other hand, C4 plants utilize MC to produce either malate or asparate as the first hydrocarbon molecule. CAM plants employ the same vegetable cells as C3 route; however, the carbon fixation occurs at night. Not only C3 but also CAM plants employ bundle sheath cells to produce carbohydrates at the initial stage of Calvin cycle but at different times (Maddalwar et al., 2021; Raines, 2011; Liu et al., 2013). In literature, there are some documents specialized in photosynthesis mechanisms citing the difference between photosynthetic routes from molecular point of view (Yamori et al., 2014; Salisbury & Ross, 1991; Solomon et al., 2013).

C3 and C4 have a special distribution through the world, for each ecosystem their predominance is affected by two environmental factors; carbon dioxide concentration in the plant surrounding air and atmospheric temperature. Thus, the balance between C3 and C4 changes with air composition (Liu et al., 2015). C3 plants are identified by their own characteristic properties, among which low plant durability and low photosynthetic yield in dried and hot environments are remarkable. The energy conversion efficiency of C3 route is around 2.4%, value that is equivalent to 4.6% respect to the maximum hypothetical yield based on the total solar irradiance (Taylor et al., 2023). Since Rubisco performs the oxidation reaction for C3 plants in the first stage of Calvin cycle, there is an energy loss of almost 25% attributed to photorespiration process which releases carbon dioxide into the atmosphere. Photorespiration process is another distinctive characteristic of C3 plants (Shaikh et al., 2021).

As was cited previously by Wang et al. (Wang et al., 2012), among the three types of plants, C4 showed the highest photosynthetic efficiency storing a major amount of carbohydrates in the roots, thus the concentration of root exudates produced by them is higher than those produced by the other plants. This fact suggests that C4 plants are a beneficial option for energy purposes. C4 plants have anatomical and structural alterations that affect photosynthetic routes; promote high solar energy conversion efficiency and increase the rhizodeposition of carbohydrates. It has been determined that C4 pathway reaches an energy conversion efficiency of around 3.7%, which corresponds to 6% of the theoretical maximum yield (Taylor et al., 2023). MCs employ phosphoenolpyruvate carboxylase enzyme to fix carbon inside its structure and to transform it into oxaloacetate which is then metabolized into malate and asparate; the first carbon molecule produced in

Calvin cycle. In agreement with Deng et al. (Deng et al., 2012), C4 plants have important advantages over C3 species; the amount of rhizodeposits is proportional to photosynthates and thus, more amount of root exudates are available for microbes' metabolism, they also show high photosynthetic efficiency and can well perform on dry and hot sites, this last one is a major disadvantage for C3 plants. Examples of C4 plants are monocots, grass plants, and some cultivar plants such as maize, sorghum, and sugarcane (Chiranjeevi et al., 2019).

Finally, CAM plants fix carbon dioxide during the night employing bundle sheath cells, this feature inhibits water loss and allows to producing of 3-PGA as the initial carbon molecule. The main characteristic of CAM plants such as sedum, pineapple, snake plant, and Opuntia is their ability to survive in arid environments and in drought conditions, moreover, their growth is very slow, and biomass production is lesser than those observed by C3 and C4 plants (Nitisoravut & Regmi, 2017). According to Weiss et al. (Weiss et al., 2010), CAM plants such as *Hylocereus undatus* and *Selenicereus megalanthus* increased their growing rate at high carbon dioxide loadings increasing the size of the fruits and the transport of nutrients to the roots. The work made by Weiss and co-workers serves as a starting point for further studies focusing on energy production by CAM plants in cities where carbon dioxide concentration is high. Table 5.1 summarizes the main characteristics of C3, C4, and CAM plants.

Table 5.1 Characteristics of the different plant species.

C3	C4	CAM
3-Phosphoglicerate as the first synthesized sugar monomer	Malate or asparate as the first synthesized sugar monomer	Malate or asparate as the first synthesized sugar monomer
Comprises around 85% of the plants	Comprises around 5% of the plants.	Comprises around 10% of the plants.
Low yield in dry and hot environments	High yield in dry and hot environments	High yield in dry and hot environments
Fixation process during the day	Fixation process during the day	Fixation process during the night
Employs bundle sheath cells	Employs mesophilic cells	Employs bundle sheath cells
Induces photorespiration process	Avoids photorespiration process	Avoids photorespiration process
There is a loss of water during metabolism process	There is a loss of water during metabolism process	Avoids water loss
Best adapted on cool and wet environments	Best adapted on hot and sunny environments	Best adapted on hot and dry environments
Photosynthetic efficiency is as follows: C4 > C3 > CAM		

There are some important factors affecting the selection of plants for PMFC construction and operation such as type of soil, bioaccumulation properties, adaptability, density of roots, tolerance to soil salinity, local availability, and plant distribution. The abundance of each plant worldwide is another important factor to take into account, among the three kinds of plants, C3 is the most abundant in nature reaching almost 85% of the total plant species in the world. Only 5% of the known plants follow C4 route, while the remaining percentage is associated with CAM species (Yamori et al., 2014).

Since almost 85% of the major plants follow C3 photosynthetic route, there is an enormous abundance of these plants worldwide. It has been reported that around 89% of the vascular plants are related to C3 pathway. Vascular plants are characterized by the possession of xylem and phloem, intercellular air spaces, stomata, and cuticle. From these, xylem and phloem are, respectively, tissues that carry out water and food conduction from roots to the rest of the plant (Yamori et al., 2014; Raven, 1984). Some examples of C3 plants include some grasses, most of trees, dicots, and cultivars such as cotton, wheat, rice, barley, and peanuts. C4 comprises monocots, grasses, and cultivars such as maize, sorghum, and sugarcane. Plants with the capability to survive in arid and hot sites such as sedum, agaves, yucca, and orchids are CAM plants. Table 5.2 resumes some common families associated with C3, C4, and CAM plants.

The faculty of plants to produce green energy is directly related to the photosynthetic efficiency to take advantage of the fixed carbon dioxide and the subsequent transformation into carbohydrate molecules during Calvin cycle. From this point

Table 5.2 Examples of common families of C3, C4, and CAM plants.

C3	C4	CAM
Fabaceae	*Arecaceae*	*Agavaceae*
Solanaceae	*Asteraceae*	*Orchidaceae*
Poaceae	*Poaceae*	*Cactaceae*
Liliaceae	*Euphorbiaceae*	*Crassulaceae*
Rosaceae	*Amaranthaceae*	*Liliaceae*
Magnoliaceae	*Zingiberaceae*	*Asteraceae*
Brassicaceae	*Bromeliaceae*	*Euphorbiaceae*
Fagaceae	*Juncaceae*	*Asparagaceae*
Myrtaceae	*Araceae*	*Bignoniaceae*
Malvaceae	*Musaceae*	
Caricaceae		
Rutaceae		

of view, the capability to produce green energy of C4 plants must be higher than those expected for C3 and CAM, i.e. the energy production performance using living plants is as follows: C4 > C3 > CAM.

The first PMFC was constructed with a C3 plant; *Glyceria maxima* colloquially known as reed mannagrass. This pioneer cell operated for around 120 days with- no plant damage reaching their maximum voltages of 250 mV after 75 days. One interesting point of this paper is that the carbon-based electrodes; graphite felt, do not negatively affect the vitality of the reed mannagrass and thus allow exploring PMFCs in the long term (Strik et al., 2008). Years after Timmers and co-workers, used the same plant to study microbe's behavior and their distribution on PMFC, observing a major predominance on graphite felt anode and in the plant roots surronding area. Moreover, denitrifiers and methanogens were identified by the presence of oxygen and nitrate in the soil. The maximum power density calculated in this paper was 80 mW/m^2 (Timmers et al., 2012). In 2013, the same research group designed and constructed a novel tubular PMFC with graphite felt as anode and reed mannagrass as a plant; in this case, the maximum power density was 60 mW/m^2 (Timmers et al., 2013).

Another pioneer plant used in PMFC was rice, which was used for the first time by Kaku et al., in late 2008 (El Gamal et al., 2018). To evaluate green energy production, Kaku and co-workers used 5 m^2 of the rice paddy field during the cropping season and installed graphite felt electrodes. Microbial analyses were performed to determine the microorganism's diversity involved in the energy production, the bacterial predominance was composed of *Natronocella acetinitrilica*, *Beijerinckiaceae*, *Rhizobiales*, and *Rhodobacter gluconicum*. However, despite the presence of EABs, the maximum power output obtained by them was only 6 mW/m^2. The effect of cathode modification, anode position, and the external electrical load on power generation in a PMFC operated in rice paddy field was also evaluated. Placing the anode at greater depth coupled with the modification of cathode with platinum metal, increased power output to a maximum of 14 mW/m^2 (Takanezawa et al., 2010). The use of biocathodes on PMFC was also reported in previous works. As example, a blue-green algal biocathode was adapted in a PMFC constructed in a paddy field employing two-chamber configurations, Nafion proton exchange membrane as separator and graphite granules as anode. The cathode was made by platinum-coated carbon cloth and submerged in BG11 nutrient solution where *Cyanobacterium phromidium* growth. At these conditions, the highest volumetric power density was 29.8 mW/m^3 (Srivastava et al., 2019). In addition, arsenic metal has been found in rice grains due to biomagnification process induced by the presence of arsenic in water and soil of paddy field. The simultaneous treatment of this metal and energy production was evaluated in a soil MFC constructed with carbon felt electrodes and operated in a greenhouse. Arsenic content decreased considerably and the maximum potential achieved was 22.2 mW/m^2 (Gustave et al., 2019). Finally, the addition of iron powder particles on paddy field fuel cell has been a promising technique to increase power density (Matsumoto et al., 2020). The effect of compost addition on energy efficiency was also studied before (Moqsud et al., 2015).

Arundo donax is another plant investigated in the early years of PMFC history. This plant follows C3 photosynthetic route and exhibits higher energy production efficiency than conventional C4 plants (Apollon et al., 2021). The first work reporting Arundo donax PMFC was made by Helder et al. (Helder et al., 2010), who constructed a cylindrical PMFC coupling to cation exchange membrane and carbon-based electrodes. Despite the efforts made by them, the system broke down and did not produce energy and any stable output. Years after, different wetland plants were evaluated in constructed wetland MFC to produce green energy, among the plants used in this study, Arundo donax showed the highest energy production efficiency reaching a maximum of 12.82 mW/m^2 (Zhou et al., 2018).

Phragmites australis is another interesting C3 plant commonly used in constructed wetland PMFC. This macrophyte has shown great potential to produce a high amount of energy during wastewater treatment and remotion of organic load in sediment sludges. The potential of this plant in BES is such that reviews of works focused on the role of *Phragmites australis* were reported before (Jingyu et al., 2020; Yang et al., 2021). There are other kinds of C3 plants used in BES to produce energy. Table 5.3 summarizes the most important plants but literature is not limited to.

For C4, *Eichhornia crassipes*, *Spartina anglica*, and *Arundinella anomala* were the pioneer plants adapted into MFC (Helder et al., 2010; Mohan et al., 2011). As can be seen in Table 5.3, in most cases the power output was higher for C4 than that observed for C3 and CAM plants. Moreover, most of the C4 plants are related to monocots. Plants were classified into C3, C4, and CAM according to previous reports and the family at which each plant corresponds. An interesting fact is observed for Poaceae family, most of these plants follow C3 route; however, a few of them share C3 and C4 characteristics, and thus they were classified as C4. *Triticum aestivum* also known as wheat is a member of the Poacea family, despite this, its own photosynthetic characteristics are more related to C4 plants. This specie generated a maximum current density of 291.29 mA/m^2 when used in a PMFC operated with bioanode and stainless steel cathode (Guan et al., 2019). Bougainvillea plant analyzed in a PMFC operated with metal electrodes reached a maximum voltage of 0.98 V, which was boosted to 2.56 V by using an electronic device composed of DC–DC converter (Prasad & Tripathi, 2018). C4 aquatic plants such as *Hydrilla verticillata* and *Echinochloa glabrescens* were used to produce energy using carbon-based electrodes as final electron acceptor and organic waste as carbon source. The voltage produced by Hydrylla plant was 585 mV (Shen et al., 2018), while the highest power output achieved with *Echinochloa*

Table 5.3 Summary of some plant species used in PMFC, operation conditions, electrode materials, and power outputs.

Plant	Family	Substrate	Anode	Cathode	Power density	References
C3						
Glyceria maxima	*Poaceae*	Hoagland nutrient solution and potassium acetate	Graphite felt and graphite granules	Graphite felt	67 mW/m^2	Strik et al. (2008)
Glyceria maxima		Hoagland nutrient solution/glucose	Graphite felt	Graphite felt	80 mW/m^2	Timmers et al. (2012)
Glyceria maxima	*Poaceae*	Hoagland nutrient solution	Graphite felt	Graphite felt	60 mW/m^2	Timmers et al. (2013)
Oryza sativa		Soil and rice straw	Graphite felt	Graphite felt	6 mW/m^2	El Gamal et al. (2018)
Oryza sativa	*Poaceae*	Rice paddy field	Graphite felt	Platinum-doped graphite felt	14 mW/m^2	Takanezawa et al. (2010)
Oryza sativa		Graphite granules and vermiculite	Graphite felt interwoven with carbon rod	Graphite felt interwoven with carbon rod	61.7 mW/m^2	Arends et al. (2014)
Oryza sativa		Compost	Carbon fiber	Carbon fiber	39.20 mW/cm^2	Moqsud et al. (2015)
Oryza sativa		Urea fertilizer	NR	NR	153.7 mW/cm^2	Putranto et al. (2018)
Oryza sativa	*Poaceae*	Hoagland nutrient solution	Graphite granules	Platinum-coated carbon cloth plus blue-green algae	29.8 mW/m^3	Srivastava et al. (2019)
Oryza sativa		Fertilizer	Carbon felt	Carbon felt	22.2 mW/m^2	Gustave et al. (2019)

(Continued)

Table 5.3 (Continued)

Plant	Family	Substrate	Anode	Cathode	Power density	References
Oryza sativa		Urea	Carbon felt	Carbon felt	$9.6\,mW/m^2$	Sudirjo et al. (2019)
Oryza sativa		Soil	Graphite felt	Platinum-doped graphite felt	$130\,mW/m^2$	Matsumoto et al. (2020)
Oryza sativa	*Poaceae*	Commercial vermicompost	Plain graphite felt	Plain graphite felt	9.1 for horizontal configuration 16.8 for vertical configuration	Kumar et al. (2020) Sudirjo et al. (2019)
Arundo donax		Hoagland nutrient solution	Graphite rod in graphite grains	Graphite felt	Unstable system	Helder et al. (2010)
Arundo donax		Nutrient solution	Graphite felt	Graphite felt	$12.82\,mW/m^2$	Zhou et al. (2018)
Arundo donax	*Poaceae*	Loamy sandy soil	Graphite	Graphite	145.2 mV	Benhabylès et al. (2021)
Phragmites australis	*Poaceae*	Dewatered alum sludges	Packed granular graphite	Packed granular graphite	$0.28\,W/m^3$	Doherty et al. (2015)
Phragmites australis		Fen peat soil	Graphite felt	Graphite felt	$0.52\,W/m^2$ $2.9\,W/m^3$	Wetser et al. (2015)
Phragmites australis		Root exudates/ nutritive solution	Graphite felt	Graphite felt	$21.87\,mW/m^2$	Salinas-Juárez et al. (2016)
Phragmites australis		Sand and ceramite	Titanium mesh/ activated carbon	Titanium mesh	$3.7\,W/m^2$	Xu et al. (2018)
Phragmites australis	*Poaceae*	Dewatered alum sludges	Stainless steel mesh filled with graphite gravel	Stainless steel mesh filled with graphite gravel	$21.7\,mW/m^2$	Yang et al. (2020)

Typha latifolia		Synthetic wastewater	Activated carbon	Activated carbon	$93\,mW/m^3$	Oon et al. (2016)
Typha latifolia	*Typhaceae*	Gravels/synthetic wastewater	Carbon felt	Carbon felt	$1.58\,mW/m^2$	Oon et al. (2020)
Typha latifolia		Synthetic wastewater and Hoagland medium	Graphite	Magnesium cathode	$13.4\,mW/m^2$	Di et al. (2020)
Typha orientalis		Glucose/nitrobenzene	Graphite felt	Graphite felt	$62.8\,mW/m^2$	Di et al. (2020)
Typha angustifolia	*Typhaceae*	Synthetic wastewater and Hoagland medium	Graphite	Magnesium cathode	$18.1\,mW/m^2$	Saz et al. (2018)
Chlorophytum comosum	*Liliaceae*	Soil	Graphite	Graphite	$18\,mW/m^2$	Azri et al. (2018)
Chasmanthe floribunda	*Iridaceae*	Soil	Graphite	Graphite	$0.21\,mW/m^2$	Azri et al. (2018)
Vigna radiata	*Fabaceae*	Garden soil	Stainless steel	Stainless steel	$0.35\,mW/m^2$	Pamintuan and Sanchez (2019)
C4						
Eichhornia crassipes		Wastewater	Graphite discs	Graphite discs	$224.9\,mW/m^2$	Mohan et al. (2011)
Eichhornia crassipes	*Pontede-riaceae*	Wastewater	Granular activated carbon	Granular activated carbon	$45.46\,mW/m^3$	Teoh et al. (2020)
Eichhornia crassipes		Soil and tap water	Zinc	Graphite carbon felt	$100.2\,mW/m^2$	Widharyanti et al. (2021)

(Continued)

Table 5.3 (Continued)

Plant	Family	Substrate	Anode	Cathode	Power density	References
Bougainvillea	*Nyctaginaceae*	Soil	Copper	Zinc	0.98 V boosted to 2.56 with DC–DC converter	Prasad and Tripathi (2018)
Lemna minuta	*Araceae*	Sunlight and water	Carbon felt	Graphite granules	380 mW/m^2	(Hubenova and Mitov (2012)
Cyperus alternifolius		Domestic wastewater and glucose	Coconut shell-activated carbon	Coconut shell-activated carbon	47.8 mW/m^3	Zhang et al. (2021)
Cyperus papyrus nanus	*Cyperaceae*	Sludge of anaerobic digester	Granular activated carbon	Platinum-coated carbon paper	510 mW/m^3	Gulamhussein and Randall (2020)
Wachendorfia thyrsiflora	*Haemodor-aceae*	Sludge of anaerobic digester	Granular activated carbon	Platinum-coated carbon paper	1036 mW/m^3	Gulamhussein and Randall (2020)
Hydrilla verticillata	*Hydrochar-itaceae*	Sludge/nutritional solution	Carbon fibre brush	Carbon fibre brush	558.5 mV	Shen et al. (2018)
Canna indica		Synthetic wastewater composition	Graphite felt	Platinum-coated carbon cloth	31.04 mW/m^3	Srivastava et al. (2017)
Canna indica	*Cannaceae*	Concentrated activated sludge	Foamed nickel	Foamed nickel	5.11 mW/m^3	Wang et al. (2017)
Canna indica		Wastewater	Carbon fiber felt	Carbon fiber felt	6.74 mW/m^2	Ge et al. (2020)
Pennisetum setaceum		Red soil	Graphite plate	Graphite plate	163 mW/m^2	Chiranjeevi et al. (2019)
Arundinella anomala	*Poaceae*	Hoagland nutrient solution	Graphite rod in graphite grains	Graphite felt	22 mW/m^2	Helder et al. (2010)

Spartina anglica		Hoagland nutrient solution	Graphite rod in graphite grains	Graphite felt	222 mW/m^2	Helder et al. (2010)
Spartina anglica	*Poaceae*	Hoagland nutrient solution	Graphite felt	Graphite felt	79 mW/m^2	Timmers et al. (2013)
Papyrus diffusus	*Cyperaceae*	Soil	Graphite	Graphite	1.08 mW/m^2	Azri et al. (2018)
Epipremnum aureum	*Araceae*	Cow dung and garden soil	Carbon fiber brush	Carbon fiber cloth	12.78 mW/m^2	Sarma and Mohanty (2018)
Epipremnum aureum		Buffalo dung and garden soil	Carbon rod coupled corn cob	Steel	6.75 mW/m^2	Sonu et al. (2023)
Amaranthus viridis	*Amaran-thaceae*	Soil	Bioanode	Stainless steel alligator clips	230.25 mA/m^2	Guan et al. (2019)
Triticum aestivum	*Poaceae*	Soil	Bioanode	Stainless steel alligator clips	291.29 mA/m^2	Guan et al. (2019)
Agapanthus africanus	*Amaryllid-aceae*	Compost	Carbon fiber textile	Carbon fiber Textile	15.55 mW/m^2	Gómora-Hernández et al. (2020)
Echinochloa glabrescens	*Poaceae*	Comercial compost	Carbon cloth	Stainless steel	0.09 mW/m^2	Bombelli et al. (2013)
Pennisetum alopecuroides	*Poaceae*	Soil	Carbon felt	Carbon felt	667.9 mV	Guan and Yu (2021)
Cyperus prolifer	*Cyperaceae*	Anaerobic sludge	Copper wires	Cooper wires	229 mW/m^3	Oodally et al. (2019)
CAM						
Dracaena braunii	*Asparagaceae*	Cow dung and garden soil	Carbon fiber brush	Carbon fiber cloth	15.38 mW/m^2	Sarma and Mohanty (2018)
Sedum album					0.0024 μW/m^2	

(Continued)

Table 5.3 (Continued)

Plant	Family	Substrate	Anode	Cathode	Power density	References
Sedum hybridum					0.092 μW/m^2	
Sedum kamischaticum					0.001 μW/m^2	
Sedum reflexum					0.001 μW/m^2	
Sedum rupestre		Sand, silt, clay, and organic matter	Graphite fiber felt	Graphite fiber felt	0.015 μW/m^2	Tapia et al. (2017)
Sedum sexangulare					0.0084 μW/m^2	
Sedum spurium	*Crassulaceae*				0.001 μW/m^2	
Opuntia albicarpa					5.47 mW/m^2	
Opuntia ficus indica					6.60 mW/m^2	
Opuntia joconostle		Mosque plants and organic matter	Graphite felt	Zinc	5.76 mW/m^2	Apollon et al. (2020)
Opuntia robusta	*Cactaceae*				2.01 mW/m^2	
Opuntia joconostle	*Cactaceae*	Garden soil and mesquite bushes	NR	NR	145 mW/m^3	Apollon et al. (2022)

NR, not reported.

glabrescens was 0.09 mW/m^2 (Bombelli et al., 2013). *Eichhornia crassipes* and *Canna indica* are other usual C4 plants evaluated in BES.

Finally, only a few papers have reported the use of CAM plants to produce energy by PMFC. As expected, the power outputs obtained by these plants are low. The lowest energy yield was obtained by Tapia et al., who designed and constructed sedum PMFC for the first time reaching around 0.001 $\mu W/m^2$ as power output (Tapia et al., 2017). Mexican Opuntia species were evaluated in a subsequent work employing graphite felt as anode and a cathode made by zinc. The substrate employed to promote microbes' growth was mosque plants organic matter. Among the evaluated plants, Opuntia ficus indica reached the highest power density; 6.60 mW/m^2 (Apollon et al., 2020).

5.7 The Role of Wetland and Drought-resistant Plants in PMFC

The capability of some C3 and C4 plants to survive in wetlands has attracted a great attention in the last years, and thus bifunctional PMFCs have been constructed not only to produce green energy but also to eliminate different pollutants from wastewater.

For example, Di et al. (Di et al., 2020) studied power generation and removal of nitrobenzene in a constructed wetland PMFC coupled to *Typha orientalis* and *Scirpus validus*. This last one showed a high-power output (19.5 mW/m^2) and a nitrobenzene removal efficiency of almost 94%. Other toxic pollutants such as heavy metals (Cr VI) (Cabezas et al., 2015; Mu et al., 2021; Cheng et al., 2019), hospital laundry wastewater (Lutterbeck et al., 2020), organic dyes (Oon et al., 2018), typical medicines (Wen et al., 2020; Dai et al., 2021; Xu et al., 2021), and fuels (Wei et al., 2014) were partially eliminated in wetland PMFC's producing green energy at the same time. This fact shows the great potential of C3 and C4 wetland plants as energy carriers in BES eliminating emergent pollutants in a promissory and sustainable way.

Drought-resistant plants usually follow CAM route. Currently, there is scarcely information reporting these plants for energy purposes by BES. However, few papers related to this topic allow us to consider CAM plants as potential biological sources for bioenergy production in water scarcity environments. Previous reports showed that plants such as *Opuntia* and *Sedum* are able to generate electricity in noncontrolling environments (Tapia et al., 2017; Apollon et al., 2020), and most recently that the energy output can be boosted by addition of ammonium nitrate on soil (Apollon et al., 2022). Research studies on energy production from CAM plants are needed in order to apply this technology on semi-arid and arid environments.

5.8 Trends and Future Perspectives

PMFCs not only remove contaminants from wastewater but also treat contaminated soil. In recent years, petroleum-derived compounds were removed employing a C4 plant; *Aglaonema commutatum* in the presence of surfactants. The hydrocarbons degradation was around 82% generating energy throughout the process (Zhao et al., 2019). Treatment of natural soils contaminated with organic pollutants and hydrocarbons is another possible application of PMFC technology (Zhang et al., 2022).

Nowadays, the use of trees has started. In our knowledge, Rusyn (Rusyn, 2014) proposed for the first time the use of trees in an urban park area. Years after, Lu et al. (Lu et al., 2020) used stems of the trees *Pachira macrocarpa* and *Populus alba* to produce stable and continuous bioelectricity at least during 40 days, reaching a power output of 7.61 mW/m^2. Since trees have long life time, these works make us think the possible production of bioenergy in a continuous way over years and even decades.

The use of ornamental plants as either indoor plants or located at the roofs of buildings focused on creating green inhabited environments has been investigated too. In previous work, Rusyn (Rusyn, 2021) summarized the energy generation and the possible application of PMFCs constructed with water logged and indoor plants to power up electronic devices and remove organic matter and pollutants from water. Interesting research was reported by Shen et al. (Shen et al., 2018). In this work, ammonium ion (NH_4^+) was efficiently removed employing the aquatic decorative plant *Hydrilla verticillata*, generating at the same time around 558.50 mV in an enclosed anode configuration. For indoor plants, a succulent-based bio-battery was developed by Apollon et al. (Apollon et al., 2020) and employed to power up an LED spotlight and digital clock in noncontrolling environment and unsaturated water conditions, the constructed system showed around 3.66 Wh/m^2 power output. *Agapanthus africanus* is another ornamental plant applied in PMFCs with energetic potential. In a previous work, 3 *Agapanthus*-PMFCs were connected in series to an electronic device composed of 1.2 V batteries in order to partially charge a mobile phone (Gómora-Hernández et al., 2020). The previous reports using ornamental plants in BES have shown the potential of these kinds of plants to produce energy. Is it necessary to take advantage of this natural energy source to power up low-demand electronic devices like LED lamps which can be applied to illuminate outdoor environments? Preliminary tests revealed that one LED spotlight is continuously turned on when two *Agapanthus africanus* PMFCs in noncontrolling environment were connected in series without any electronic boosting system (Figure 5.5). The connection of PMFCs in parallel or series configuration can be an interesting research area for further studies. It has been observed that the parallel or serial connection of PMFC's

Figure 5.5 LED-spotlight turned on using PMFC connected in series.

proportionally increases current and the generated voltage according to the number of fuel cells, i.e. major number of PMFC's generates major voltage or current, allowing to achieve the required output for the desired application (Rusyn et al., 2021; Bataillou et al., 2022; Osorio-de-la-Rosa et al., 2021).

The application of electronic principles is necessary to better take advantage of the energy produced by PMFC and to design data acquisition systems that allow to study the effect of either biotic or abiotic factors on power generation (Prasad & Tripathi, 2018; Koffi & Okabe, 2022). Finally, recent trends suggest the use of plant food in PMFC technology as a contribution toward sustainability (Guan et al., 2019; Pamintuan & Sanchez, 2019; Kuleshova et al., 2022).

5.9 Conclusions

PMFC is a novel technology able to produce green energy from plant roots chemical energy, the performance of this biological cell is directly related to biotic and abiotic factors, from which photosynthetic route of the plant species is considerably important. Despite C4 plants showed high photosynthetic efficiency, some C3 plants have shown great potential to produce energy by BES. PMFC is a universal technology, since any plant can be used to produce energy in different environments and conditions. In fact, nowadays, an enormous diversity of plant species from ornamental and aquatic to drought resistant and succulents have been used for energy purposes. An important characteristic of PMFCs is their capability to remove organic matter from either wastewater or soil, and even to remove recalcitrance pollutants, dyes, aromatic compounds, drugs, and among others from polluted sites. This last one allows us to consider this technology as the most suitable and sustainable for green energy production, however, the optimization of the systems, exploration of other plants, and taking advantage of the generated energy are necessary to achieve a social impact application.

The potential of PMFCs can be such that in the near future, outdoor environments such as gardens, parks, rural communities, and outer hallways can be illuminated continuously and in a long term employing the chemical energy of plants and threes with the simultaneous removal of organic matter and pollutants.

Acknowledgments

J. C. Gómora-Hernández thanks the division of Environmental Engineering for providing the facilities to carry out this work.

References

Al-Mamun, A., Ahmad, W., Baawain, M.S. et al. (2018). A review of microbial desalination cell technology: configurations, optimization and applications. *Journal of Cleaner Production* 183: 458–480. https://doi.org/10.1016/j.jclepro.2018.02.054.

Apollon, W., Kamaraj, S.K., Silos-Espino, H. et al. (2020). Impact of opuntia species plant bio-battery in a semi-arid environment: demonstration of their applications. *Applied Energy* 279: 115788. https://doi.org/10.1016/j.apenergy.2020.115788.

Apollon, W., Luna-Maldonado, A.I., Kamaraj, S.K. et al. (2021). Progress and recent trends in photosynthetic assisted microbial fuel cells: a review. *Biomass and Bioenergy* 148: 106028. https://doi.org/10.1016/j.biombioe.2021.106028.

Apollon, W., Valera-Montero, L.L., Perales-Segovia, C. et al. (2022). Effect of ammonium nitrate on novel cactus pear genotypes aided by biobattery in a semi-arid ecosystem. *Sustainable Energy Technologies and Assessments* 49: 101730. https://doi.org/10.1016/j.seta.2021.101730.

Arends, J.B.A., Speeckaert, J., Blondeel, E. et al. (2014). Greenhouse gas emissions from rice microcosms amended with a plant microbial fuel cell. *Applied Microbiology and Biotechnology* 98: 3205–3217. https://doi.org/10.1007/s00253-013-5328-5.

Arulmani, S.R.B., Gnanamuthu, H.L., Kandasamy, S. et al. (2021). Sustainable bioelectricity production from *Amaranthus viridis* and *Triticum aestivum* mediated plant microbial fuel cells with efficient electrogenic bacteria selections. *Process Biochemistry* 107: 27–37. https://doi.org/10.1016/j.procbio.2021.04.015.

Arun, S., Sinharoy, A., Pakshirajan, D., and Lens, P.N.L. (2020). Algae based microbial fuel cells for wastewater treatment and recovery of value-added products. *Renewable and Sustainable Energy Reviews* 132: 110041. https://doi.org/10.1016/j.rser.2020.110041.

Azri, Y. M., Toy, I., Sadi, M., Benhabyles, L. (2018). Bioelectricity generation from three ornamental plants: *Chlorophytum comosum*, *Chasmanthe floribunda* and *Papyrus diffusus*. *International Journal of Green Energy* 15:254-263. https://doi.org/10.1080/15435075.2018.1432487.

Bais, H.P., Weir, T.L., Perry, L.G. et al. (2006). The role of root exudates in rhizosphere interactions with plants and other organisms. *Annual Review of Plant Biology* 57: 233–266. https://doi.org/10.1146/annurev.arplant.57.032905.105159.

Bataillou, G., Haddour, N., and Vollaire, C. (2022). Bioelectricity production of PMFC using Lobelia Queen Cardinalis in individual and shared soil configurations. *E3S Web of Conferences*. 334:08001. https://doi.org/10.1051/e3sconf/202233408001.

Benhabylès, L., Azri, Y.M., Tou, I., and Sadi, M. (2021). Bioelectricity production from Arundo donax MFC and Chlorophytum Comosum. In: *Advances in Renewable Hydrogen and Other Sustainable Energy Carrier*, vol. 1 (ed. A. Khellaf). Singapore: Springer Chapter 37.

Bidwell, R.G.S. (1979). *Plant Physiology*. USA: MacMillan Publishing Co. Inc.

Bombelli, P., Iyer, D.M.R., Covshoff, S. et al. (2013). Comparison of power output by rice (*Oryza sativa*) and an associated weed (*Echinochloa glabrescens*) in vascular plant bio-photovoltaic (VP-BPV) systems. *Applied Microbiology and Biotechnology* 97: 429–438. https://doi.org/10.1007/s00253-012-4473-6.

Cabezas, A., Pommerenke, B., Boon, N., Friedrich, M. W. (2015). Geobacter, Anaeromyxobacter and Anaerolineae populations are enriched on anodes of root exudate-driven microbial fuel cells in rice field soil. *Environmental Microbiology Reports* 7:489-497. https://doi.org/10.1111/1758-2229.12277.

Cao, Y., Mu, H., Liu, W. et al. (2019). Electricigens in the anode of microbial fuel cells: pure cultures versus mixed communities. *Microbial Cell Factories* 18: 39. https://doi.org/10.1186/s12934-019-1087-z.

Cardemil, L. and Squeo, F.A. (2007). *Plant Physiology*. Chile: Serena University, Universidad de La Serena.

Cardoso, J.A., Rincón, J., De la Cruz Jiménez, J. et al. (2013). Morpho-anatomical adaptations to waterlogging by germplasm accessions in a tropial forage grass. *AoB Plants* 5: 047. https://doi.org/10.1093/aobpla/plt047.

Cheng, C., Hu, Y., Shao, S. et al. (2019). Simultaneous Cr(VI) reduction and electricity generation in plant-sediment microbial fuel cells (P-SMFCs): synthesis of non-bonding Co_3O_4 nanowires onto cathodes. *Environmental Pollution* 247: 647–657. https://doi.org/10.1016/j.envpol.2019.01.084.

Chicas, S.D., Sivasankar, V., Omine, K. et al. (2018). Plant microbial fuel cell technology: developments and limitations. In: *Microbial fuel Cell Technology for Bioelectricity*, vol. vol. 1 (ed. V. Sivasankar, P. Mylsamy, and K. Omine). Cham, Switzerland: Springer Chapter 3.

Chiranjeevi, P., Yeruva, D.K., Kumar, A.K. et al. (2019). Plant-microbial fuel cell technology. In: *Microbial Electrochemical Technology*, vol. vol. 1 (ed. S.V. Mohan, S. Varjani, and A. Pandey). Amsterdam, Netherlands: Elsevier Chapter 3.8.

Dai, M., Zhang, Y., Wu, Y. et al. (2021). Mechanism involved in the treatment of sulfamethoxazole in wastewater using a constructed wetland microbial fuel cell system. *Journal of Environmental Chemical Engineering* 9: 106193. https://doi.org/10.1016/j.jece.2021.106193.

Deng, H., Chen, Z., Zhao, F. (2012). Energy from plants and microorganisms: progress in plant-microbial fuel cells. *ChemSusChem* 5:1006-011. https://doi.org/10.1002/cssc.201100257.

Di, L., Li, Y., Nie, L. et al. (2020). Influence of plant radial oxygen loss in constructed wetland combined with microbial fuel cell on nitrobenzene removal from aqueous solution. *Journal of Hazardous Materials* 394: 122542. https://doi.org/10.1016/j.jhazmat.2020.122542.

Doherty, L., Zhao, Y., Zhao, X. et al. (2015). A review of a recently emerged technology: constructed wetland – microbial fuel cells. *Water Research* 85: 38–45.

Doherty, L., Zhao, Y., Zhao, X., and Wang, W. (2015). Nutrient and organic removal from swine slurry with simultaneous electricity generation in an alum sludge-based constructed wetland incorporating microbial fuel cell technology. *Chemical Engineering Journal* 266: 74–81. https://doi.org/10.1016/j.cej.2014.12.063.

El Gamal, M., Mousa, H.A., El-Naas, M.H. et al. (2018). Bio-regeneration of activated carbon: a comprehensive review. *Separation and Purification Technology* 197: 345–359. https://doi.org/10.1016/j.seppur.2018.01.015.

Ge, X., Cao, X., Song, X. et al. (2020). Bioenergy generation and simultaneous nitrate and phosphorus removal in a pyrite based constructed wetland microbial fuel cell. *Bioresource Technology* 2020 (296): 122350. https://doi.org/10.1016/j.biortech.2019.122350.

Gómora-Hernández, J.C., Serment-Guerrero, J.H., Carreño-De-León, M.C., and Flores-Álamo, N. (2020). Voltage production in a plant microbial fuel cell using *Agapanthus africanus*. *Revista Mexicana de Ingenieria Quimica* 19: 227–237. https://doi.org/10.24275/rmiq/IA542.

Guan, C.Y. and Yu, C.P. (2021). Evaluation of plant microbial fuel cells for urban green roofs in a subtropical metropolis. *Science of the Total Environment* 765: 142786. https://doi.org/10.1016/j.scitotenv.2020.142786.

Guan, C.Y., Hu, A., and Yu, C.P. (2019). Stratified chemical and microbial characteristics between anode and cathode after long-term operation of plant microbial fuel cells for remediationof metal contaminated soils. *Science of the Total Environment* 670: 585–594. https://doi.org/10.1016/j.scitotenv.2019.03.096.

Gulamhussein, M. and Randall, D.G. (2020). Design and operation of plant microbial fuel cells using municipal sludge. *Journal of Water Process Engineering* 38: 101653. https://doi.org/10.1016/j.jwpe.2020.101653.

Gustave, W., Yuan, Z.F., Ren, Y.X. et al. (2019). Arsenic alleviation in rice by using paddy soil microbial fuel cells. *Plant and Soil* 441: 111–127. https://doi.org/10.1007/s11104-019-04098-7.

Hamelers, H.V., Ter Heijne, A., Sleutels, T.H. et al. (2010). New applications and performance of bioelectrochemical systems. *Applied Microbiology and Biotechnology* 85: 1673–1685. https://doi.org/10.1007/s00253-009-2357-1.

He, L., Du, P., Chen, Y. et al. (2017). Advances in microbial fuel cells for wastewater treatment. *Renewable and Sustainable Energy Reviews* 71: 388–403. https://doi.org/10.1016/j.rser.2016.12.069.

Helder, M., Strik, D.P.B.T.B., Hamelers, H.V.M. et al. (2010). Concurrent bio-electricity and biomass production in three plant-microbial fuel cells using *Spartina anglica*, *Arundinella anomala* and *Arundo donax*. *Bioresource Technology* 101: 3541–3547. https://doi.org/10.1016/j.biortech.2009.12.124.

Hirose, A., Kasai, T., Koga, R. et al. (2019). Understanding and engineering electrochemically active bacteria for sustainable biotechnology. *Bioresources and Bioprocessing* 6: 10. https://doi.org/10.1186/s40643-019-0245-9.

Hohmann-Marriorr, M. F., Blankenship, R. E. (2011). Evolution of photosynthesis. *Annual Review of Plant Biology* 62:515-548. https://doi.org/10.1146/annurev-arplant-042110-103811.

Hu, Y., Han, X., Shi, L., and Cao, B. (2022). Electrochemically active biofilm-enabled biosensors: current status and opportunities for biofilm engineering. *Electrochimica Acta* 428: 140917. https://doi.org/10.1016/j.electacta.2022.140917.

Hubenova, Y. and Mitov, M. (2012). Conversion of solar energy into electricity by using duckweed in direct photosynthetic plant fuel cell. *Bioelectrochemistry* 87: 185–191. https://doi.org/10.1016/j.bioelechem.2012.02.008.

Jingyu, H., Miwornunyuie, N., Ewusi-Mensah, D., and Koomson, D.A. (2020). Assessing the factors influencing the performance of constructed wetland-microbial fuel cell integration. *Water Science and Technology* 81: 631–643. https://doi.org/10.2166/wst.2020.135.

Jung, S., Lee, J., Park, Y.K., and Kwon, E.E. (2020). Bioelectrochemical systems for a circular bioeconomy. *Bioresource Technology* 300: 122748. https://doi.org/10.1016/j.biortech.2020.122748.

Kabutey, F.T., Zhao, Q., Wei, L. et al. (2019). An overview of plant microbial fuel cells (PMFCs): configurations and applications. *Renewable and Sustainable Energy Reviews* 110: 402–414. https://doi.org/10.1016/j.rser.2019.05.016.

Kadier, A., Kalil, M.S., Abdeshahian, P. et al. (2016). Recent advances and emerging challenges in microbial electrolysis cells (MECs) for microbial production of hydrogen and value-added chemicals. *Renewable and Sustainable Energy Reviews* 61: 501–525. https://doi.org/10.1016/j.rser.2016.04.017.

Kaku, N., Yonezawa, N., Kodama, Y., and Watanabe, K. (2008). Plant/microbe cooperation for electricity generation in a rice paddy field. *Applied Microbiology and Biotechnology* 79: 43–49. https://doi.org/10.1007/s00253-008-1410-9.

Katuri, K.P., Ali, M., and Saikaly, P.E. (2019). The role of microbial electrolysis cell in urban wastewater treatment: integration options, challenges and prospects. *Current Opinion in Biotechnology* 57: 101–110. https://doi.org/10.1016/j.copbio.2019.03.007.

Koffi, N.J. and Okabe, S. (2022). High electrical energy harvesting performance of an integrated microbial fuel cell and low voltage booster-rectifier system treating domestic wastewater. *Bioresource Technology* 359: 127455. https://doi.org/10.1016/j.biortech.2022.127455.

Kuleshova, T., Rao, A., Bhadra, S. et al. (2022). Plant microbial fuel cells as an innovative vesatile agro-technology for green energy generation combined with wastewater treatment and food production. *Biomass and Bioenergy* 167: 106629. https://doi.org/10.1016/j.biombioe.2022.106629.

Kumar, V.K., Mohan, K.M., Manangath, S.P. et al. (2020). Resource recovery from paddy field using plant microbial fuel cell. *Process Biochemistry* 99: 270–281. https://doi.org/10.1016/j.procbio.2020.09.015.

Lara, M.V., Drincovich, M.F., and Andreo, C.S. (2010). Transiciones metabólicas en la fijación fotosintética del carbono en plantas del género Portulaca. In: *C4 y CAM, características generales y uso en programas de desarrollo de tierras áridas y semiáridass* (ed. J.L. González and A. Chueca). Madrid, España: Consulado Superior de Investigaciones Científicas Chapter 3.

Li, Y., Duan, Y., Luebke, D., and Morreale, B. (2013). Advances in CO_2 capture technology: a patent review. *Applied Energy* 102: 1439–1447. https://doi.org/10.1016/j.apenergy.2012.09.009.

Li, S., Cheng, C., and Thomas, A. (2017). Carbon-based microbial fuel cell electrodes: from conductive supports to active catalysts. *Advanced Materials* 29: 1602547. https://doi.org/10.1002/adma.201602547.

Liu, Z., Sun, N., Yang, S. et al. (2013). Evolutionary transition from C3 to C4 photosynthesis and the route to C4 rice. *Biologia* 68: 577–586. https://doi.org/10.2478/s11756-013-0191-5.

Liu, X., Bo, Y., Zhang, J., and He, Y. (2015). Classification of C3 and C4 vegetation types using MODIS and ETM+ blended high spatio-temporal resolution data. *Remote Sensing* 7: 15244–15268. https://doi.org/10.3390/rs71115244.

Logan, B.E. and Regan, J.M. (2006). Electricity-producing bacterial communities in microbial fuel cells. *Trends in Microbiology* 14: 512–518. https://doi.org/10.1016/j.tim.2006.10.003.

Logan, B.E., Rossi, R., Ragab, A., and Saikaly, P.E. (2019). Electroactive microorganisms in bioelectrochemical systems. *Nature Reviews Microbiology* 17: 307–319. https://doi.org/10.1038/s41579-019-0173-x.

Lu, Z., Yin, D., Chen, P. et al. (2020). Power-generation trees: direct bioelectricity production from plants with microbial fuel cells. *Applied Energy* 268: 115040. https://doi.org/10.1016/j.apenergy.2020.115040.

Lutterbeck, C.A., Machado, E.L., Sánchez-Barrios, A. et al. (2020). Toxicity evaluation of hospital laundry wastewaters treated by microbial fuel cells and constructed wetlands. *Science of the Total Environment* 729: 138816. https://doi.org/10.1016/j.scitotenv.2020.138816.

Maddalwar, S., Nayak, K.K., Kumar, M., and Singh, L. (2021). Plant microbial fuel cell. Opportunities, challenges and prospects. *Bioresource Technology* 341: 125772. https://doi.org/10.1016/j.biortech.2021.125772.

Martins, F., Felgueiras, C., Smitkova, M., and Caetano, N. (2019). Analysis of fossil fuel energy consumption and environmental impacts in European countries. *Energies* 12: 964. https://doi.org/10.3390/en12060964.

Matsumoto, A., Nagoya, M., Tsuchiya, M. et al. (2020). Enhanced electricity generation in rice paddy-field microbial fuel cells supplemented with iron powders. *Bioelectrochemistry* 136: 107625. https://doi.org/10.1016/j.bioelechem.2020.107625.

Mohan, S.V., Mohanakrishna, G., and Chiranjeevi, P. (2011). Sustainable power generation from floating macrophytes based ecological microenvironment through embedded fuel cells along with simultaneous wastewater treatment. *Bioresource Technology* 102: 7036–7042. https://doi.org/10.1016/j.biortech.2011.04.033.

Moqsud, M. A., Yoshitake, J., Bushra, Q. S., Hyodo, M., Omine, K., Strik, D. (2015). Compost in plant microbial fuel cell for bioelectricity generation. *Waste Management* 36:63-69. https://doi.org/10.1016/j.wasman.2014.11.004.

Mu, C., Wang, L., and Wang, L. (2021). Removal of Cr(VI) and electricity production by constructed wetland combined with microbial fuel cell (CW-MFC): influence of filler media. *Journal of Cleaner Production* 320: 128860. https://doi.org/10.1016/j.jclepro.2021.128860.

Narasimhan, K., Basheer, C., Bajic, V. B., Swarup, S. (2003). Enhancement of plant – microbe interactions using a rhizosphere metabolomics – driven approach and its application in the removal of polychlorinated biphenyls. *Plant Physiology* 132:146-153. https://doi.org/10.1104/pp.102.016295.

Nitisoravut, R. and Regmi, R. (2017). Plant microbial fuel cells: a promising biosystems engineering. *Renewable and Sustainable Energy Reviews* 76: 81–89. https://doi.org/10.1016/j.rser.2017.03.064.

Oodally, A., Gulamhussein, M., and Randall, D.G. (2019). Investigating the performance of constructed wetland microbial fuel cells using three indigenous South African wetland plants. *Journal of Water Process Engineering* 32: 100930. https://doi.org/10.1016/j.jwpe.2019.100930.

Oon, Y.L., Ong, S.A., Ho, L.N. et al. (2016). Synergistic effect of up-flow constructed wetland and microbial fuel cell for simultaneous wastewater treatment and energy recovery. *Bioresource Technology* 203: 190–197. https://doi.org/10.1016/j.biortech.2015.12.011.

Oon, Y.L., Ong, S.A., Ho, L.N. et al. (2018). Up-flow constructed wetland-microbial fuel cell for azo dye, saline, nitrate remediation and bioelectricity generation: from waste to energy approach. *Bioresource Technology* 266: 97–108. https://doi.org/10.1016/j.biortech.2018.06.035.

Oon, Y.L., Ong, S.A., Ho, L.N. et al. (2020). Construct wetland-microbial fuel cell for azo dyes degradation and energy recovery. Influence of molecular structure, kinetics, mechanisms and degradation pathways. *Science of the Total Environment* 720: 137370. https://doi.org/10.1016/j.scitotenv.2020.137370.

Osorio-de-la-Rosa, E., Vázquez-Castillo, J., Castillo-Atoche, A. et al. (2021). Arrays of plant microbial fuel cells for implementing self-sustainable wireless sensor networks. *IEEE Sensors Journal* 21: 1965–1974. https://doi.org/10.1109/JSEN.2020.3019986.

Pamintuan, K.R.S. and Sanchez, K.M. (2019). Power generation in a plant microbial fuel cell assembly with graphite and stainless steel electrodes growing *Vigna Radiata*. *IOP Conference Series: Materials Science and Engineering* 703: 012037.

Pant, D., Van Bogaert, G., Porto-Carrero, C. et al. (2011). Anode and cathode materials characterization for a microbial fuel cell in half cell configuration. *Water Science and Technology* 63: 2457–2461. https://doi.org/10.2166/wst.2011.217.

Potter, M.C. (1911). Electrical effects accompanying the decomposition of organic compounds. *Proceedings of the Royal Society B* 84: 260–276. https://doi.org/10.1098/rspb.1911.0073.

Prasad, P.N. and Kalla, S. (2021). Plant-microbial fuel cells: a bibliometric analysis. *Process Biochemistry* 111: 250–260. https://doi.org/10.1016/j.procbio.2021.10.001.

Prasad, J. and Tripathi, R.K. (2018). Plant microbial fuel cell energy harvesting boost converter with/without the super capacitor. *Majlesi Journal of Mechatronic Systems* 7: 7–13.

Putranto, A.W., Sugiarto, A.W., Kusumarini, N. et al. (2018). The effect of urea fertilizer and electrode gaps toward the voltage of plant microbial fuel cell based on *Oryza sativa*. *Jurnal Teknologi Pertanian 19*: 43–50. https://doi.org/10.21776/ub.jtp.2018.019.01.5.

Rabaey, K., Van de Sompel, K., Maignien, L. et al. (2006). Microbial fuel cells for sulfide removal. *Environmental Science & Technology* 40: 5218–5224. https://doi.org/10.1021/es060382u.

Raines, C. A. (2011). Increasing photosynthetic carbon assimilation in C3 plants to improve crop yield: current and future strategies. *Plant Physiology* 155:36-42. https://doi.org/10.1104/pp.110.168559.

Raven, J. A. (1984). Physiological correlates of the morphology of early vascular plants. *Botanical Journal of the Linnean Society* 88:105-126. https://doi.org/10.1111/j.1095-8339.1984.tb01566.x.

Raya-Pérez, J.C. and Aguirre-Mancilla, C.L. (2008). Aparición y evoluación de la fotosíntesis C4. *Revista Chapingo Serie Ciencias Forestales y del Ambiente* 14: 45–50.

Reece, J.B., Urry, L.A., Cain, M.L., and Wasserman, S.A. (2014). *Campbell Biology*. New York, NY, USA: Pearson.

Regmi, R., Nitisoravut, R., and Ketchaimongkol, J. (2018). A decade of plant-assisted microbial fuel cells: looking back and moving forward. *Biofuels* 9: 605–612. https://doi.org/10.1080/17597269.2018.1432272.

Rusyn, I.B. (2014). Bioelectricity of plant-microbe associations of urban soi in a park areas. In: *Proceedings of the 1st International Academic Congress. Fundamental and Applied Studies in the Pacific and Atlantic Oceans Countries*, vol. vol. 2, 75–78. Tokyo: University Press.

Rusyn, I. (2021). Role of microbial community and plant species in performance of plant microbial fuel cells. *Renewable and Sustainable Energy Reviews* 152: 111697. https://doi.org/10.1016/j.rser.2021.111697.

Rusyn, I.B., Medvediev, O.V., and Valko, B.T. (2021). Enhancement of bioelectric parameters of multi-electrode plant-microbial fuel cells by combining of serial and parallel connection. *International Journal of Environmental Science and Technology* 18: 1323–1334. https://doi.org/10.1007/s13762-020-02934-3.

Saeed, T., Majed, N., Yadav, A.K. et al. (2022). Constructed wetlands for drained wastewater treatment and sludge stabilization: role of plants, microbial fuel cell and earthworm assistance. *Chemical Engineering Journal* 430: 132907. https://doi.org/10.1016/j.cej.2021.132907.

Salinas-Juárez, M.G., Roquero, P., Del Durán-Domínguez-de-Bazúa, M., and C. (2016). Plant and microorganisms support media for electricity generation in biological fuel cells with living hydrophytes. *Bioelectrochemistry* 112: 145–152. https://doi.org/10.1016/j.bioelechem.2016.02.007.

Salisbury, F.B. and Ross, C.W. (1991). *Plant Physiology*. Beverly, MA, USA: Wadsworth Publishing Company.

Santoro, C., Arbizzani, C., Erable, B., and Ieropoulos, I. (2017). Microbial fuel cells: from fundamentals to applications. A review. *Journal of Power Sources* 356: 225–244. https://doi.org/10.1016/j.jpowsour.2017.03.109.

Sarma, P.J. and Mohanty, K. (2018). *Epipremnum aureum* and *Dracaena braunii* as indoor plants for enhanced bio-electricity generation in a plant microbial fuel cell with electrochemically modified carbno fiber brush anode. *Journal of Bioscience and Bioengineering* 126: 404–410. https://doi.org/10.1016/j.jbiosc.2018.03.009.

Sathe, S.M., Chakraborty, I., Dubey, B.K., and Ghangrekar, M.M. (2022). Microbial fuel cell coupled Fenton oxidation for the cathodic degradation of emerging contaminants from wastewater: applications and challenges. *Environmental Research* 204: 112135. https://doi.org/10.1016/j.envres.2021.112135.

Saz, C., Türe, C., Türker, O.C., and Yakar, A. (2018). Effect of vegetation type on treatment performance and bioelectric production of constructed wetland modules combined with microbial fuel cell (CW-MFC) treating synthetic wastewater. *Environmental Science and Pollution Research* 25: 8777–8792. https://doi.org/10.1007/s11356-018-1208-y.

Shaikh, R., Rizvi, A., Quraishi, M. et al. (2021). Bioelectricity production using plant-microbial fuel cell: present state of the art. *South African Journal of Botany* 140: 393–408. https://doi.org/10.1016/j.sajb.2020.09.025.

Shen, X., Zhang, J., Liu, D. et al. (2018). Enhance performance of microbial fuel cell coupled surface flow constructed wetland by using submerged plants and enclosed anodes. *Chemical Engineering Journal 351*: 312–318. https://doi.org/10.1016/j.cej.2018.06.117.

Solomon, E.P., Berg, L.R., and Martin, D.W. (2013). *Biology*. USA: Cengage Learning.

Sonu, K., Sogani, W., and Sharma, M.K. (2023). Performance evaluation of plant microbial fuel cell with Epipremnum aureum plant using composite anode made of corn cob and carbon rod. *International Journal of Environmental Studies* 80: 8–19. https://doi.org/10.1080/00207233.2021.1957620.

Srivastava, P., Dwivedi, S., Kumar, N. et al. (2017). Performance assessment of aeration and radial oxygen loss assisted cathode based integrated constructed wetland microbial fuel cell system. *Bioresource Technology* 244: 1178–1182. https://doi.org/10.1016/j.biortech.2017.08.026.

Srivastava, P., Gupta, S., Garaniya, V. et al. (2019). Up to 399 mV bioelectricity generated by a rice paddy-planted microbial fuel cell assisted with a blue-green algal cathode. *Environmental Chemistry Letters* 17: 1045–1051. https://doi.org/10.1007/s10311-018-00824-2.

Strik, D.P.B.T.B., Hamelers, H.V.M., Snel, J.F.H., and Buismas, C.J.N. (2008). Green electricity production with living plants and bacteria in a fuel cell. *International Journal of Energy Research* 32: 870–876. https://doi.org/10.1002/er.1397.

Sudirjo, E., Jager, P., Buisman, C.J.N., and Strik, D.P.B.T.B. (2019). Performance and long distance data acquisition via LoRa technology of a tubular plant microbial fuel cell located in a paddy field in west Kalimantan, Indonesia. *Sensors* 19: 4647. https://doi.org/10.3390/s19214647.

Sun, J., Xu, W., Yang, P. et al. (2019). Enhancing the performance of photo-bioelectrochemical fuel cell using grapheme oxide/cobalt/polypyrrole composite modified photo-biocathode in the presence of antibiotic. *International Journal of Hydrogen Energy* 44: 1919–1929. https://doi.org/10.1016/j.ijhydene.2018.11.111.

Takanezawa, K., Nishio, K., Kato, S. et al. (2010). Factors affecting electric output from rice paddy microbial fuel cells. *Bioscience, Biotechnology, and Biochemistry* 74: 1271–1273. https://doi.org/10.1271/bbb.90852.

Tapia, N.F., Rojas, C., Bonilla, C.A., and Vargas, I.T. (2017). Evaluation of Sedum as driver for plant microbial fuel cells in a semi-arid green roof ecosystem. *Ecological Engineering* 108: 203–210. https://doi.org/10.1016/j.ecoleng.2017.08.017.

Taylor, G., Garassino, F., Aarts, M.G.M., and Harbinson, J. (2023). Improving C3 photosynthesis by exploiting natural genetic variation: *Hirschfeldia incana* as a model species. *Food and Energy Security* 12: e420. https://doi.org/10.1002/fes3.420.

Teoh, T.P., Ong, S.A., Ho, L.N. et al. (2020). Up-flow constructed wetland-microbial fuel cell: influence of floating plant, aeration and circuit connection on wastewater treatment performance and bioelectricity generation. *Journal of Water Process Engineering* 36: 101371. https://doi.org/10.1016/j.jwpe.2020.101371.

Timmers, R.A., Rothballer, M., Strik, D.P.B.T.B. et al. (2012). Microbial community structure elucidates performance of *Glycera maxima* plant microbial fuel cell. *Applied Mirobiology and Biotechnology* 94: 537–548. https://doi.org/10.1007/s00253-012-3894-6.

Timmers, R.A., Strik, D.P.B.T.B., Hamelers, H.V.M., and Buisman, C.J.N. (2013). Electricity generation by a novel design tubular plant microbial fuel cell. *Biomass and Bioenergy* 51: 60–67. https://doi.org/10.1016/j.biombioe.2013.01.002.

Treesubsuntorn, C. and Thiravetyan, P. (2021). Suitable application of *Echinodorus Cordifolius* microbial fuel cells inoculated with *Bacilus Thuringiensis*. *Waste and Biomass Valorization* 12: 2237–2247. https://doi.org/10.1007/s12649-020-01024-2.

Waller, A.D. (1903). *Eight Lectures on the Signs of Life from Their Electrical Aspect*. USA: Murray.

Wang, C., Guo, L., Li, Y., and Wang, Z. (2012). Systematic comparison f C3 and C4 plants based on metabolic network analysis. *BMC Systems Biology* 6: S2–S9. https://doi.org/10.1186/1752-0509-6-S2-S9.

Wang, J., Song, X., Wang, Y. et al. (2017). Effects of electrode material and substrate concentration on the bioenergy output and wastewater treatment in air-cathode microbial fuel cell integrating with constructed wetland. *Ecological Engineering* 99: 191–198. https://doi.org/10.1016/j.ecoleng.2016.11.015.

Wei, J., Liu, X., Zhang, X. et al. (2014). Rhizosphere effect of *Scirpus triqueter* on soil microbial structure during phytoremediation of diesel-contaminated wetland. *Environmental Technology* 35: 514–520. https://doi.org/10.1080/09593330.2013.836549.

Weiss, I., Mizrahi, Y., and Raveh, E. (2010). Effect of elevated CO_2 on vegetative and reproductive growth characteristics of the CAM plants *Hylocereus undatus* and *Selenicereus megalanthus*. *Scientia Horticulturae* 123: 531–536. https://doi.org/10.1016/j.scienta.2009.11.002.

Wen, H., Zhu, H., Yan, B. et al. (2020). Treatment of typical antibiotics in constructed wetlands integrated with microbial fuel cells: roles of plant and circuit operation mode. *Chemosphere* 250: 126252. https://doi.org/10.1016/j.chemosphere.2020.126252.

Wetser, K., Liu, J., Buisman, C., and Strik, D. (2015). Plant microbial fuel cell applied in wetlands: spatial, temporal and potential electricity generation of *Spartina anglica* salt marshes and *Phragmites australis* peat soils. *Biomass and Bioenergy* 83: 543–550. https://doi.org/10.1016/j.biombioe.2015.11.006.

Widharyanti, I.D., Hendrawan, M.A., and Christwardana, M. (2021). Membraneless plant microbial fuel cell using water hyacinth (*Eichhornia crassipes*) for green energy generation and biomass production. *International Journal of Renewable Energy Development* 10: 71–78. https://doi.org/10.14710/ijred.2021.32403.

Xiao, X., Yu, H. Q. (2020). Molecular mechanisms of microbial transmembrane electron transfer of electrochemically active bacteria. *Current Opinion in Chemical Biology* 59:104-110. https://doi.org/10.1016/j.cbpa.2020.06.006.

Xu, L., Zhao, Y., Tang, C., and Doherty, L. (2018). Influence of glass wool as separaton on bioelectricity generation in a constructed wetland-microbial fuel cell. *Journal of Environmental Management* 207: 116–123. https://doi.org/10.1016/j.jenvman.2017.11.035.

Xu, H., Song, H.L., Singh, R.P. et al. (2021). Simultaneous reduction of antibiotics leakage and methane emission from constructed wetland by integrating microbial fuel cell. *Bioresource Technology* 320: 124285. https://doi.org/10.1016/j.biortech.2020.124285.

Yaday, A.K. (2010). Design and development of novel constructed wetland cum microbial fuel cell for electricity production and wastewater treatment. In: *Proceedings of 12th International Conference on Wetland Systems for Water Pollution Control (IWA)*, vol. vol. 1, Venice, Italy (ed. A.M. Cerioni and R. Motta), 1085–1089.

Yamori, W., Hikosaka, K., and Way, D.A. (2014). Temperature response of photosynthesis in C3, C4 and CAM plants: temperature acclimation and temperature adaptation. *Photosynthesis Research* 119: 101–117. https://doi.org/10.1007/s11120-013-9874-6.

Yang, X. and Chen, S. (2021). Microorganisms in sediment microbial fuel cells: ecological niche, microbial response, and environmental function. *Science of the Total Environment* 756: 144145. https://doi.org/10.1016/j.scitotenv.2020.144145.

Yang, Y., Zhao, Y., Tang, C., Xu, L., Morgan, D., Liu, R. (2020). Role of macrophyte species in constructed wetland-microbial fuel cell for simultaneous wastewater treatment and bioenergy generation. *Chemical Engineering Journal* 392:123708. https://doi.org/10.1016/j.cej.2019.123708.

Yang, Y., Zhao, Y., Tang, C. et al. (2021). Dual role of macrophytes in constructed wetland-microbial fuel cells using pyrrhotite as cathode material: a comparative assessment. *Chemosphere* 263: 128354. https://doi.org/10.1016/j.chemosphere.2020.128354.

Zhang, K., Wu, X., Wang, W. et al. (2021). Effects on plant location on methane emission, bioelectricity generation, pollutant removal and related biological processes in microbial fuel cell constructed wetland. *Journal of Water Process Engineering* 43: 102283. https://doi.org/10.1016/j.jwpe.2021.102283.

Zhang, H., Chao, B., Gao, X. et al. (2022). Effect of starch-derived organic acids on the removal of polycyclic aromatic hydrocarbons in an aquaculture-sediment microbial fuel cell. *Journal of Environmental Management* 311: 114783. https://doi.org/10.1016/j.jenvman.2022.114783.

Zhao, L., Deng, J., Hou, H. et al. (2019). Investigation of PAH and oil degradation along with electricity generation in soil using an enhanced plant-microbial fuel cell. *Journal of Cleaner Production* 221: 678–683. https://doi.org/10.1016/j.jclepro.2019.02.212.

Zhi, W., Ge, Z., He, Z., and Zhang, H. (2014). Methods for understanding microbial community structures and functions in microbial fuel cells: a review. *Bioresource Technology* 171: 461–468. https://doi.org/10.1016/j.biortech.2014.08.096.

Zhou, Y., Xu, D., Xiao, E. et al. (2018). Relationship between electrogenic performance and physiological change of four wetland plants in constructed wetland-microbial fuel cells during non-growing seasons. *Journal of Environmental Sciences* 70: 54–62. https://doi.org/10.1016/j.jes.2017.11.008.

Part II

The Diversity of Photosynthesis-Assisted Power Generation

The part presents an overview of the diversity of photosynthetic organisms such as anoxygenic photosynthetic bacteria and also oxygenic photosynthetic organisms such as cyanobacteria, microalgae, bryophytes, and other plants exploited in photosynthesis-assisted power generation. Electrogenic microalgae and the effects of biomass weight and light intensity, temperature, and pH on the performance of photosynthetic microbial fuel cells are considered. The dry-surface biofilms of purple photosynthetic bacteria are clarified. Recent studies of bryophytes as a promising object of photosynthesis-assisted power generation technology due to their high stress resistance and survival in a wide range of temperature conditions are presented. The effect of plant species on the efficiency of plant microbial fuel cell (PMFC) and the suitability of wetland and drought-resistant plants to photosynthesis-assisted power generation are discussed.

6

Insights on Algae-based Microbial Fuel Cells

Nivedha Jayaseelan[1], *Vennila Lakshmanan*[1,*], *Kanimozhi Kaliyamoorthi*[1], *Olikkavi Subashchandrabose*[1], *Tani Carmel Raj*[1], *and Sathish-Kumar Kamaraj*[2]

[1] *Department of Biochemistry and Biotechnology, Faculty of Science, Annamalai University, Chidambaram, Tamil Nadu, India*
[2] *Instituto Politécnico Nacional (IPN)-Centro de Investigación en Ciencia Aplicada y Tecnología Avanzada, Unidad Altamira (CICATA-Altamira), Altamira, Tamps., México*

6.1 Introduction

Fossil fuels account for 86% of global energy production, yet their usage degrades the environment (Arutyunov and Lisichkin 2017). Solar, wind, and bioenergy are sustainable and carbon-neutral, but further research is required to reduce operational costs and installation and also to increase efficiency and yields (Schenk et al. 2008). Alternative sustainable systems without producing carbon dioxide with high energy conversion efficiencies, such as photosynthetic algal microbial fuels (PAMFCs) and microbial fuel cells (MFCs), must be developed (Arun et al. 2020; Shukla and Kumar 2018).

Algae is an aquatic species that can perform photosynthesis. Because of the chlorophyll pigment, which gives certain algae their distinctive green color, they are referred to as "green algae" (Saad and Atia 2014). Green algae create more than half of the planet's photosynthetic product (Chapman 2013). It may be further separated into microalgae and macroalgae depending on their cell and size type (unicellular or multicellular) (Mashjoor et al. 2016). Algal fuel cells (AFCs)

[*]Corresponding author

Photosynthesis-Assisted Energy Generation: From Fundamentals to Lab Scale and In-Field Applications, First Edition. Edited by Sathish-Kumar Kamaraj and Iryna Rusyn.

are typically MFCs that employ algae as a biocatalyst. Using algae in a fuel cell has some advantages because it produces biomass that can be converted into biofuels in addition to power (Kannan and Donnellan 2021). Like bacterial MFCs, microalgae serve as biocatalysts in the cathode chamber and provide high power densities at small reactor volumes; however, these systems cannot produce economically viable power densities at larger volumes (Nookwam et al. 2022).

MFCs are favored for bioelectricity generation because they can sustainably generate power over the long term (Verma et al. 2021). The generation of biohydrogen and other value-added products may be made possible by the wastewater treatment potential of these inexpensive, compact electrical devices known as MFCs (Kumar et al. 2019). The aeration costs in the cathodic compartment comprise about half of the total operating expenses, which limits the MFC's performance (del Campo et al. 2013). The O_2 supply provided by the algae in the system boosts power output, allows for simultaneous CO_2 removal and strong cathode polarization resistance, and lowers aeration costs (Saba et al. 2017). PAMFCs include additional electron acceptors (especially nitrate and sulfate in the cathode compartment), and bioelectricity is expected to be created throughout the night. During the day, algae absorb CO_2 and ammonium and produce O^2 through photosynthesis (Helder 2012). Algal bioelectrochemical systems can use microalgae as biocatalysts to extract phosphate, nitrogen, and organic materials from wastewater (Bhambri and Karn 2020).

Because of algal photosynthesis, oxygen is generated, and during the daytime, there is a concurrent absorption of phosphorus (P), carbon dioxide (CO_2), and nitrogen (N) for cathodic reactions. Additionally, fresh or pretreated algal biomass is employed as a substrate in the anode compartment (Reddy et al. 2019). The cost of mechanical aeration in such bioelectrochemical systems can be significantly reduced. Algae-based MFC technologies have a number of advantages over conventional MFC, which include transferring electrons directly via electro-active protein, producing oxygen (electron acceptor) for cathodic reactions through photosynthesis, capturing carbon dioxide in situ, and removing P, C, and N from SMFC at a low cost (Kumar et al. 2021; Capodaglio et al. 2021). Biofilm production and low-cost harvesting techniques to produce value-added compounds such as lipids and pigments may reduce process costs (Siddiki et al. 2022; Devi et al. 2022). Our grasp of the several algae-based MFCs and their potential applications are significantly lacking due to the absence of in-depth research on the subject.

Therefore, this chapter provides an overview of the application of MFCs based on algae for bioelectricity generation and their potential for manufacturing other important things. Many algae-based MFCs have production limits set by the oxygen available in the cathode section. Other parameters to consider for improving the economics of the process include the kind of biocatalyst, MFC operation and installation, nutritional supply, light intensity, electrode size, membrane type, scalability, and maximum power density (Palanisamy et al. 2019). Moreover, we take a closer look at the efficiency of several algae-based MFCs and the factors that affect them. The designs and performance features of algae-based MFCs will be discussed in detail, highlighting future research directions in this field.

6.2 Algae-based Microbial Fuel Cells (AMFCs)

6.2.1 Microbial Carbon Capture Cells (MCCs)

Carbon capture is possible by transforming the MFC into an MCC (Jadhav et al. 2017). This presents a potential approach to remove organic matter from wastewater in an anodic chamber, recover energy in the form of electricity, and utilize CO_2 to produce algal biomass in a cathodic section, which can subsequently be collected and used as a feedstock for the manufacture of biodiesel (Gajda et al. 2015). The MCC combines MFC with several algae grown in cathodic chambers can capture CO_2 and supply oxygen to support cathodic reduction processes (Figure 6.1). Using algal species in cathodic chambers lowers the expense of external aeration, which is required in aqueous cathode MFC since they form oxygen during photosynthesis and make it available for cathodic reduction (Shukla and Kumar 2018).

In MCC with MFC, the useable algal biomass with photosynthetic activity is developed by utilizing the CO_2 produced during the oxidative breakdown of organic matter in the anodic chamber (Patel et al. 2021). This allows for simultaneous power generation, CO_2 sequestration, wastewater treatment, and biomass production. To effectively reduce CO_2 emissions while also providing the opportunity to generate electricity concurrently without external aeration, this method of introducing anodic off gases created during the oxidation of organic matter into the cathodic chamber encourages the growth of algae (Wang et al. 2010). The biological events that occur in an MCC's anodic and cathodic chambers are shown in Eq. (6.1) through Eq. (6.4).

In the anodic chamber:

$$CH_3COO^- + 2H_2O \rightarrow 2CO_2 + 7H^+ + 8e^- \tag{6.1}$$

Two reactions, light-dependent and light-independent, occur in an MCC's cathodic chamber, which is illustrated in Eqs. (6.2), (6.3), and (6.4), respectively.

Light-dependent reaction in the cathodic chamber:

$$nCO_2 + nH_2O \rightarrow \left(CH_2O\right)_n + nO_2 \tag{6.2}$$

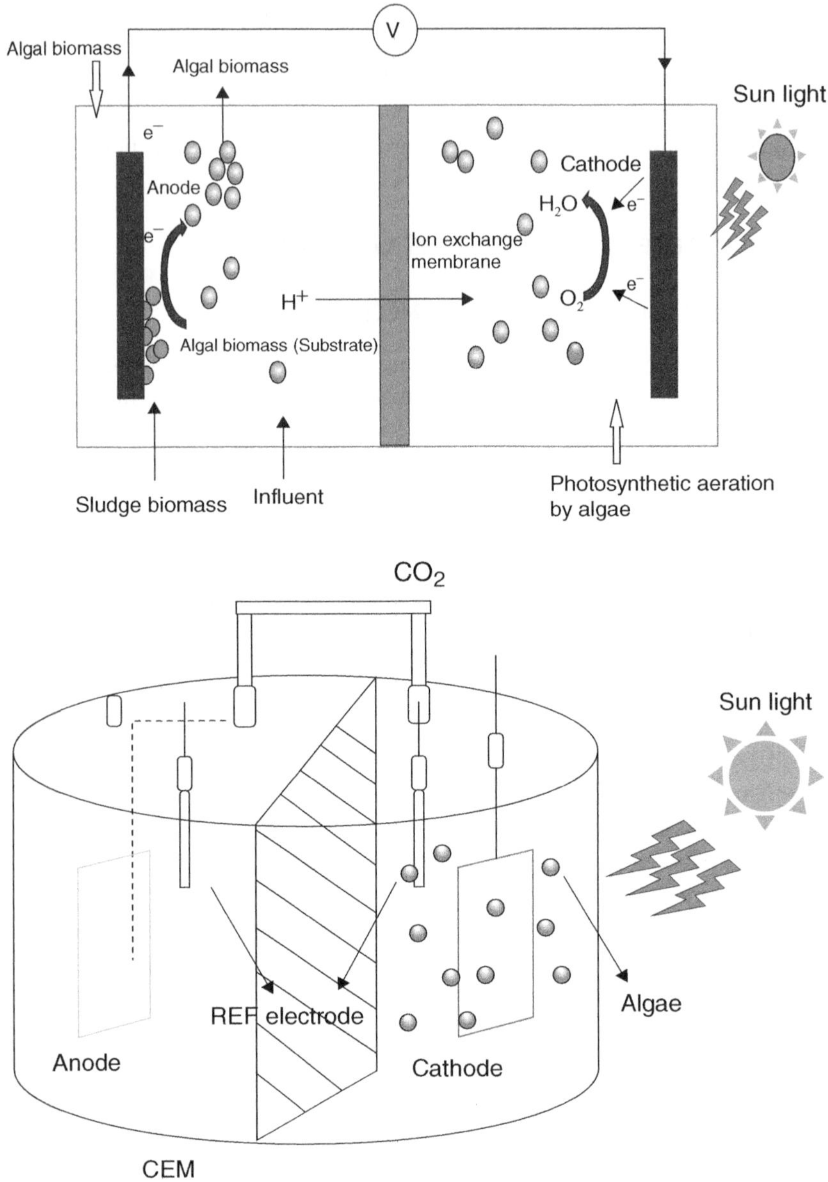

Figure 6.1 Schematic representation of flat plate MFC and MCCs with CO_2 exchange from the anode compartment.

$$2O_2 + 8e^- + 8H^+ \rightarrow 4H_2O \quad (6.3)$$

Light-independent reaction in the cathodic chamber:

$$C_2H_4O_2 + 2O_2 \rightarrow 2CO_2 + 2H_2O \quad (6.4)$$

6.2.2 Sediment Microbial Fuel Cells (SMFC)

Long-term water pollution may originate from the sediments found in aquatic environments (Xue et al. 2020). Animal and plant waste, the breakdown of dead bacteria and feces, plankton, and man-made organic substances all contribute to the formation of soil and sediments (Zhou et al. 2009; Round 1984). Organic carbon levels (26%) in sediments can be between 0.4 and 2.2 wt. Therefore, in some areas, the organic carbon content of sediments may be considered an adequate energy source (Xu et al. 2015). Sediment remediation has often utilized a wide variety of physicochemical techniques, many of which are both expensive and resource-intensive (Ahmed et al. 2021).

Nonetheless, exoelectrogens, which carry electrons beyond the cell, can metabolize this organic carbon. Sediment bioremediation using sediment microbial fuel cells (SMFCs) was recently created because of its low cost and environmental friendliness (Zabihallahpoor et al. 2015; Zhang et al. 2022). SMFCs are membraneless MFCs with a distinct capacity to process sediments free of organic molecules. It is a basic MFC design that harnesses the power of water. Low-power SMFCs can generate bioelectricity (He et al. 2017). In addition to electricity generation, SMFCs improve the oxidation of organic molecules in buried soil (De Schamphelaire et al. 2008). An anode electrode is buried in anaerobic silt, and a cathode electrode is hung above the water's surface, completing an electrical circuit that generates electricity. Dissolved oxygen (DO) gradient throughout the water depth and membranes is not necessary for proton transmission in SMFCs (Abbas et al. 2017). Similar MFCs do not have an anaerobic anode as SMFCs do. The studies conducted by Reimers and Tender around ten years ago served as inspiration for the development of the first fully working SMFCs. Various organic-rich aquatic sediments have been used in research, including river, marine, freshwater, paddy, rice, and other underwater sediments. Benthic microbial fuel cells (BMFCs) and marine microbial fuel cells (MMFCs) are two types of SMFCs (Zabihallahpoor et al. 2015).

Many crucial processes control microorganisms' involvement in SMFCs at the anode and cathode. Microorganisms from the *Geobacteracea* family and the *Desulfucapsa* or *Desulfobulbus* species dominate the anode biofilm. *Desulfucapsa* and *Desulfobulbus* species oxidize the S_0 generated by the anode to SO_4^{2-}, whereas

Geobacteracea directly oxidizes acetate in the sediment to reduce the anode (Abbas and Rafatullah 2021). Anaerobic bacteria in sediment produce acetate through the fermentation of organic matter (e.g. clostridium). Another reaction at the anode was the oxidation of S^{2-} to S^0. MnO_4, O_2, SO_4^{2-}, and Fe_2O_3 are reduced in that order between the sediment surface and the anode when organic material is oxidized (Yadav et al. 2022). Increasing sediment depth results in a buildup of more powerful reductants in each successive sediment layer (Egger et al. 2017). Microorganisms abundant on anode surface or dissolved and solid-phase comprised of reduced ions in the soil can transfer electrons generated by active biocatalysts (e.g. sulfides in marine sediment) (Gambino et al. 2021; Hennebel et al. 2015). In certain MFCs, an electron mediator is used to shuttle produced electrons to the anode surface, while a proton exchange membrane channel produces protons to the cathode area. Several uses for SMFCs are not possible with more conventional energy sources. High efficiency, low cost, low operating temperatures, simple construction, less frequent maintenance requirements (e.g. periodic replacement), abundant and inexpensive fuel resources, the absence of toxic byproducts, and the ability to be placed in remote areas are the most important benefits of direct conversion of organic matter into electricity by SMFCs (Algar et al. 2020). Low working voltage, nonlinear scaling up, difficulty in producing continuous power, and low cell potential are some problems with SMFCs (Figure 6.2).

6.3 The Implementation of Algae in MFCs

6.3.1 MFCs with algae-assisted cathodic compartment

About 50% of MFC running costs may be attributed to the price of mechanical aeration (Arun et al. 2020). Microalgae-based MFCs may have a higher energy efficiency due to the algal photosynthesis in the cathodic compartment, which may provide oxygen for the cathodic processes (Elshobary et al. 2021) (Figure 6.3a and b). Using this symbiotic relationship, which has the benefit of capturing CO_2 and providing power at the same time, might pave the way for the sustainable development of MFCs. The surplus algal biomass is employed as an electron donor in the anodic compartment, while the CO_2 is utilized as an inorganic carbon source for algae growth (Mekuto et al. 2020). Figure 6.3 depicts the many pathways by which electrons might be delivered from the electrode to the algae in the cathodic compartment. The direct passage of electrons from the cathode to the algae is shown in Figure 6.3a, with carbon dioxide serving as the acceptor (Saba et al. 2017). Extracellular electron transport is facilitated by mediators that do not require an external energy source, such as the c-type cytochromes in the outer

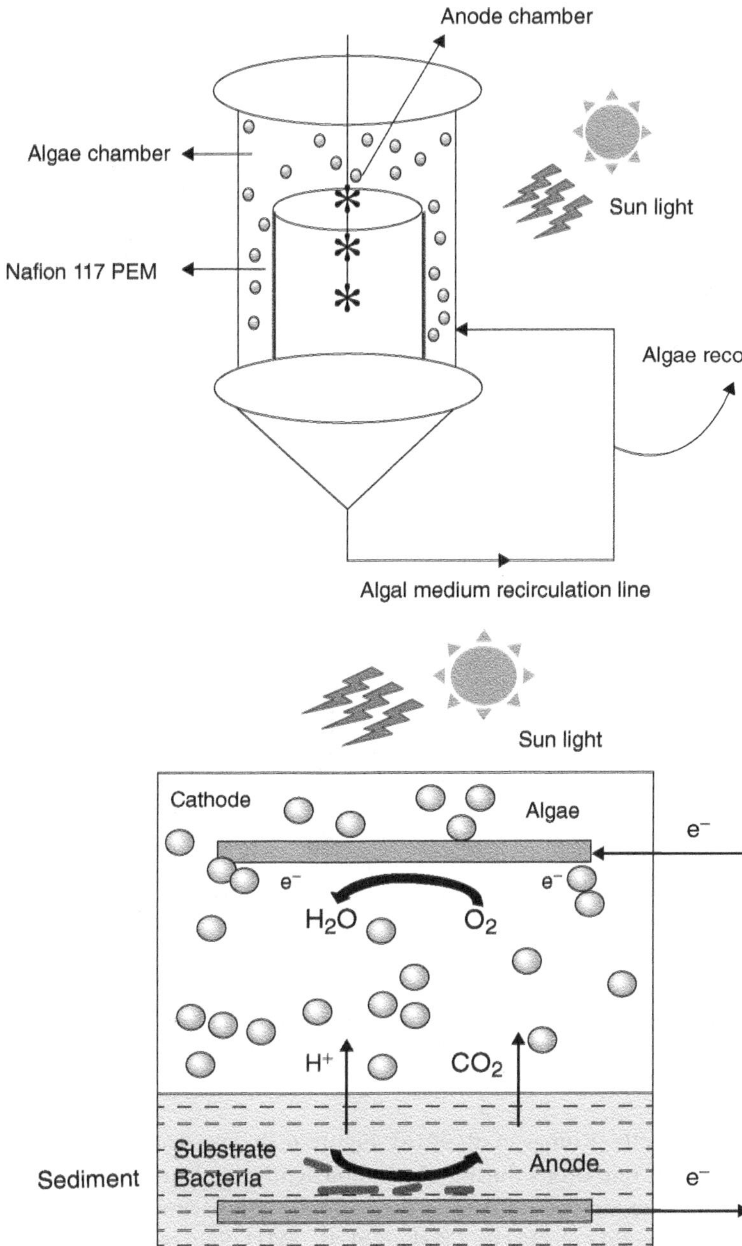

Figure 6.2 Schematic representation of tubular PMFC and sediment type PMFC.

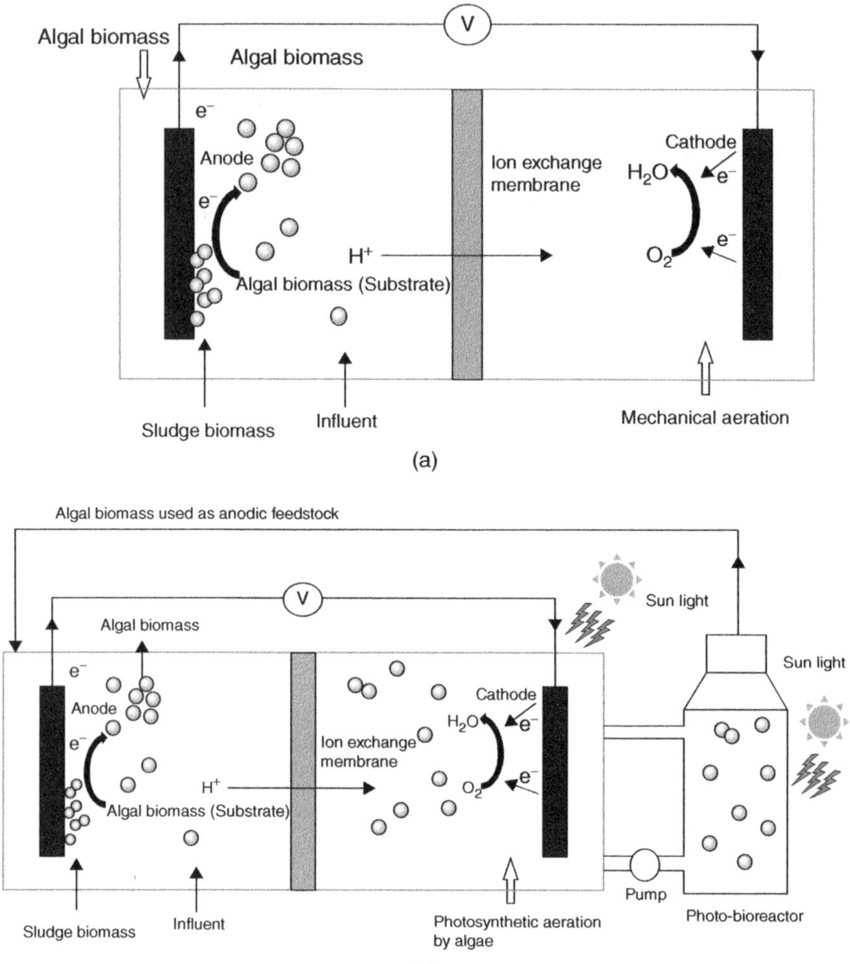

Figure 6.3 Algae-based MFCs with (a) direct utilization of algal biomass as anodic substrate and (b) harvested biomass from the cathodic compartment used as feedstock for the anodic compartment.

membrane of algal cells (Kumar et al. 2016). Figure 6.3b demonstrates how the oxygen consumption rate drops during algal photosynthesis. Numerous studies have shown that algae may be utilized in the cathodic compartment of a membrane fuel cell to generate electricity (Taşkan et al. 2021). Successful electricity generation in MFCs was achieved with an algae-supported cathode at a maximum power density of 21 mW/m^2, which was improved to 38 mW/m^2 by altering the cathode (Table 6.1) (covering it with a carbon nanotube) (Dyantyi et al. 2020; Jaswal et al. 2020).

Wang et al. (2010) looked at the CO_2 removal and power output values of algae-based MFCs. They measured the absorption at 658 nm for a variety of algal concentrations, from 0.21 to 0.85 OD_{658}. Algal photosynthesis appears to have been successful, as the DO level dropped from 7.6 to 0.9 mg/l (Lin et al. 2002). This indicates that it participated in a cathodic process involving an abundance of inorganic carbon as an electron acceptor (Rosenbaum et al. 2011). By utilizing algae in both the anode (lipid-extracted algal [LEA] biomass was used as an electron donor for the anodic reaction) and the cathode (oxygen produced by algal photosynthesis was used as an electron acceptor for the cathodic reaction), algae-based MFCs were able to achieve a maximum power density of 2.7 W/m^3 (Khandelwal et al. 2021).

6.3.2 Biomass-derived Algae as a Substrate in the Anodic Compartment

Algal biomass is a complex organic substance that may be utilized successfully in the anode's half-cell to produce bioelectricity. Therefore, there is no requirement to prepare it before utilizing it as a substrate in MFCs (Greenman et al. 2019; Obileke et al. 2021). The procedures used in MFCs that use wastewater as the substrate for the anodic compartment are comparable to those used in MFCs that feed algae for the goal of producing bioelectricity from wastewater (Figures 6.2b and 6.3). The maximum power outputs have been recorded to be 1.78, 0.85, and 0.56 W/m^2 for dry algal biomass concentrations of 5, 2, and 1 g/l, respectively (Myers 2015) (Table 6.1). This proves the viability of the algal-fed MFC process and shows that the bioelectricity produced is equal to that produced by using other organic substrates, such as acetate, which produces a maximum power density of 0.661 W/m^2 (Kakarla and Min 2014). In the anodic compartment of MFCs, *Chlorella vulgaris* and *Ulva lactuca* were fed by Velasquez et al. in a dry powder form. Compared to the *U. lactuca* MFC, which produced 2.0 kW h/kg dry weight, the *C. vulgaris* MFC produced 2.5 kWh/kg dry weight (Velasquez-Orta et al. 2009). The output of the *C. vulgaris* MFC was 2.5 kWh/kg dry weight, but the *U. lactuca* MFC had a maximum power density of 0.98 W/m^2. Several concentrations of algal biomass (0.25, 0.39, 0.56, and 0.72 g/l) were fed to MFCs by Gajda et al. (2015), and the values for power output that were obtained varied depending on the concentration of the algal biomass and were 10, 17, 11, and 17 μW, respectively (Gajda et al. 2015) (Table 6.1).

Walter et al. examined the usage of fresh algal biomass in a cascade of 9 MFCs. The fresh algal culture with 6×10^5 cells/ml had a maximum output of 42 W/m^3 (Walter et al. 2015). Using *C. vulgaris* and *Microcystis aeruginosa* from an algal bloom in an aquatic environment, Wang et al. (2010) recorded maximum volumetric power densities of 3.70 W/m^3 (*C. vulgaris*) and 4.14 W/m^3 (*M. aeruginosa*). Watts per cubic meter were used to measure these densities (Buckley et al. 2015).

Table 6.1 Power densities of various algal microbial fuel cells.

Organism	Anode|Cathode	Reactor setup|anolyte|catholyte	Power or cell voltage	References
C. vulgaris	Carbon felt | Carbon fiber cloth Bubbling	PAMFC | Anaerobic sludge with phosphate buffer supplemented with glucose | BG-11	1108.9 mW/m^3	(Li et al. 2019)
C. vulgaris	Graphite | Sheet copper	DCMFC | Anaerobic sludge | Aerated BG-11	19,151 mW/m^3	(Hou et al. 2016)
Mixed Algal Culture	Carbon Fiber Brush (CFB) | CFB	DCMFC | Leachate wastewater (WW) | Leachate WW	330 mV	(Nguyen and Min 2020)
Scenedesmus sp.	Graphite plate | Graphite plate	DCMFC | Diluted food WW | BG-11	28.5 ± 0.3 mW/m^2	(Naina Mohamed et al. 2019)
C. vulgaris	Nickel foam|Graphene	DCMFC | Activated sludge | BG-11	36.4 mW/m^2	(Zhang et al., 2018)
Mixed Algal Culture	CFB | Carbon cloth	PAMFC | Anaerobic sludge | Anolyte effluent anaerobic sludge	268 mW/m^2	(Jiang, 2017)
C. pyrenoidosa	Carbon felt | Carbon felt	DCMFC | Mixed anaerobic sewage sludge in synthetic WW | BG-11	6400 mW/m^3	(Jadhav et al. 2017)
S. obliquus	Toray carbon paper | Toray carbon paper	DCMFC | Municipal wastewater replaced later with BBM | Ferricyanide	951 mW/m^3	(Kondaveeti et al. 2014)
Mixed Algal Culture	Carbon brush | Carbon cloth	PAMFC | Anaerobic sludge | Anolyte effluent anaerobic sludge	3300 mW/m^3	(Tse et al. 2016)
Mixed Algal Culture	CFB | CFB	DCMFC | Anaerobic sludge | Leachate WW	11 mV	(Nguyen et al. 2017)
S. obliquus	Plain carbon paper | Platinum-coated carbon paper	DCMFC | Municipal wastewater | Bold's Basal Medium (BBM)	153 mW/m^2	(Kakarla and Min 2014)
S. quadricauda SDEC-8	Carbon cloth woven with Ti | Carbon cloth woven with Ti	SCMFC | Domestic sewer WW | Domestic sewer WW	62.93 mW/m^2	(Yang et al. 2018)
Golenkinia sp. SDEC-16	Graphite | Sheet copper	SCMFC | Kitchen waste anaerobically digested effluent (KWADE) | KWADE	6150 mW/m^3	(Hou et al. 2016)

The cathode compartment's LEA may also be used as a substrate for the anodic half-cell. The maximum power density of this substrate capable of handling is 67.07 mW/m^2 (Rashid et al. 2013). Bacteria that reside in the anode chamber of MFCs are typically unable to fully use the algal biomass because of its thick cell wall. The solution to this issue is first to pretreat the algal biomass by utilizing a variety of methods, including sonication, thermal treatment, or microwave therapy (Reddy et al. 2019). For instance, heating and sonicating 5 g/l of dry algal biomass produced a maximum power density of 1.78 W/m^2 (Mirsiaghi 2016). The *Laminaria* brown algae biomass was used as the substrate in the anodic compartment of MFCs by autoclaving and microwave irradiating it (Gadhamshetty et al. 2013). The power density energy generated through autoclave-based pretreatment (218 mW/m^2) significantly outperformed both microwave irradiation (118 mW/m^2) and the absence of pretreatment (86 mW/m^2) (Rathinavel et al. 2018). The findings of these studies suggest that a number of pretreatment techniques can effectively increase the usage of algal biomass in MFCs (Figure 6.4).

6.4 The Wastewater Treatment Using Algae-assisted MFCs (AMFCs)

6.4.1 COD Reduction in Algae-based MFC with the Use of Algal Biomass

A wide range of algal biomass products generated in the cathode compartment can be used in algae-based solid-fuel cells (Chia et al. 2022). Some products include fresh, untreated algal biomass recovered from lipids, pretreatment algal biomass, algal powder, and pre-digested algal biomass (Lakaniemi et al. 2012). For a better understanding of the viability of using algal biomass as chemical oxygen demand (COD) in the anode compartment to increase the efficiency of producing bioelectricity and removing COD, prior studies assessed the performance of algae-based MFCs with a range of biomass concentrations (1–5 g COD/l), in order to understand the possibility of employing algal biomass as COD in the anode compartment (Yuan et al. 2011). The current and the power density were impacted by how much biomass was present in the algae. In a recent study, the COD of pretreated *Scenedesmus obliquus* algal biomass was analyzed to see whether employing it as the substrate for the anode in MFCs would be feasible on a total COD (tot) and soluble COD (sol) level (Singh et al. 2019).

However, processed algal material produced CODtot and CODsol removal efficiency of 59.2% and 66.4%, respectively, compared to raw algal biomass, which is readily biodegradable (Yuan et al. 2011). The processed and unprocessed algal biomass possess a huge difference. The greatest COD removal efficiency for MFCs

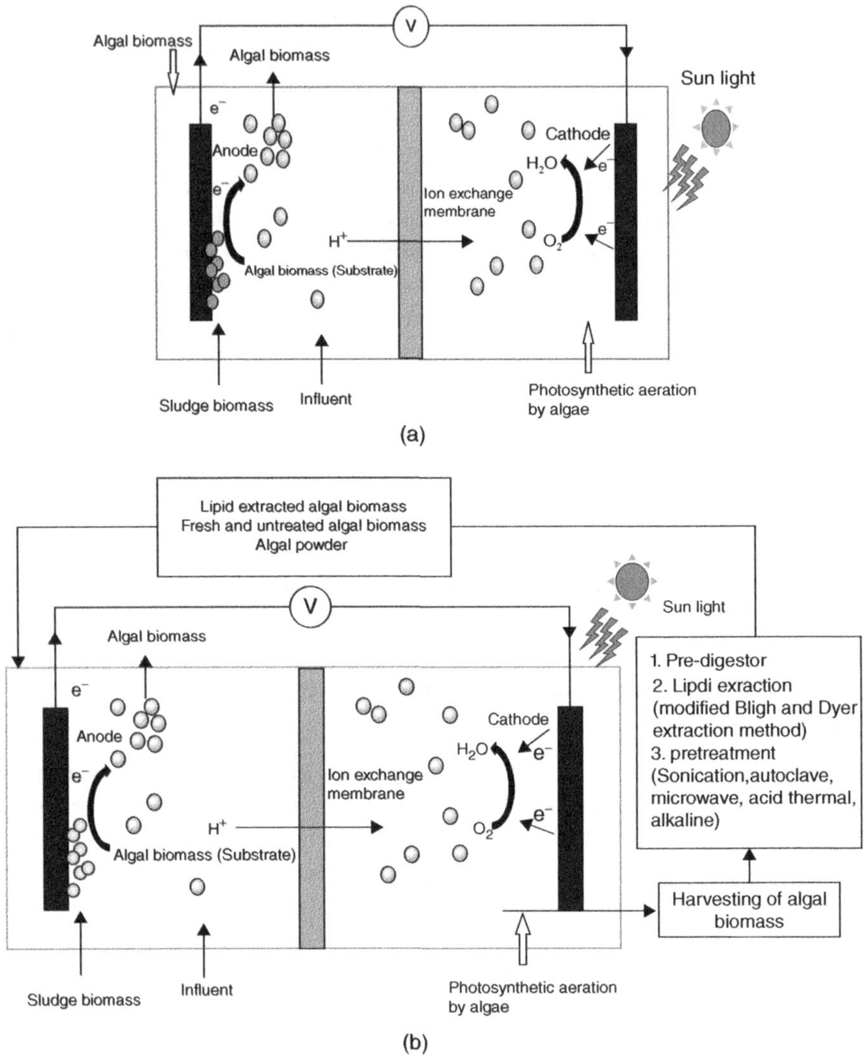

Figure 6.4 Algae-based MFCs: (a) direct use of algae as an electron acceptor for cathodic reaction biomass and (b) harvested biomass from cathodic compartment followed by different pretreatment techniques used as feedstock for the anodic compartment (electron donor).

fed with LEA biomass and fruit pulp waste, respectively, was determined to be 70.8% and 62.3% by Khandelwal et al. (2021). For LEA biomass, the rates of COD removal were 0.29 (0.11) kg/m^{3}d, while for waste from fruit pulp, they were 0.28 (0.04) kg/m^{3}d (Khandelwal et al. 2021).

6.4.2 The Removal of Nitrogen and Phosphorus Utilizing AMFCs at the Cathode

In recent years, microalgae have developed a reputation as appealing organisms that may be used in bioprocesses to remove phosphate and nitrogen (Cuellar-Bermudez et al. 2017; Oliveira et al. 2021). They work well at removing N, P, and COD from wastewater and producing oxygen, which may be utilized as an electron acceptor in algae-supported cathodes. Furthermore, they have the ability to generate oxygen on their own. *Chlamydomonas reinhardtii* and *Pseudokirchneriella subcapitata* used an integrated photo biochemical system (MFC inside an algal reactor) to reduce nitrogen (98% as ammonium) and phosphorus (82% as phosphate) from municipal wastewater (Azarpour et al. 2022). In the anode and cathode chambers, nitrogen and phosphorus were both removed. Fifty-five percent of the phosphate was eliminated by algal activity in the cathode chamber, while the remaining 27% was eliminated by bacterial activity in the anode chamber (Tao et al. 2014).

An integrated system consisting of a photo-bioreactor and an up-flow membrane-less MFC has a substantial level of potential for removing nitrogen (58.7% ammonium) and phosphate (99% phosphate) (Ghangrekar et al. 2020). It has been used successfully to fix CO_2 from flue gas produced during fermentation and burning fossil fuels and treat digestate from anaerobic digestion (AD) of food waste with a low COD/N ratio (Uggetti et al. 2014). Additionally, an integrated system has treated digestate from AD of food waste with a high COD/N ratio.

6.4.3 The Recuperation of Compounds with Value-added Microorganisms Based on Algae

6.4.3.1 Formation of Photosynthetic Biofilm on the Cathode

An alternate strategy for reducing the expenses related to biomass collection is the production of a biofilm on the cathode electrodes in photosynthetic algal-assisted MFCs (PAMFCs), also known as algae cellulose, algae wire, or algae string (Don and Babel 2021). With this technique, the algae may be removed from the growing media by simply scraping. The quantity of biofilm adhesion on the electrode surface impacts the amount of oxygen transport, MFC performance, corrosion, and light limitation in this approach (Anam et al. 2021). In MFCs with algae, algae act as a mediator for direct oxygen supply to the electrode. They do so up to a particular thickness of the biofilm present on the electrode surface (Kakarla and Min 2014). The extracellular polymeric substance (EPS) matrix created on the electrode surface may be thoroughly studied in photosynthetic algae biofilms on cellulose (Logroño et al. 2017).

It has been demonstrated that the type of cathode (surface texture, roughness, or porosity) and the kind of material can significantly impact the formation of biofilms on cathode electrodes (better attachment) (Santoro et al. 2014).

Algal biomass developed on the surfaces of algae wire more slowly than on algae string and algae cellulose surfaces due to corrosion brought on by stainless steel (Ghirardi et al. 1996). All three electrodes encouraged algae growth on the cathode electrodes' surface to boost electricity output. The development of the biofilm was influenced by the pH and CO_2 concentrations, the quantity of light, and the mass transfer conditions (Rincon et al. 2017). According to recent research, Scenedesmus obliquus may successfully build a biofilm on a platinized cathode by utilizing the oxygen produced during this process as an electron acceptor (Gaur et al. 2018). A significant exchange of current density for oxygen reduction on the electrode was found in this particular research. This was caused by algal biofilms making direct oxygen transport easier. The algal biomass trapped inside the biofilm system was removed by scraping. The development of an algal biofilm looks to be a viable technique for harvesting biomass, even if it is unclear how it may be used in MFCs for the production of energy or the collecting of microalgae for the synthesis of other compounds with added value. However, additional study on the development of algal biofilm is required for increased O_2 transfer, improved MFC efficiency, high algal biomass production, and effective harvesting from the cathode compartment (Elmaadawy et al. 2020).

6.4.4 Carbon Dioxide (CO_2) Removal by the Use of MFCs with Algae-assisted Cathodes

Carbon dioxide (CO_2) is a source of energy that microalgae may use to power their autotrophic development (Doucha et al. 2005). As a result, when the microalgae are exposed to light, they may absorb CO_2 from the anode compartment and turn it into oxygen for the cathodic process. The quantity of CO_2 emitted by the anode chamber and the amount of CO_2 absorbed by microalgae in the cathode chamber in these systems determine the amount of net CO_2 removal (Kim et al. 2021). The catholyte completely removed CO_2 from the anodic chamber in microbial carbon capture cells (MCCs) with a maximum power density of $5.6\,W/m^3$ (Konur 2015). Through wastewater treatment without mechanical aeration, this method may be used to produce electricity, create biodiesel, and trap carbon emissions all at once. An anoxic MFC was employed for CO_2 removal in the presence of light. This biocathode's anticipated average CO_2 reduction was 0.28 (0.02) mol bicarbonate per mole electron (Cao et al. 2009).

6.5 Photosynthetic Algae Microbial Fuel Cell (PAMFC)

The algae in the cathodic chamber of an AMFC serve as terminal electron acceptors for reduction. Therefore, unlike bacterial MFCs, mechanical aeration is no longer required since oxygenic photosynthesis generates enough DO (Yadav et al. 2020). Like higher plants, algae include chloroplasts and stacked discs known as thylakoids (Gu et al. 2022). The two protein-pigment photosynthetic systems, PSI and PSII, are present in the thylakoid membranes and are named after the order in which they were discovered. Chlorophyll and carotenoids are found in PSI, but PSII also contains xanthophyll, phycobilin, and chlorophyll. Chlorophyll absorbs light in the red range (650–750 nm), while carotenoids do so in the blue zone (450–500 nm) (Masojídek et al. 2013).

The light-dependent phase of photosynthesis begins at the light-harvesting complex (LHC) in PSII, even though the light is absorbed concurrently in PSI and PSII (LHC). The pigment molecules in PSII exchange their light energy with one another. Due to its photobleaching characteristics at 680 nm, this energy is then focused in the P680 reaction center, where it excites electrons to a higher energy state and induces a charge separation. The pool of the plastoquinone membrane then receives the electron (Germano et al. 2004).

PSII is also known as water-plastoquinone oxidoreductase because of the simultaneous events that occur throughout this process and the splitting of a water molecule (the terminal electron donor) to create electrons (Renger and Renger 2008). The electron from the water molecule replaces the lower energy electron from the LHC after being absorbed by an electron transporter. As a result, inside the cell, free O_2 is created. High-energy electrons lose energy through the electron transport chain and help H^+ ions chemiosmose across the membrane to create ATP (Nicholls and Ferguson 2002). A type of chlorophyll known as P700 in PSI experiences a comparable energy concentration, and the electron from PSII substitutes for the missing electron in the excited P700 electrons. It then generates $NADP^+$ and moves down the electron transport chain, where it is reduced to NADPH and produces ATP. The Calvin Cycle, the light-independent phase of algal photosynthesis when glucose is finally generated, cannot function without NADPH and ATP (Dorlon 2021).

6.5.1 Reactor Design – PAMFCs

Algal growth is possible in a variety of bioreactor designs (Duan and Shi 2014). PAMFCs may either add a bubble column to an existing dual-chamber MFC (DCMFC) cathode chamber or combine an algal photobioreactor with DCMFC. Numerous studies have discovered that switching to a PAMFC from a traditional DCMFC may enhance the amount of power produced and the amount of COD eliminated from a system (Kannan and Donnellan 2021). When there is no oxygen availability, Archaea employs AD to use the fermentative method to generate volatile fatty acids (VFA). The biogas output rose by 62% and 104%, respectively, by combining AD with bacterial MFCs and AFCs (Logroño et al. 2017).

Increased nitrogen and organic compound removal rates, enhanced pH stability, and increased AD output are just a few advantages of combining the AD and PAMFC systems. With a high potential of 490 mV reaching during the process from a starting potential of 270 mV, the AD-AMFC system still produced relatively little power. As a result, rather than producing fuel cell energy, the AD-PAMFC systems are utilized to increase the effectiveness of the AD process. Other reactor connection designs must be taken into account in order to enhance the latter.

6.5.2 Photosynthesis-related Factors Influencing Reactor Performance

6.5.2.1 Effect of Light on AMFCs

Since photosynthesis is necessary for an AMFC, it is critical to understand the significance of light's action and intensity. Any visible spectrum light source, such as fluorescent lights, LEDs, halogen lamps, and even the sun, produces algae at the cathode. According to the most efficient light source for algae growth, the mixture of red and blue LED light and fluorescent light may increase biomass output by 102% (Chen et al. 2011). The light source has no obvious effect since the organism, and the culture conditions govern the light affecting the biomass development (Yeh et al. 2010). Neither research has provided any data on the power the AFC system generates under various lighting situations. Carvalho et al. (2011) and Kommareddy and Anderson (2004) found that light with a wavelength between 400 and 500 nm is ideal for photosynthesis under the absorption of pigments involved in the process. This enables the use of light with a range of visible spectrum wavelengths. Red LED light (650–670 nm) promotes the quickest development of biomass (Tsai et al. 2017). While red LED light is being utilized, it is advantageous to have blue LED light. It is recommended to have very tiny amounts of white LED light to address any demands unrelated to photosynthesis. Similar to the studies mentioned above, further study is needed to determine the amount of power created in the system by these light sources. A power output of 62.7 mW/m^2 from the 96 $E/(m^2 s)$ high-intensity fluorescent white light was generated, which is ten times more than from the 26 $E/(m^2 s)$ low-intensity light (Gouveia et al. 2014).

However, it was also demonstrated that employing fluorescent light with intensities of more than 125 mW/cm^2 might potentially decrease the system's capacity to generate power (Juang et al. 2012). To promote mixotrophic development for the generation of biofuels or other bioproducts and maintain the health of the photosynthetic organism, algal culture in a light/dark period is crucial (Khan et al. 2018). One of the primary arguments in favor of selecting AFCs over bacteria MFCs is the fact that CO_2 produced in the system during the dark phase of photosynthesis is transformed into beneficial products like sugars and biomass. Because certain algae normally experience darkness and use the catholyte for development, they may adapt. The dark phase continues to be the rate-limiting stage in algal development. Suppose an AMFC is made to run to produce bioproducts; in that case, the suitable lighting time must be chosen because prior research has indicated that little to no power is created during the dark cycle (Xiao et al. 2012). The ideal light intensity for an AFC containing *C. vulgaris* was 5000 lx of white fluorescent light (Zhang et al. 2018). Even though the system reached these conclusions after 24 hours of nonstop light, it is still unclear how the algal cultures fared throughout the test. A 16 hours/8 hours light/dark regime only utilized 11% less electricity than continuous illumination, although being more advantageous economically. A distinct finding is that the type of substrate utilized for algal growth controls the impact of the lighting conditions more effectively than just light intensity (Patel et al. 2021). Together, the two experiments demonstrated that continuous light exposure is not required because it had less effect on biomass production than a light/dark cycle. It is eventually undesirable to use the extra energy to illuminate algae continually. The algal strain and system's intended output ultimately decide the most effective method to grow algae in an AFC, regardless of whether the system's intended output is predominantly energy production or biomass growth.

6.5.2.2 Genetic Modification of Algae for Improved Photosynthesis

Genetic modification is also necessary to perform research to increase photosynthesis without the constant requirement for light by altering biochemical procedures. Since photosynthesis is directly related to the production of required power. *Synechosystis* algae were genetically modified to boost the quantity of electrogenic transmission by a factor of five (Saar et al. 2018). The techniques used in this work increased the PSII gene expression, yielding a power density of $0.5\,W/m^2$ (Williams 1988). Algae have been attempted to be genetically altered to improve their ability to fix CO_2 through the process of photosynthesis as a proof of concept. To boost CO_2 absorption, the light-independent Calvin cycle was altered (Anto et al. 2020). Long-term costs are lower for fourth-generation algal biofuels, which use genetically altered algae to improve lipid assimilation and CO_2 absorption due to enhanced lipid absorption (Logroño et al. 2017). These genetic modifications boost the benefits of growing algae for carbon sequestration and the development of biofuels; however, their impact on the production of electricity is questionable and should be the subject of further inquiry.

6.6 Conclusion

In order to improve power generation, biomass control and biomass recovery are necessary to understand and control the mechanisms of electron transfer and CO_2 uptake in the cathode. Algae-assisted MFCs are capable of producing power more effectively. CO_2 and nutrient (N and P) removal are sought for commercial applications to recover value-added goods or biofuels. Under these circumstances, energy savings and benefits from resource recovery will outweigh the additional expenses related to gathering algal biomass, developing AMFCs, and supplying CO_2. Additional research is needed to scale up existing algae-based MFCs and develop new algae-based MFCs. However, more groundwork has to be done to identify the number of large-scale applications, including the examination of the effect of the reactor's parts and the amount of electricity produced by an AMFC system, tackling the AMFC system's economic and performance-based scale-up, using genetic engineering principles to enhance power production, and creating innovative ways to operate AMFC systems under direct natural sunlight.

References

Abbas, S.Z. and Rafatullah, M. (2021). Recent advances in soil microbial fuel cells for soil contaminants remediation. *Chemosphere* 272: 129691.

Abbas, S.Z., Rafatullah, M., Ismail, N., and Syakir, M.I. (2017). A review on sediment microbial fuel cells as a new source of sustainable energy and heavy metal remediation: mechanisms and future prospective. *International Journal of Energy Research* 41 (9): 1242–1264.

Ahmed, S.F., Mofijur, M., Nuzhat, S. et al. (2021). Recent developments in physical, biological, chemical, and hybrid treatment techniques for removing emerging contaminants from wastewater. *Journal of Hazardous Materials* 416: 125912.

Algar, C.K., Howard, A., Ward, C., and Wanger, G. (2020). Sediment microbial fuel cells as a barrier to sulfide accumulation and their potential for sediment remediation beneath aquaculture pens. *Scientific Reports* 10 (1): 1–12.

Anam, M., Gomes, H.I., Rivers, G. et al. (2021). Evaluation of photoanode materials used in biophotovoltaic systems for renewable energy generation. *Sustainable Energy & Fuels* 5: 4209–4232.

Anto, S., Mukherjee, S.S., Muthappa, R. et al. (2020). Algae as green energy reserve: technological outlook on biofuel production. *Chemosphere* 242: 125079.

Arun, S., Sinharoy, A., Pakshirajan, K., and Lens, P.N. (2020). Algae based microbial fuel cells for wastewater treatment and recovery of value-added products. *Renewable and Sustainable Energy Reviews* 132: 110041.

Arutyunov, V.S. and Lisichkin, G.V. (2017). Energy resources of the 21st century: problems and forecasts. Can renewable energy sources replace fossil fuels. *Russian Chemical Reviews* 86 (8): 777.

Azarpour, A., Zendehboudi, S., Mohammadzadeh, O. et al. (2022). A review on microalgal biomass and biodiesel production through co-cultivation strategy. *Energy Conversion and Management* 267: 115757.

Bhambri, A. and Karn, S.K. (2020). Biotechnique for nitrogen and phosphorus removal: a possible insight. *Chemistry and Ecology* 36 (8): 785–809.

Buckley, A., Ferrando, J., Lloyd, S. et al. (2015). LHAPDF6: parton density access in the LHC precision era. *The European Physical Journal C* 75 (3): 1–20.

Cao, X., Huang, X., Liang, P. et al. (2009). A completely anoxic microbial fuel cell using a photo-biocathode for cathodic carbon dioxide reduction. *Energy & Environmental Science* 2 (5): 498–501.

Capodaglio, A.G., Bolognesi, S., and Cecconet, D. (2021). Sustainable, decentralized sanitation and reuse with hybrid nature-based systems. *Water* 13 (11): 1583.

Carvalho, A.P., Silva, S.O., Baptista, J.M., and Malcata, F.X. (2011). Light requirements in microalgal photobioreactors: an overview of biophotonic aspects. *Applied Microbiology and Biotechnology* 89 (5): 1275–1288.

Chapman, R.L. (2013). Algae: the world's most important "plants"—an introduction. *Mitigation and Adaptation Strategies for Global Change* 18 (1): 5–12.

Chen, C.Y., Yeh, K.L., Aisyah, R. et al. (2011). Cultivation, photobioreactor design and harvesting of microalgae for biodiesel production: a critical review. *Bioresource Technology* 102 (1): 71–81.

Chia, S.R., Nomanbhay, S.B.H.M., Chew, K.W. et al. (2022). Algae as potential feedstock for various bioenergy production. *Chemosphere* 287: 131944.

Cuellar-Bermudez, S.P., Aleman-Nava, G.S., Chandra, R. et al. (2017). Nutrients utilization and contaminants removal. A review of two approaches of algae and cyanobacteria in wastewater. *Algal Research* 24: 438–449.

De Schamphelaire, L., Rabaey, K., Boeckx, P. et al. (2008). Outlook for benefits of sediment microbial fuel cells with two bio-electrodes. *Microbial Biotechnology* 1 (6): 446–462.

del Campo, A.G., Cañizares, P., Rodrigo, M.A. et al. (2013). Microbial fuel cell with an algae-assisted cathode: a preliminary assessment. *Journal of Power Sources* 242: 638–645.

Devi, N.D., Chaudhuri, A., and Goud, V.V. (2022). Algae biofilm as a renewable resource for production of biofuel and value-added products: a review. *Sustainable Energy Technologies and Assessments* 53: 102749.

Don, C.D.Y.A. and Babel, S. (2021). Comparing the performance of microbial fuel cell with mechanical aeration and photosynthetic aeration in the cathode chamber. *International Journal of Hydrogen Energy* 46 (31): 16751–16761.

Dorlon, J. (2021). Metabolisms. *Microreviews in Cell and Molecular Biology* 8 (2).

Doucha, J., Straka, F., and Lívanský, K. (2005). Utilization of flue gas for cultivation of microalgae *Chlorella* sp. in an outdoor open thin-layer photobioreactor. *Journal of Applied Phycology* 17 (5): 403–412.

Duan, Y. and Shi, F. (2014). Bioreactor design for algal growth as a sustainable energy source. In: *Reactor and process Design in Sustainable Energy Technology*, 27–60. Elsevier.

Dyantyi, N., Mekuto, L., and Olowolafe, A.V. (2020). Microalgae as a biocathode and feedstock in anode chamber for a self-sustainable microbial fuel cell technology: a review. *South African Journal of Chemical Engineering 31* (1): 7–16.

Egger, M., Hagens, M., Sapart, C.J. et al. (2017). Iron oxide reduction in methane-rich deep Baltic Sea sediments. *Geochimica et Cosmochimica Acta* 207: 256–276.

Elmaadawy, K., Hu, J., Guo, S. et al. (2020). Enhanced treatment of landfill leachate with cathodic algal biofilm and oxygen-consuming unit in a hybrid microbial fuel cell system. *Bioresource Technology* 310: 123420.

Elshobary, M.E., Zabed, H.M., Yun, J. et al. (2021). Recent insights into microalgae-assisted microbial fuel cells for generating sustainable bioelectricity. *International Journal of Hydrogen Energy* 46 (4): 3135–3159.

Gadhamshetty, V., Belanger, D., Gardiner, C.J. et al. (2013). Evaluation of Laminaria-based microbial fuel cells (LbMs) for electricity production. *Bioresource Technology* 127: 378–385.

Gajda, I., Greenman, J., Melhuish, C., and Ieropoulos, I. (2015). Self-sustainable electricity production from algae grown in a microbial fuel cell system. *Biomass and Bioenergy* 82: 87–93.

Gambino, E., Chandrasekhar, K., and Nastro, R.A. (2021). SMFC as a tool for the removal of hydrocarbons and metals in the marine environment: a concise research update. *Environmental Science and Pollution Research* 28 (24): 30436–30451.

Gaur, N., Narasimhulu, K., and PydiSetty, Y. (2018). Recent advances in the bio-remediation of persistent organic pollutants and its effect on environment. *Journal of Cleaner Production* 198: 1602–1631.

Germano, M., Gradinaru, C.C., Shkuropatov, A.Y. et al. (2004). Energy and electron transfer in photosystem II reaction centers with modified pheophytin composition. *Biophysical Journal* 86 (3): 1664–1672.

Ghangrekar, M.M., Das, S., and Tiwari, B.R. (2020). Integration of bioelectrochemical systems with other existing wastewater treatment processes. In: *Integrated Microbial Fuel Cells for Wastewater Treatment*, 229–248. Butterworth-Heinemann.

Ghirardi, M., Markov, S., and Seibert, M. (1996). Development of an efficient algal H2-producing system. In: *Proceedings of the 1996 US Department of Energy Hydrogen Program Review*, vol. 1, 285–302.

Gouveia, L., Neves, C., Sebastião, D. et al. (2014). Effect of light on the production of bioelectricity and added-value microalgae biomass in a photosynthetic alga microbial fuel cell. *Bioresource Technology* 154: 171–177.

Greenman, J., Gajda, I., and Ieropoulos, I. (2019). Microbial fuel cells (MFC) and microalgae; photo microbial fuel cell (PMFC) as complete recycling machines. *Sustainable Energy & Fuels* 3 (10): 2546–2560.

Gu, L., Grodzinski, B., Han, J. et al. (2022). Granal thylakoid structure and function: explaining an enduring mystery of higher plants. *New Phytologist* 236 (2): 319–329.

He, L., Du, P., Chen, Y. et al. (2017). Advances in microbial fuel cells for wastewater treatment. *Renewable and Sustainable Energy Reviews* 71: 388–403.

Helder, M. (2012). *Design Criteria for the Plant-Microbial Fuel Cell: Electricity Generation with Living Plants: from Lab Tot Application*. Wageningen University and Research.

Hennebel, T., Boon, N., Maes, S., and Lenz, M. (2015). Biotechnologies for critical raw material recovery from primary and secondary sources: R&D priorities and future perspectives. *New Biotechnology* 32 (1): 121–127.

Hou, Q., Nie, C., Pei, H. et al. (2016). The effect of algae species on the bioelectricity and biodiesel generation through open-air cathode microbial fuel cell with kitchen waste anaerobically digested effluent as substrate. *Bioresource Technology* 218: 902–908.

Jadhav, D.A., Jain, S.C., and Ghangrekar, M.M. (2017). Simultaneous wastewater treatment, algal biomass production and electricity generation in clayware microbial carbon capture cells. *Applied Biochemistry and Biotechnology* 183 (3): 1076–1092.

Jaswal, V., Rani, G., and Yogalakshmi, K.N. (2020). Photosynthetic microbial fuel cells: from fundamental to potential applications. In: *Bioelectrochemical Systems*, 1–19. Singapore: Springer.

Jiang, H. (2017). Combination of microbial fuel cells with microalgae cultivation for bioelectricity generation and domestic wastewater treatment. *Environmental Engineering Science* 34: 489–495.

Juang, D.F., Lee, C.H., and Hsueh, S.C. (2012). Comparison of electrogenic capabilities of microbial fuel cell with different light power on algae grown cathode. *Bioresource Technology* 123: 23–29.

Kakarla, R. and Min, B. (2014). Photoautotrophic microalgae *Scenedesmus obliquus* attached on a cathode as oxygen producers for microbial fuel cell (MFC) operation. *International Journal of Hydrogen Energy* 39 (19): 10275–10283.

Kannan, N. and Donnellan, P. (2021). Algae-assisted microbial fuel cells: a practical overview. *Bioresource Technology Reports* 15: 100747.

Khan, M.I., Shin, J.H., and Kim, J.D. (2018). The promising future of microalgae: current status, challenges, and optimization of a sustainable and renewable industry for biofuels, feed, and other products. *Microbial Cell Factories* 17 (1): 1–21.

Khandelwal, A., Dhindhoria, K., Dixit, A., and Chhabra, M. (2021). Superiority of activated graphite/CuO composite electrode over Platinum based electrodes as cathode in algae assisted microbial fuel cell. *Environmental Technology and Innovation* 24: 101891.

Kim, G.Y., Heo, J., Kim, K. et al. (2021). Electrochemical pH control and carbon supply for microalgae cultivation. *Chemical Engineering Journal* 426: 131796.

Kommareddy, A. and Anderson, G. (2004). Study of light requirements of a photobioreactor. In: *North Central ASAE/CSAE Conference Presentation*, MB04–MB111. USA: Winnipeg.

Kondaveeti, S., Choi, K.S., Kakarla, R., and Min, B. (2014). Microalgae Scenedesmus obliquus as renewable biomass feedstock for electricity generation in microbial fuel cells (MFCs). *Frontiers of Environmental Science & Engineering* 8 (5): 784–791.

Konur, O. (2015). Current state of research on algal bioelectricity and algal microbial fuel cells. In: *Marine Bioenergy: Trends and Developments*, 527–556. CRC Press.

Kumar, K., Gunaseelan, K., and Gajalakshmi, S. (2021). Insights into the role of bioinspiration, photosynthetic organisms, and biomass-derived electrodes/membranes in the development of bioelectrochemical systems. *Sustainable Energy Technologies and Assessments* 48: 101570.

Kumar, R., Singh, L., and Zularisam, A.W. (2016). Exoelectrogens: recent advances in molecular drivers involved in extracellular electron transfer and strategies used to improve it for microbial fuel cell applications. *Renewable and Sustainable Energy Reviews* 56: 1322–1336.

Kumar, S.S., Kumar, V., Kumar, R. et al. (2019). Microbial fuel cells as a sustainable platform technology for bioenergy, biosensing, environmental monitoring, and other low power device applications. *Fuel* 255: 115682.

Lakaniemi, A.M., Tuovinen, O.H., and Puhakka, J.A. (2012). Production of electricity and butanol from microalgal biomass in microbial fuel cells. *Bioenergy Research* 5 (2): 481–491.

Li, M., Zhou, M., Luo, J. et al. (2019). Carbon dioxide sequestration accompanied by bioenergy generation using a bubbling-type photosynthetic algae microbial fuel cell. *Bioresource Technology* 280: 95–103.

Lin, Y.F., Jing, S.R., Lee, D.Y., and Wang, T.W. (2002). Removal of solids and oxygen demand from aquaculture wastewater with a constructed wetland system in the start-up phase. *Water Environment Research* 74 (2): 136–141.

Logroño, W., Pérez, M., Urquizo, G. et al. (2017). Single chamber microbial fuel cell (SCMFC) with a cathodic microalgal biofilm: a preliminary assessment of the generation of bioelectricity and biodegradation of real dye textile wastewater. *Chemosphere* 176: 378–388.

Mashjoor, S., Yousefzadi, M., Esmaeili, M.A., and Rafiee, R. (2016). Cytotoxicity and antimicrobial activity of marine macro algae (*Dictyotaceae* and *Ulvaceae*) from the Persian Gulf. *Cytotechnology* 68 (5): 1717–1726.

Masojídek, J., Torzillo, G., and Koblížek, M. (2013). Photosynthesis in microalgae. In: *Handbook of Microalgal Culture: Applied Phycology and Biotechnology*, 21–36. Wiley.

Mekuto, L., Olowolafe, A.V., Huberts, R. et al. (2020). Microalgae as a biocathode and feedstock in anode chamber for a selfsustainable microbial fuel cell technology: a review. *South African Journal of Chemical Engineering* 31 (1): 7–16.

Mirsiaghi, M. (2016). Bioconversion of lipid-extracted algal biomass into ethanol. Doctoral dissertation, Colorado State University.

Myers, S.G. (2015). A Proposed Treatment System for the Remediation of Landfill Leachate Using Algae.

Naina Mohamed, S., Jayabalan, T., and Muthukumar, K. (2019). Simultaneous bioenergy generation and carbon dioxide sequestration from food wastewater using algae microbial fuel cell. *Energy Sources, Part A: Recovery, Utilization, and Environmental Effects* 45 (1): 1–9.

Nguyen, H.T. and Min, B. (2020). Leachate treatment and electricity generation using an algae-cathode microbial fuel cell with continuous flow through the chambers in series. *Science of the Total Environment* 723: 138054.

Nguyen, H.T., Kakarla, R., and Min, B. (2017). Algae cathode microbial fuel cells for electricity generation and nutrient removal from landfill leachate wastewater. *International Journal of Hydrogen Energy* 42 (49): 29433–29442.

Nicholls, D.G. and Ferguson, S.J. (2002). *Bioenergetics*, vol. 3. Gulf Professional Publishing.

Nookwam, K., Cheirsilp, B., Maneechote, W. et al. (2022). Microbial fuel cells with photosynthetic-cathodic chamber in vertical cascade for integrated Bioelectricity, biodiesel feedstock production and wastewater treatment. *Bioresource Technology* 346: 126559.

Obileke, K., Onyeaka, H., Meyer, E.L., and Nwokolo, N. (2021). Microbial fuel cells, a renewable energy technology for bio-electricity generation: a mini-review. *Electrochemistry Communications* 125: 107003.

Oliveira, C.Y.B., Oliveira, C.D.L., Prasad, R. et al. (2021). A multidisciplinary review of Tetradesmus obliquus: a microalga suitable for large-scale biomass production and emerging environmental applications. *Reviews in Aquaculture* 13 (3): 1594–1618.

Palanisamy, G., Jung, H.Y., Sadhasivam, T. et al. (2019). A comprehensive review on microbial fuel cell technologies: Processes, utilization, and advanced developments in electrodes and membranes. *Journal of Cleaner Production* 221: 598–621.

Patel, A., Arkatkar, A., Singh, S. et al. (2021). Physico-chemical and biological treatment strategies for converting municipal wastewater and its residue to resources. *Chemosphere* 282: 130881.

Rashid, N., Cui, Y.F., Rehman, M.S.U., and Han, J.I. (2013). Enhanced electricity generation by using algae biomass and activated sludge in microbial fuel cell. *Science of the Total Environment* 456: 91–94.

Rathinavel, L., Jothinathan, D., Sivasankar, V. et al. (2018). Algal microbial fuel cells—nature's perpetual energy resource. In: *Microbial Fuel Cell Technology for Bioelectricity*, 81–116. Cham: Springer.

Reddy, C.N., Nguyen, H.T., Noori, M.T., and Min, B. (2019). Potential applications of algae in the cathode of microbial fuel cells for enhanced electricity generation with simultaneous nutrient removal and algae biorefinery: current status and future perspectives. *Bioresource Technology* 292: 122010.

Renger, G. and Renger, T. (2008). Photosystem II: the machinery of photosynthetic water splitting. *Photosynthesis Research* 98 (1): 53–80.

Rincon, S.M., Romero, H.M., Aframehr, W.M., and Beyenal, H. (2017). Biomass production in *Chlorella vulgaris* biofilm cultivated under mixotrophic growth conditions. *Algal Research* 26: 153–160.

Rosenbaum, M., Aulenta, F., Villano, M., and Angenent, L.T. (2011). Cathodes as electron donors for microbial metabolism: which extracellular electron transfer mechanisms are involved. *Bioresource Technology* 102 (1): 324–333.

Round, F.E. (1984). The ecology of algaee. CUP Archive.

Saad, A. and Atia, A. (2014). Review on freshwater blue-green algae (*Cyanobacteria*): occurrence, classification and toxicology. *Biosciences, Biotechnology Research Asia* 11 (3): 1319–1325.

Saar, K.L., Bombelli, P., Lea-Smith, D.J. et al. (2018). Enhancing power density of biophotovoltaics by decoupling storage and power delivery. *Nature Energy* 3: 75–81.

Saba, B., Christy, A.D., Yu, Z., and Co, A.C. (2017). Sustainable power generation from bacterio-algal microbial fuel cells (MFCs): an overview. *Renewable and Sustainable Energy Reviews* 73: 75–84.

Santoro, C., Guilizzoni, M., Baena, J.C. et al. (2014). The effects of carbon electrode surface properties on bacteria attachment and start up time of microbial fuel cells. *Carbon* 67: 128–139.

Schenk, P.M., Thomas-Hall, S.R., Stephens, E. et al. (2008). Second generation biofuels: high-efficiency microalgae for biodiesel production. *Bioenergy Research* 1 (1): 20–43.

Shukla, M. and Kumar, S. (2018). Algal growth in photosynthetic algal microbial fuel cell and its subsequent utilization for biofuels. *Renewable and Sustainable Energy Reviews* 82: 402–414.

Siddiki, S.Y.A., Mofijur, M., Kumar, P.S. et al. (2022). Microalgae biomass as a sustainable source for biofuel, biochemical and biobased value-added products: an integrated biorefinery concept. *Fuel* 307: 121782.

Singh, H.M., Pathak, A.K., Chopra, K. et al. (2019). Microbial fuel cells: a sustainable solution for bioelectricity generation and wastewater treatment. *Biofuels* 10 (1): 11–31.

Tao, Q., Luo, J., Zhou, J. et al. (2014). Effect of dissolved oxygen on nitrogen and phosphorus removal and electricity production in microbial fuel cell. *Bioresource Technology* 164: 402–407.

Taşkan, B., Bakır, M., and Taşkan, E. (2021). Enhanced power generation from algal biomass using multi-anode membrane-less sediment microbial fuel cell. *International Journal of Energy Research* 45 (2): 2011–2022.

Tsai, C.C., Huang, C.F., Chen, C.F., and Yue, C.F. (2017). The study of LED light source illumination conditions for ideal algae cultivation. In: *Smart Photonic and Optoelectronic Integrated Circuits*, vol. 10107, 153–158. SPIE.

Uggetti, E., Sialve, B., Trably, E., and Steyer, J.P. (2014). Integrating microalgae production with anaerobic digestion: a biorefinery approach. *Biofuels, Bioproducts and Biorefining* 8 (4): 516–529.

Velasquez-Orta, S.B., Curtis, T.P., and Logan, B.E. (2009). Energy from algae using microbial fuel cells. *Biotechnology and Bioengineering* 103 (6): 1068–1076.

Verma, P., Daverey, A., Kumar, A., and Arunachalam, K. (2021). Microbial fuel cell–a sustainable approach for simultaneous wastewater treatment and energy recovery. *Journal of Water Process Engineering* 40: 101768.

Walter, X.A., Greenman, J., Taylor, B., and Ieropoulos, I.A. (2015). Microbial fuel cells continuously fuelled by untreated fresh algal biomass. *Algal Research* 11: 103–107.

Wang, X., Feng, Y., Liu, J. et al. (2010). Sequestration of CO2 discharged from anode by algal cathode in microbial carbon capture cells (MCCs). *Biosensors and Bioelectronics* 25 (12): 2639–2643.

Williams, J.G. (1988). Construction of specific mutations in photosystem II photosynthetic reaction center by genetic engineering methods in Synechocystis 6803. *Methods in Enzymology* 167: 766–778.

Xiao, L., Young, E.B., Berges, J.A., and He, Z. (2012). Integrated photo-bioelectrochemical system for contaminants removal and bioenergy production. *Environmental Science & Technology* 46 (20): 11459–11466.

Xu, X., Li, W., Fujibayashi, M. et al. (2015). Predominance of terrestrial organic matter in sediments from a cyanobacteria-blooming hypereutrophic lake. *Ecological Indicators* 50: 35–43.

Xue, B., Zhang, L., Li, R. et al. (2020). Underestimated microplastic pollution derived from fishery activities and "hidden" in deep sediment. *Environmental Science & Technology* 54 (4): 2210–2217.

Yadav, A., Kumar, P., Rawat, D. et al. (2022). Microbial fuel cells for mineralization and decolorization of azo dyes: recent advances in design and materials. *Science of the Total Environment* 826: 154038.

Yadav, G., Sharma, I., Ghangrekar, M., and Sen, R. (2020). A live bio-cathode to enhance power output steered by bacteria-microalgae synergistic metabolism in microbial fuel cell. *Journal of Power Sources* 449: 227560.

Yang, Z., Pei, H., Hou, Q. et al. (2018). Algal biofilm-assisted microbial fuel cell to enhance domestic wastewater treatment: nutrient, organics removal and bioenergy production. *Chemical Engineering Journal* 332: 277–285.

Yeh, K.L., Chang, J.S., and Chen, W.M. (2010). Effect of light supply and carbon source on cell growth and cellular composition of a newly isolated microalga *Chlorella vulgaris* ESP-31. *Engineering in Life Sciences* 10 (3): 201–208.

Yuan, Y., Chen, Q., Zhou, S. et al. (2011). Bioelectricity generation and microcystins removal in a blue-green algae powered microbial fuel cell. *Journal of Hazardous Materials* 187 (1-3): 591–595.

Zabihallahpoor, A., Rahimnejad, M., and Talebnia, F. (2015). Sediment microbial fuel cells as a new source of renewable and sustainable energy: present status and future prospects. *RSC Advances* 5 (114): 94171–94183.

Zhang, H., Chao, B., Wang, H., and Li, X. (2022). Effects of carbon source on electricity generation and PAH removal in aquaculture sediment microbial fuel cells. *Environmental Technology* 43 (26): 4066–4077.

Zhang, Y., He, Q., Xia, L. et al. (2018). Algae cathode microbial fuel cells for cadmium removal with simultaneous electricity production using nickel foam/graphene electrode. *Biochemical Engineering Journal* 138: 179–187.

Zhou, Q., Li, K., Jun, X., and Bo, L. (2009). Role and functions of beneficial microorganisms in sustainable aquaculture. *Bioresource Technology* 100 (16): 3780–3786.

7

An Overview of Photosynthetic Bacteria-Based Microbial Fuel Cells

Kuppurangan Gunaseelan[1], *Moogambigai Sugumar*[2], *and Selvaraj Gajalakshmi*[1]

[1] *Sustainable Fuel Cells Technology Lab, Centre for Pollution Control and Environmental Engineering, Pondicherry University, Puducherry, India*
[2] *Department of Medical Biotechnology and Integrative Physiology, Saveetha School of Engineering, Saveetha Institute of Medical and Technical Sciences, Thandalam, Chennai, India*

7.1 Introduction

Microbial fuel cell (MFC), a developing technology, employs microorganisms to decompose organic matter in wastewater, producing electricity (Ieropoulos et al. 2010; Logan et al. 2007).

Exoelectrogenic bacteria present in the substrate derive energy from organics via anodic respiration and transfer electrons to the anode in MFCs, based on the principle of oxidation and reduction (Gunaseelan et al. 2021; Sathish-Kumar 2018). MFCs can handle wastewater from different sources, such as households, tanneries, slaughterhouses, piggeries, and pharmaceutical and dye industries as substrates (Kamaraj et al. 2015; Kiran Kumar et al. 2018).

Microbes serve as biocatalysts in the anodic chamber, oxidizing organic compounds and producing electrons that reduce the anode and generate electricity via an external connection to reach the cathode (Pant et al. 2010).

Solar energy is transformed into chemical energy by photosynthetic prokaryotes using bioenergetic mechanisms (Strik et al. 2011). The most recent technique

Photosynthesis-Assisted Energy Generation: From Fundamentals to Lab Scale and In-Field Applications, First Edition. Edited by Sathish-Kumar Kamaraj and Iryna Rusyn.

for generating electricity from solar energy is the photosynthetic microbial fuel cell (PhFC) or photoMFC (Nadzri et al. 2023). Photo-MFCs use phototrophic microorganisms and the photobioelectrochemical method to generate bioelectricity from captured sunlight (Chandra et al. 2017). This interdisciplinary renewable energy solution combines physical, chemical, and life science (McCormick et al. 2015; Xiao et al. 2014) and has attracted significant concern in both theory and practice owing to its renewable and sustainable nature (Barone et al. 2023).

PhFCs/photo-MFCs have simultaneously treated the wastewater and generated bioenergy by converting light into chemical energy through photosynthesis by phototrophic prokaryotes. The dark fermentation reaction in an anodic compartment is recognized in this bioelectrochemical system (BES) as a crucial metabolic process of microorganisms. Depending on their photosynthetic mechanism, PhFCs or biophotovoltaic systems (BPVs) use oxygenic or anoxygenic photosynthetic prokaryotes (Rosenbaum et al. 2010). Phototropic prokaryotes are divided into oxygenic and anoxygenic phototrophic bacteria (OPB and APB) based on the presence of protein pigments like chlorophyll and carotenoid (George et al. 2020). Cyanobacteria are a type of OPB that utilize photosynthesis to eliminate carbon dioxide from the atmosphere and are considered traditional photoautotrophs (George et al. 2020).

APB are capable of anaerobic growth and the process of photosynthesis without oxygen production. Various types of bacteriochlorophylls (BChls), including a, b, c, d, and g, are present in APB such as heliobacteria, purple bacteria, and green bacteria, which depend on the various BChls (Qi et al. 2018). Hydrogen sulfide is used as an electron donor by anaerobic purple sulfur bacteria (PSB), and it is subsequently oxidized to elemental sulfur. APB was utilized in various research fields, including but not limited to wastewater treatment, aquaculture, medical applications, and bioelectricity generation, depending on their unique properties (Sasaki et al. 2005).

A PMFC generates more electricity as the light and microbes work together to increase the cell's voltage (Khan et al. 2022; Sathe et al. 2022).

Several microorganisms, namely, blue-green algae, cyanobacteria, and anaerobic PSB that utilize photosynthesis, have been employed in PMFCs as anodic microbes (Mekuto et al. 2020; Zafar et al. 2023). In contrast, microalgae work on the cathodes of the PMFC and moreover generate bioelectricity, they have also created biofuels, O_2, proteins, carbohydrates, and carotenoids (Gupta et al. 2021; Khan et al. 2021).

Electrochemically active microorganisms and microalgae in PMFC are superior to MFC since they do not rely on redox mediators (Ahirwar et al. 2023) and can generate free oxygen through photosynthesis (Chandra et al. 2012; Venkata Mohan et al. 2016).

Strik et al. (2008) conducted a study to compare the power generation of MFC and algae-assisted PMFC. According to the findings, the maximum amount of power produced by the algae-assisted PMFC was 110 mW/m^2, which was more than double the output of the MFC (67 mW/m^2). Considering the already reported works and their impacts, this comprehensive book chapter covers the ecology and metabolism of photosynthetic prokaryotes followed by the need for light optimization, governing factors, and bottlenecks of photosynthetic bacteria-based MFCs.

7.2 Ecology, Metabolism, and Extracellular Electron Transport in OPB and APB

Photoautotrophs exhibit diverse nutritional behaviors, including photoautotrophic, photoheterotrophic, and chemoheterotrophic modes, and they also undergo respiration. The enhanced electrogenic activity observed in the existence of light is the result of a combination of factors, such as oxygenic and anoxygenic photosynthesis, as well as chemo-heterotrophic and autotrophic nutritional behaviors of the phototrophic bacteria (Navaneeth et al. 2015).

In the lack of light, the photosynthetic activity of bacteria stops, but their respiratory and chemo-heterotrophic behaviors persist. As a consequence, they generate electricity even in complete darkness (Navaneeth et al. 2015).

Phototrophic prokaryotes are categorized into OPB and APB. OPB includes cyanobacteria, the major photoautotrophs that utilize atmospheric carbon dioxide for carbon fixation (George et al. 2020). Photoautotropes vary in terms of their photosynthetic machinery, the type of chlorophyll and carotenoid pigments they contain, and the electron donor they use for photosynthesis (George et al. 2020). Oxygenic photosynthesis is sometimes recognized as the most significant microbial innovation. The development of oxygen-evolving photosynthesis, supported by geologic processes, is credited with tipping the balance from a decreasing early Earth to an oxygenated environment that eventually gave rise to complex life (Hamilton 2019).

APB have the ability to conduct photosynthesis without releasing oxygen. They also can thrive in anaerobic environments, but oxygen can interfere with their photosynthetic machinery and pigments in most groups of these bacteria (George et al. 2020). Most anoxygenic phototrophs can be adapted to low light and also reducing environments similar to those present on early Earth and have evolved to evade the harmful effects of oxygen. However, the evolution of anoxygenic photosynthesis is still unclear. Geobiologists are studying the physiology of the first phototrophs on Earth and their significance in the planet's evolution by utilizing omics techniques and characterizing transitional taxa (Hamilton 2019). A recent study examined various strains of APB and their performance in bioelectricity generation in photo-MFC. Analysis revealed that the APB in photo-MFC included *Rhodospirillum*, *Rhodopseudomonas*, *Rhodobacter*, *Chlorobium*, *Rhodovulum*, and mixed cultures. *Rhodopseudomonas* is the commonly used APB and is considered a model bacterium for wastewater treatment (Qi et al. 2018).

Growing evidence suggests that anoxygenic photosynthesis played a crucial role in the development of oxygenic photosynthesis, and detoxification mechanisms against reactive oxygen species (ROS) in pre-GOE Cyanobacteria. However, the vast taxonomic and physiological diversity of the anoxygenic phototrophs makes it challenging to study their response to oxygen, and they appear to have developed fewer and simpler defenses against oxygen and ROS (Hamilton 2019). The process of intact bacterial photobioelectrocatalysis, known as photo-induced extracellular electron transfer (PEET), includes the transmission of electrons from a microorganism's photosynthetic system to an electrode surface. Research is dedicated to understanding these electron transfer processes in BESs of intact photosynthetic bacteria (Torquato and Grattieri 2022).

Since oxygenic and anoxygenic photosynthetic bacteria have limited direct PEET, researchers are investigating altering the electron transfer mechanism artificially. Although the carbon-based and indium tin oxide-based electrodes used in the early research were unaltered, it has been studied whether altering these materials will improve the PEET process (Torquato and Grattieri 2022). Furthermore, researchers have worked on creating synthetic methods to improve PEET, which has led to the development of exciting applications in photobioelectrocatalysis, making it a growing area of interest for scientists (Torquato and Grattieri 2022).

7.2.1 Oxygenic Photoautotrophs-Based Microbial Fuel Cells

Many studies have explored PhFCs using photosynthetic species to generate electrons through the photocatalytic reaction of the water. The research began with *Synechocystis* and *Anabaena* cyanobacterial species, which resulted with an increased and decreased power output under light and dark conditions respectively (Chandra et al. 2018). Cyanobacteria are a diverse group of light-harvesting microorganisms with various shapes, physiology, and metabolic abilities that allow them to thrive in different environments. Figure 7.1 represents the schematics of oxygenic phototroph's electron transport mechanism and metabolism. Considering they have both respiratory and photosynthetic electron transport chains in the thylakoid, which allow them to control the flow of electron and energy metabolism in accordance with their needs, their metabolism is more complex than that of other photosynthetic bacteria (Sarma et al. 2016).

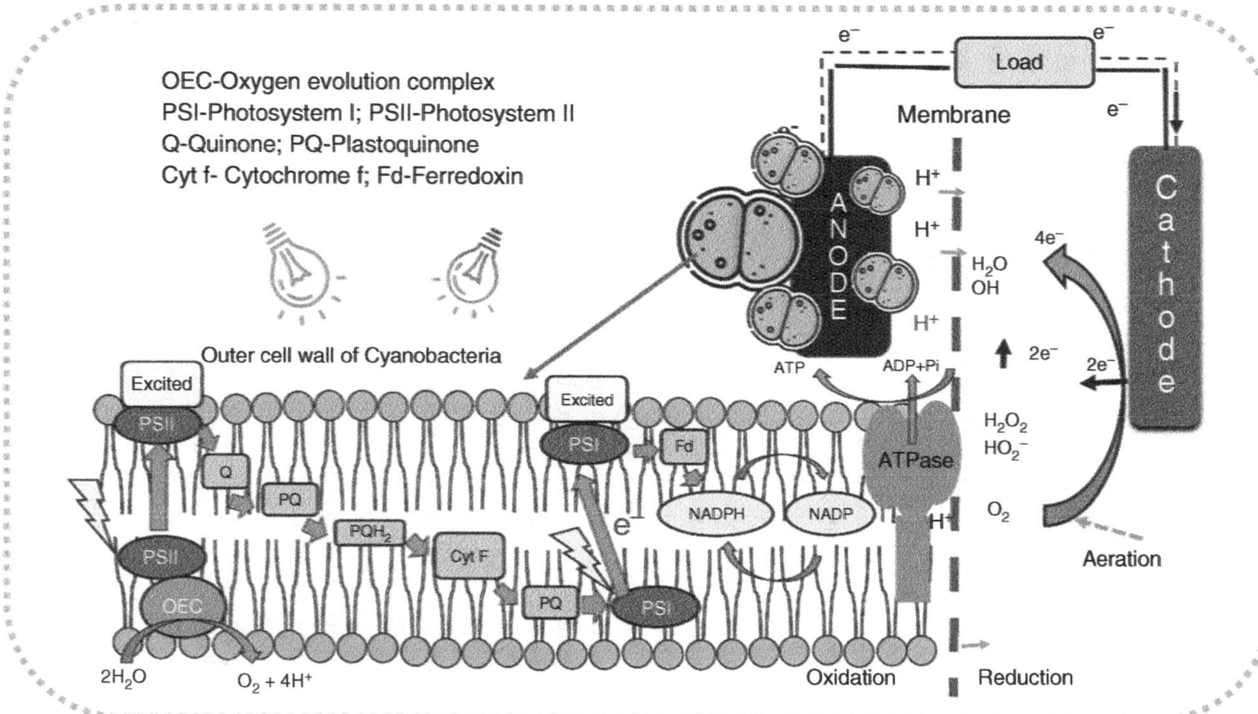

Figure. 7.1 Oxygenic Phototroph's electron transport mechanism and metabolism.

Unlike heterotrophs, cyanobacteria rely on light for electricity generation by giving electrons to an acceptor in the outer membrane under the presence of light. It is not known, nevertheless, whether cytochrome bd quinol oxidase moves the electrons directly to the external acceptor or whether other substances are involved further down the line (Chandra et al. 2018). Placing a phototrophic biofilm on the anode of a PhFC will allow it to produce electricity from solar energy. Oxygenic photosynthetic species start the process of charge separation by delivering electrons and protons to a series of reactions termed photophosphorylation. Three membrane-bound protein complexes are involved in oxygenic photosynthesis, including PSI, PSII, and cytochrome bf complex, and interact to transport an electron from H_2O to NADP for oxygen generation. As a result, in an established MFC system, it is possible to link the production of bioelectricity with oxygenic photosynthesis (Chandra et al. 2018).

Plastoquinone (PQ) and plastocyanin (PC), two mobile compounds, transfer electrons between large protein complexes. PQ is essential to the cyanobacteria's electrogenic process. These compounds are able to transport electrons over long distances and are essential for photosynthetic energy conversion (Chandra et al. 2018). According to Pisciotta et al. 2010 [13], under light conditions, cyanobacteria release extracellular electrons via the photosynthetic electron transfer chain (PETC), and the photolysis of water by PSII. PQ and bd quinol oxidase are found to be important electron transporters to the extracellular medium, supporting the release of electrons from H_2O, rather than organic matter.

These systems function through oxygenic photosynthesis, which leads to the release of H^+ and e^- as well as O_2 via photolysis of H_2O. The release of the electrons, oxygen, and H^+ occurs simultaneously, with O_2 neutralizing H^+ and the electrons due to the high electronegativity (Chandra et al. 2018). Dissolved oxygen reduces the electrogenic activity by consuming electrons and protons generated by cyanobacterial metabolism. While the synthesis of reducing equivalents has a positive impact on electrogenic activity, the production of oxygen has a negative impact on total activity. Previous studies on oxygenic phototroph-based PhotoMFC by other authors are given in Table 7.1. As a mediator in a PhFC to transmit electrons from microorganisms to the anode, studies have used 2-hydroxy-1,4-naphthoquinone. The use of unsustainable and environmentally hazardous nonnatural redox mediators, however, impedes the practical application of PhFCs and the advancement of scientific research (Chandra et al. 2018).

Microalgae MFCs are a unique type of photosynthetic MFCs with similar properties to photosynthetic bacteria MFCs. Microalgae can function on both anodic and cathodic compartments of the MFC system, making them suitable for improving MFC performance (Xu et al. 2021). Chemical energy found in algae can be utilized to produce sustainable fuels and goods. Photosynthesis and MFC integration can produce bioelectricity from sunlight in an economical and sustainable way (Shukla and Kumar 2018).

Table. 7.1 Oxygenic phototroph-based PhotoMFC.

Oxygenic phototrophic bacteria (OPB)	System design	Membrane	Power output (mW/m^2)	References
Oscillatoria limnetica	Multichannel	Nafion	0.02	(Bombelli et al. 2012)
Synechocystis sp.	Single chamber	Nafion	0.2	(Bradley et al. 2013)
Spirulina platensis	Single chamber	Membrane less	10	(Lin et al. 2013)
Chlamydomonas reinhardtii	Dual chamber	Nafion	13	(Lan et al. 2013)
Synechocystis sp.	Microfluidic	Nafion	105	(Bombelli et al. 2015)
Cladophora sp.	Single chamber	Membrane less	619.1	(Taşkan and Taşkan 2022)
Chlorella vulgaris	Dual chamber	CEM	19.8	(Khazraee Zamanpour et al. 2017)
Synechococcus sp.	Dual chamber	Nafion	95.6	(Lakshmidevi et al. 2020)
Chlorella vulgaris	Dual chamber	Nafion	89	(Khandelwal et al. 2020)
Chlorella sp. G29-5	Dual chamber	CMI	505.6	(J. Y. Wu et al. 2021)

7.2.2 Anoxygenic Photoautotrophic Bacteria-Based Microbial Fuel Cells

APB comprises species that are distinct from OPB, the most conserved species, in both phylogenetics and photosynthesis due to two shared characteristics. Instead of oxidizing water, they reduce photosynthetic output by using organic compounds, hydrogen sulfide, or other comparable electron donors as their photopigment (Imhoff 2008). Figure 7.2 represents the schematics of anoxygenic phototroph's electron transport mechanism and metabolism.

Bacteria in ABP employ either type I or II reaction center (RC) for the process of photosynthesis. Type I RC is utilized by species in Firmicutes, Chlorobi, and Acidobacteria phyla, whereas Proteobacteria, Chloroflexi, and Gemmatimonadetes phyla use type II RC (George et al. 2020).

APBs with varying characteristics can coexist in related environments due to their ability to gather light in the range of 740 to 1020 nm. *Thermomicrobium roseum*, a nonmotile chemotroph, and *Herpetosiphon aurantiacus*, a closely related filamentous chemotroph, are referred to as "green non-sulfur bacteria" because of their high bacteriochlorophyll C concentration (Hanada 2016).

Filamentous APBs are distinct from the green sulfur bacteria in both phylogenetics and physiology. The major pigment-protein complexes in APB include

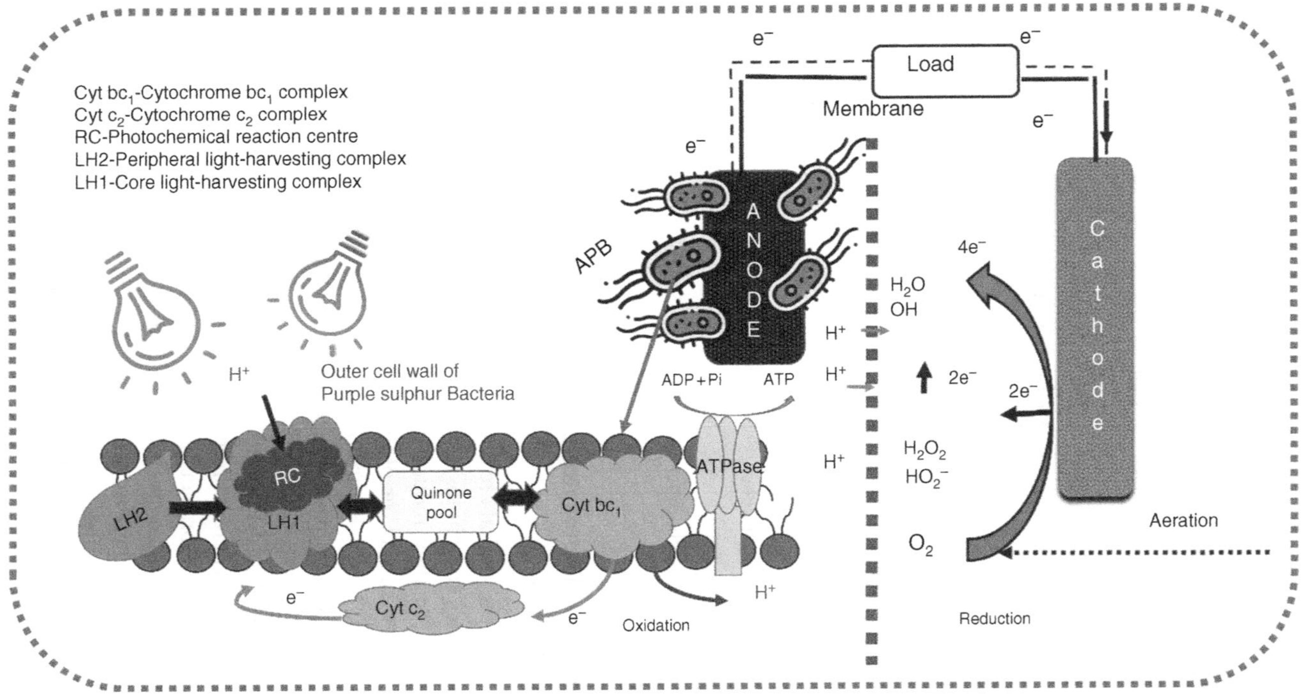

Figure. 7.2 Anoxygenic phototroph's electron transport mechanism and metabolism.

BChl, apoprotein, and carotenoid (CD), as well as photosynthetic components like core and peripheral light-harvesting complexes (LH1 and LH2) and photochemical RC (Qi et al. 2018). APB includes purple, green, and heliobacteria with different bacteriochlorophylls (BChl a, b, c, d, and g, respectively). In APB, oxygen is not produced throughout the metabolic process as a byproduct. APB's distinctive properties make it useful in various fields such as aquaculture, new energy development, environmental modification, and medicine, including mycoprotein production (Hallenbeck and Liu 2016; Sasaki et al. 2005).

The initial report of bioelectricity generation through phototrophic prokaryotes was not with cyanobacteria but with APB. In 1964, Berk and his team were the first to investigate the conversion of bioelectrochemical energy by utilizing resting cells of the *Rhodospirillum rubrum*. Recently, researchers have been attempting to accomplish self-sustaining bioelectricity generation with APB via photo-MFC, as described in a study by Qi et al. (2018). Despite having been proposed for many years, photo-MFC with APB has not garnered much attention in research (Qi et al. 2018). The previous studies on oxygenic phototroph-based PhotoMFC by other authors are given in Table 7.2.

Table. 7.2 Anoxygenic Phototroph-based PhotoMFC.

Anoxygenic phototrophic bacteria (APB)	System design	Membrane	Power output (mW/m^2)	References
Rhodobacter sphaeroides	Single chamber	Membrane less	790	(Cho et al. 2008)
Rhodopseudomonas palustris	Single chamber	Membrane less	2780	(Xing et al. 2008)
Rhodopseudomonas sp.	Single chamber	Nafion 117	112.2	(Chandra et al. 2015)
Rhodopseudomonas palustris	Single chamber	Nafion	0.15	(Lai et al. 2017)
Rhodopseudomonas sp.	Dual chamber	Cation exchange membrane	221	(Zheng et al. 2017)
Rhodopseudomonas palustris	Single chamber	Nafion	0.15	(Lai et al. 2017)
Rhodospirillum rubrum	Dual chamber	CEM	125	(Gomez et al. 2014)
Rhodopseudomonas palustris	Dual chamber	Nafion 115	10.4	(Inglesby et al. 2012)
Mixed culture	Dual chamber	Ceramic	360	(Gunaseelan et al. 2022)
Mixed culture	Dual chamber	Ceramic	275	(Gunaseelan et al. 2021)

In photo MFC anodes, APB's bioelectricity generation involves two pathways. First, APB produces electrons through anoxygenic photosynthesis or endogenous respiration. Second, hydrogen generated during APB photosynthesis can be utilized as a medium for electron production (Qi et al. 2018).

The key factor for the photosynthetic and growth efficiency of APB is the light source, and optimizing it can enhance the performance of photo-MFC. APB has two primary pigment protein complexes (CD and BChl), which have an absorption spectrum of 450–550 and 715–1050 nm (Qi et al. 2018).

7.3 Advantages of the APB over Algae and Cyanobacteria

A broad group of bacteria known as APB uses organic and inorganic electron donors to execute anoxygenic photosynthesis. They have appealing properties that make them valuable for a variety of environmental biotechnology applications, including growth in anaerobic conditions, low energy requirements, low growth and maintenance costs, and most crucially varied modes of metabolism (George et al. 2020). Algae are often used in the MFCs to reduce oxygen or as bacteria substrates. However, they also can produce electric current indirectly through cytochromes at the anode in a photo-algal MFC (PAMFC). These systems are environmentally friendly, but their low efficiency makes them unfeasible (Shukla and Kumar 2018).

Therefore, it is crucial to comprehend the bioelectrochemical mechanisms that enable the electrons to transfer from algae to electrodes. The fuel cell's performance is influenced by oxygen, pH, and substrate loading rate. The mechanisms involved in the transfer of an electron from algae to electrodes and the factors that impact the functioning of PAMFC were explained by Shukla and Kumar 2018.

Compared to plants, unicellular microalgae are efficient autotrophs because of their high photosynthetic efficiency. However, the lack of coordination among light harvesting, electrophotochemical, and electron transfer processes leads to reduced productivity and efficiency (Melis 2009).

Purple nonsulfur bacteria (PNSB) are photosynthetic microorganisms that live in various aquatic habitats with sunlight and can change their metabolism depending on nutrients, oxygen, and light. They can grow in anoxic conditions and tolerate toxicants like heavy metals, making them useful in remediating contaminated sites. Their pigmentation ranges from deep red to brown (Asif et al. 2021).

Bacteria use bacteriochlorophyll (Bchl) molecules to convert sunlight into energy during anoxygenic photosynthesis, while organic waste or CO_2 serves as an electron supply. The RC of the photosystem is activated by the energy from the pigment molecules, which is directed to bacteriochlorophyll a (Bchla). This triggers a series of photochemical reactions that create a separation of positive and negative charges through the membrane without generating oxygen (Venkata Subhash et al. 2013).

The two systems showed different treatment efficiency apart from electricity generation. PhFC AX was more efficient (90%) than PhFC OX (72%), possibly due to differences in the initial COD levels of the feed. PhFC AX was fed with artificial wastewater containing a high COD level of 1000 mg/l, which was metabolically reduced by anoxygenic photosynthesis (Venkata Subhash et al. 2013).

Moreover, APB can generate bioelectricity by biodegrading slow-release substrates such as polymers, and this sustainable bioelectricity generation is driven by solar energy without the need for any additional operations (Qi et al. 2018).

7.4 Optimization of Light Source for Sustainable Electricity Production

Good illumination is crucial for the development of photosynthetic microorganisms. The requirement for an efficient light source also acts as an intrinsic barrier since a balance must be attained between the reactors' surface/volume ratios (which determine the amount of substrate per illuminated area/region) and a high enough concentration of COD to promote growth. Multiple studies have shown that increasing the intensity of illumination can greatly enhance electricity production, likely by increasing the amount of dissolved oxygen (Gouveia et al. 2014; Wu et al. 2014).

The growth and enrichment of microbes in Photo-BES can be influenced by the type and strength of available light (George et al. 2020). Studies have indicated that light quality is crucial for microalgae and may also be important for photoactive bacteria in MFCs. While there is little information specifically about the use of photoactive bacteria in MFCs, general literature suggests that certain light sources may be suitable. More information is available about the effectiveness of light on power generation in photo-MFCs using microalgae, which typically require specific visible light bands (Fischer 2018).

Some studies suggest that less intense lighting can lead to increased electricity production (Juang et al. 2012). This indicates that phototrophic organisms may act as important biocatalysts by producing oxygen, and this could be influenced by microbial species and operational conditions (Gouveia et al. 2014).

The costs of illumination for PPB seem to outweigh the potential revenue from the products they produce. It is important to note that the maximal biomass energy yield of 59 g COD/kWh only takes into account the practical energy consumed by the light source (Capson-Tojo et al. 2020). To overcome this challenge, the most apparent solution is to use solar light. Only a few studies have utilized sunlight to cultivate photosynthetic prokaryotic bacteria (PPB) thus far (Capson-Tojo et al. 2020).

While these studies indicate that it is technically possible to cultivate PPB outdoors using tubular PBRs, they were conducted using synthetic culture media and pure PPB cultures. The goal of these outdoor studies was to produce biomass, PHA, and hydrogen (Adessi et al. 2012; Carlozzi and Sacchi 2001).

7.4.1 Source of Light

The role of light sources with various wavelengths and intensities in the development of photosynthetic prokaryotes in a PMFC is essential. One of the key factors affecting the effectiveness of PMFC is light stress, which has a wide range of effects. Additionally, all of the biopolymer types produced and included in a PMFC as well as the metal contaminants derived from them are discussed in detail. Since light plays a crucial role in regulating photosynthesis, both high and low light levels can change the logarithm of algal growth (Khan et al. 2022).

The primary difference between light source results is brought on by the different photosynthetic pigment absorption peaks in PSB. The content, ratio, and species of BChls and Carotenoids (CDs) may result in small variations in absorption peak, and only a light source that matches closely the absorbance peak (Kuo et al., 2012). For light optimization, it could act as an obstacle (Chen et al. 2020). The best monochromatic light is considered to be red (605–700 nm), and blue (435–480 nm) for the most common bacteria, including *Rhodobacter sphaeroides* and *Rhodopseudomonas sphaeroides* (Lee et al., 2010; Zhou et al., 2015). The regions of blue and red light include the majority of BChls absorption peak, shown in Table 7.3. Additionally, CDs mainly absorb light in the 400–500 nm band to compensate for BChls subpar visible spectrum absorption capacity (Chen et al. 2020).

The most effective artificial light source for PPB growth is NIR-emitting LEDs since they maximize NIR emission while being energy-efficient. LEDs have higher COD removal rates than incandescent lamps due to their lower energy costs and NIR specificity (Suwan Duangkamon Chitapornpan Sukhuma 2014). Compared to traditional light sources, LEDs produce biomass energy yields that are between 10 and 100 times greater per kWh (Hülsen et al. 2016, 2018). The typical plug wall efficiency of NIR LEDs is 20–30%, and their photon conversion efficiency is a commendably high 80% (Auf Der Maur et al. 2016). The reason for this is that the mature phosphide emitter technology uses low forward voltages and ideal wavelengths. Additionally, because of its long wavelength, NIR light has comparatively low medium attenuation (Capson-Tojo et al. 2020).

The type of light source used in experiments employing pure cultures and axenic conditions is another aspect that limits the extrapolation of results. The majority of experiments used nonspecific LEDs, fluorescent, halogen, or tungsten lamps as their light sources. To reduce heat emission, halogen, fluorescent, and VIS-emitting LEDs have reduced output at NIR wavelengths. Because PPB absorb primarily above 805 nm (by BChl a and b), the accessible light spectrum for them is severely constrained (Capson-Tojo et al. 2020)

Simple incandescent bulbs and white LEDs remain to be the standard in industrial applications to meet the high light exposure requirements. Since they readily meet the illuminance requirements and emit a full spectrum of light. The choice between the two will rely on the situation, with the white LED taking preference while taking energy conservation and expenses into consideration. There is no research currently on the effects of simultaneous irradiation from monochromatic light sources. However, there are currently devices available on the market that blend red and blue light to promote plant growth (Chen et al. 2020)

Under IR irradiation, they often take on a dominant role, fully utilize their superiority, and achieve enrichment from wastewater leading to a reduced reactor start-up period. Wastewater treatment may be improved as a result of the IR radiation's potential to increase PSB reactivity. The idea for choosing IR is similar to choosing visible light, however, there have been very few studies (Chen et al. 2020). Another potential low-power light source is LED lighting, which has the advantage of having a flexible color spectrum. Investigating the possibilities of LED light with MFCs was Lan and colleagues' effort. When *Chalmydomonas reinhardtii* was used, an LED-assisted microalgae-MFC produced higher electricity with red light than with blue light (Fischer 2018).

7.4.2 Photoperiod

A longer duration of light could upsurge oxygen synthesis and thus power output because illumination is a restraining factor. However, it has been found that prolonged light periods reduce power production. This finding supports the notion that dark periods are essential for controlling organisms and healthy cyanobacteria consortia (Wu et al. 2013; Xiao et al. 2012). Although a rise in the frequency of the light/dark cycle, resulted in a decrease in electricity generation and algal biomass, other research further supported the significance of a proper dark period for a photosynthetic MFC (Xiao et al. 2012). However, a prolonged dark cycle can result in the photosynthetic system operating inefficiently, which may inhibit growth (Chen et al. 2020). There was a temperature increase in the cultivation of microalgae as a result of the intense light. In some ways, higher temperatures encourage the growth of microalgae. When employing a pure microalgae culture of *Desmodesmus* sp, an irradiation limit for microalgae-MFC power generation at its maximum level was discovered (Wu et al. 2014). The light period should therefore be longer than the duration of dark in the setting of the cycle (Chen et al. 2020).

7.4.3 Light Intensity

In accordance with the presence of light, PSB produces chromophores rapidly in the cell for photosynthesis, which converts light sources into chemical energy (Chen et al. 2020). The amount of incident radiation and the photosynthetic unit's capacity to collect light should be coordinated. Appropriate light intensity may deliver PSB more light energy and promote their biological processes, which can have an impact on the removal of organic pollutants. According to earlier research, PSB can significantly promote growth by increasing light intensity within a suitable range, but over that point, it is ineffective and repressive. The increasing light intensity would also result in a decrease in light conversion efficiency (Chen et al. 2020).

Gürgan et al. (2020) testified that when *Rhodobacter capsulatus* was exposed to 10,000 lx of light, researchers discovered that genes involved in photosynthesis were downregulated. To diffuse extra electrons, biohydrogen synthesis was increased together with an overexpression of the electron transport system genes (Gürgan et al. 2020). Excessive light may cause impairment to phototrophic organisms' cells, but on the other side, it can enhance the activity of exoelectrogen due to the degeneracy of reducing power (George et al. 2020). There are reasons to believe that *Rhodopseudomonas*, which emerges most frequently in wastewater treatment studies, reacts best to light density range between 2000 and 4000 lx. Regarding COD removal and biomass generation within this range, PSB performs well, although the difference is not significant (Zhou et al. 2014). The photosynthetic unit rises to more efficiently capture the limited amount of photons that are available as the PPB grows in low-intensity light (Muzziotti et al. 2017).

When there was a change in light intensity, photosynthetic pigment's composition and content likewise underwent significant changes. To prevent light damage and the photoinhibition effect, excessive light may stimulate the CD's light protection mechanism. The light is increased to 8000 Lux, which inhibits PSB but increases CD production (Zhou et al. 2014). In other words, the impacts of extreme natural light illumination should be considered. Additionally, by appropriately increasing light intensity, CD production can be performed (Chen et al. 2020).

7.5 Governing Factors and Bottlenecks of Photosynthetic Bacteria-Based Microbial Fuel Cells

There are several governing factors and bottlenecks in using photosynthetic phototrophs-based MFCs for the generation of electricity, resource recovery, and pollutant removal (George et al. 2020). One of the focal challenges is the low amount of photosynthetic products available for energy conversion, which also leads to low energy potential. Additionally, the anode activities may be restricted by oxygen generated during photosynthesis, and MFC reactor designs have to accommodate for illumination, which causes challenges for reactor operation (Qi et al. 2018). It is essential to evaluate the stability of an MFC system including phototrophic bacteria over an extended period of time using real wastewater. In addition, it is important to understand how phototrophic microbes interact with other microorganisms in a mixed microbial community (Xiao and He 2014). It is possible to combine phototrophic organisms with MFCs, as their ability to operate as efficient oxygenators improves the cathodic reaction of PMFCs. Moreover, microalgae are useful for removing phosphorus and nitrogen, which is crucial since MFC technology is insufficient in this area (Elmekawy et al. 2014).

PMFC technology provides a promising opportunity for energy production and carbon fixation from challenging waste materials, as well as microalgae biomass production. However, further research is necessary to scale up and enhance integrated PMFC systems, including developing more efficient coupled system designs that consider the impact of photobioreactors (PBRs) and hydraulic retention time (HRT) of MFC on photosynthetic prokaryotes growth and pollutant removal (Elmekawy et al. 2014). BES technologies follow allometric scaling laws where the increased volume of a single MFC reactor leads to decreased power yield efficacy, and an increased number of stack MFC units leads to increased power yield efficiency. The work on allometric scaling is essential because small-scale MFCs can provide more power density than large-scale MFCs because the latter has a lower electrode surface area-to-volume ratio (Sarma et al. 2016). The scale-up plan should be considered a vital issue since it greatly affects the economy. The economic aspect is also an important consideration when pushing the marketing of light-enhanced MFC systems for real-world application (Xu et al. 2021). In the final analysis, a coordinated effort is required to link genomics, proteomics, metabolic engineering, and advanced molecular and computational biology techniques in order to better comprehend these resilient organism's ability to harvest clean energy, such as bioelectricity and biofuels.

7.6 Conclusion

The research interest in intact photosynthetic bacteria photobioelectrochemistry is on the rise, as shown by several studies conducted in the last decade. This field of study is still in its early stages.

The previously discussed exceptional research has enhanced our knowledge of the extracellular electron transportation mechanism with electrodes for both oxygenic and anoxygenic photosynthetic bacteria followed by the importance of light source, governing factors, and the bottlenecks related to PMFC.

Despite significant improvements, practical applications of intact photosynthetic prokaryotes-based MFCs require further research inputs.

References

Adessi, A., Torzillo, G., Baccetti, E., and De Philippis, R. (2012). Sustained outdoor H2 production with *Rhodopseudomonas palustris* cultures in a 50L tubular photobioreactor. *International Journal of Hydrogen Energy 37* (10): 8840–8849. https://doi.org/https://doi.org/10.1016/j.ijhydene.2012.01.081.

Ahirwar, A., Das, S., Das, S. et al. (2023). Photosynthetic microbial fuel cell for bioenergy and valuable production: a review of circular bio-economy approach. *Algal Research 70* (December 2022): 102973. https://doi.org/10.1016/j.algal.2023.102973.

Asif, A., Mohsin, H., and Rehman, Y. (2021). Purple nonsulfur bacteria: an important versatile tool in biotechnology, Chap. 9. In: *Recent Advancement in Microbial Biotechnology* (ed. S. De Mandal and A.K. Passari), 309–337. Academic Press https://doi.org/https://doi.org/10.1016/B978-0-12-822098-6.00003-3.

Auf Der Maur, M., Pecchia, A., Penazzi, G. et al. (2016). Efficiency drop in green InGaN/GaN light emitting diodes: the role of random alloy fluctuations. *Physical Review Letters 116* (2): 1–5. https://doi.org/10.1103/PhysRevLett.116.027401.

Barone, G.D., Cernava, T., Ullmann, J. et al. (2023). Recent developments in the production and utilization of photosynthetic microorganisms for food applications. *Heliyon 9* (4): e14708. https://doi.org/https://doi.org/10.1016/j.heliyon.2023.e14708.

Bombelli, P., Zarrouati, M., Thorne, R.J. et al. (2012). Surface morphology and surface energy of anode materials influence power outputs in a multi-channel mediatorless bio-photovoltaic (BPV) system. *Physical Chemistry Chemical Physics 14* (35): 12221–12229. https://doi.org/10.1039/c2cp42526b.

Bombelli, P., Müller, T., Herling, T.W. et al. (2015). A high power-density, mediator-free, microfluidic biophotovoltaic device for cyanobacterial cells. *Advanced Energy Materials 5* (2): 1–6. https://doi.org/10.1002/aenm.201401299.

Bradley, R.W., Bombelli, P., Lea-Smith, D.J., and Howe, C.J. (2013). Terminal oxidase mutants of the cyanobacterium *Synechocystis* sp. PCC 6803 show increased electrogenic activity in biological photo-voltaic systems. *Physical Chemistry Chemical Physics 15* (32): 13611–13618. https://doi.org/10.1039/c3cp52438h.

Capson-Tojo, G., Batstone, D.J., Grassino, M. et al. (2020). Purple phototrophic bacteria for resource recovery: challenges and opportunities. *Biotechnology Advances 43* (May): 107567. https://doi.org/10.1016/j.biotechadv.2020.107567.

Carlozzi, P. and Sacchi, A. (2001). Biomass production and studies on *Rhodopseudomonas palustris* grown in an outdoor, temperature controlled, underwater tubular photobioreactor. *Journal of Biotechnology 88* (3): 239–249. https://doi.org/https://doi.org/10.1016/S0168-1656(01)00280-2.

Chandra, R., Venkata Subhash, G., and Venkata Mohan, S. (2012). Mixotrophic operation of photo-bioelectrocatalytic fuel cell under anoxygenic microenvironment enhances the light dependent bioelectrogenic activity. *Bioresource Technology 109*: 46–56. https://doi.org/10.1016/j.biortech.2011.12.135.

Chandra, R., Annie Modestra, J., and Venkata Mohan, S. (2015). Biophotovoltaic cell to harness bioelectricity from acidogenic wastewater associated with microbial community profiling. *Fuel 160*: 502–512. https://doi.org/10.1016/j.fuel.2015.07.090.

Chandra, R., Sravan, J.S., Hemalatha, M. et al. (2017). Photosynthetic synergism for sustained power production with microalgae and photobacteria in a biophotovoltaic cell. *Energy and Fuels 31* (7): 7635–7644. https://doi.org/10.1021/acs.energyfuels.7b00486.

Chandra, R., Castillo-Zacarias, C., Mancera-Andrade, E.I. et al. (2018). Fundamentals of biophotovoltaics for conversion of solar energy to bioelectricity. In: *Biomass, Biofuels, Biochemicals: Microbial Electrochemical Technology: Sustainable Platform for Fuels, Chemicals and, Remediation*. Elsevier B.V. https://doi.org/10.1016/B978-0-444-64052-9.00020-0.

Chen, J., Wei, J., Ma, C. et al. (2020). Photosynthetic bacteria-based technology is a potential alternative to meet sustainable wastewater treatment requirement? *Environment International 137* (November 2019): 105417. https://doi.org/10.1016/j.envint.2019.105417.

Cho, Y.K., Donohue, T.J., Tejedor, I. et al. (2008). Development of a solar-powered microbial fuel cell. *Journal of Applied Microbiology 104* (3): 640–650. https://doi.org/10.1111/j.1365-2672.2007.03580.x.

Elmekawy, A., Hegab, H.M., Vanbroekhoven, K., and Pant, D. (2014). Techno-productive potential of photosynthetic microbial fuel cells through different configurations. *Renewable and Sustainable Energy Reviews 39*: 617–627. https://doi.org/10.1016/j.rser.2014.07.116.

Fischer, F. (2018). Photoelectrode, photovoltaic and photosynthetic microbial fuel cells. *Renewable and Sustainable Energy Reviews 90* (March): 16–27. https://doi.org/10.1016/j.rser.2018.03.053.

George, D.M., Vincent, A.S., and Mackey, H.R. (2020). An overview of anoxygenic phototrophic bacteria and their applications in environmental biotechnology for sustainable resource recovery. *Biotechnology Reports 28*: e00563. https://doi.org/10.1016/j.btre.2020.e00563.

Gomez, M.V., Mai, G., Greenwood, T., and Mullins, J.P. (2014). The development and maximization of a novel photosynthetic microbial fuel cell using *Rhodospirillum rubrum*. *Journal of Emerging Investigators 3*: 1–7. https://www.emerginginvestigators.org/articles/the-development-and-maximization-of-a-novel-photosynthetic-microbial-fuel-cell-using-em-rhodospirillum-rubrum-em%0Ahttps://www.emerginginvestigators.org/articles/the-development-and-maximization-of-a-novel-ph.

Gouveia, L., Neves, C., Sebastião, D. et al. (2014). Effect of light on the production of bioelectricity and added-value microalgae biomass in a photosynthetic alga microbial fuel cell. *Bioresource Technology 154*: 171–177. https://doi.org/https://doi.org/10.1016/j.biortech.2013.12.049.

Gunaseelan, K., Jadhav, D.A., Gajalakshmi, S., and Pant, D. (2021). Blending of microbial inocula: an effective strategy for performance enhancement of clayware Biophotovoltaics microbial fuel cells. *Bioresource Technology* 323: 124564. https://doi.org/10.1016/j.biortech.2020.124564.

Gunaseelan, K., Jadhav, D.A., Pant, D., and Gajalakshmi, S. (2022). Effectiveness of biophotovoltaics system modified with fuller-clay composite separators for chromium removal. *Electrochimica Acta 426*: 140714. https://doi.org/https://doi.org/10.1016/j.electacta.2022.140714.

Gupta, A.K., Seth, K., Maheshwari, K. et al. (2021). Biosynthesis and extraction of high-value carotenoid from algae. *Frontiers in Bioscience-Landmark 26* (March): 171–190. https://doi.org/10.52586/4932.

Gürgan, M., Koku, H., Eroglu, I., and Yücel, M. (2020). Microarray analysis of high light intensity stress on hydrogen production metabolism of *Rhodobacter capsulatus*. *International Journal of Hydrogen Energy 45* (5): 3516–3523. https://doi.org/https://doi.org/10.1016/j.ijhydene.2018.12.205.

Hallenbeck, P.C. and Liu, Y. (2016). Recent advances in hydrogen production by photosynthetic bacteria. *International Journal of Hydrogen Energy 41* (7): 4446–4454. https://doi.org/10.1016/j.ijhydene.2015.11.090.

Hamilton, T.L. (2019). The trouble with oxygen: the ecophysiology of extant phototrophs and implications for the evolution of oxygenic photosynthesis. *Free Radical Biology and Medicine 140* (May): 233–249. https://doi.org/10.1016/j.freeradbiomed.2019.05.003.

Hanada, S. (2016). Anoxygenic photosynthesis — a photochemical reaction that does not contribute to oxygen reproduction. *Microbes and Environments 31* (1): 1–3. https://doi.org/10.1264/jsme2.ME3101rh.

Hülsen, T., Barry, E.M., Lu, Y. et al. (2016). Low temperature treatment of domestic wastewater by purple phototrophic bacteria: performance, activity, and community. *Water Research 100*: 537–545. https://doi.org/https://doi.org/10.1016/j.watres.2016.05.054.

Hülsen, T., Hsieh, K., Lu, Y. et al. (2018). Simultaneous treatment and single cell protein production from agri-industrial wastewaters using purple phototrophic bacteria or microalgae – a comparison. *Bioresource Technology 254*: 214–223. https://doi.org/https://doi.org/10.1016/j.biortech.2018.01.032.

Ieropoulos, I., Greenman, J., and Melhuish, C. (2010). Improved energy output levels from small-scale microbial fuel cells. *Bioelectrochemistry 78* (1): 44–50. https://doi.org/10.1016/j.bioelechem.2009.05.009.

Imhoff, J.F. (2008). *Systematics of Anoxygenic Phototrophic Bacteria*, 269–287. Springer https://doi.org/10.1007/978-1-4020-6863-8_14.

Inglesby, A.E., Beatty, D.A., and Fisher, A.C. (2012). *Rhodopseudomonas palustris* purple bacteria fed *Arthrospira* maxima cyanobacteria: demonstration of application in microbial fuel cells. *RSC Advances 2* (11): 4829–4838. https://doi.org/10.1039/c2ra20264f.

Juang, D.F., Lee, C.H., and Hsueh, S.C. (2012). Comparison of electrogenic capabilities of microbial fuel cell with different light power on algae grown cathode. *Bioresource Technology 123*: 23–29. https://doi.org/https://doi.org/10.1016/j.biortech.2012.07.041.

Kamaraj, S.K., Romano, S.M., Moreno, V.C. et al. (2015). Use of novel reinforced cation exchange membranes for microbial fuel cells. *Electrochimica Acta 176*: 555–566. https://doi.org/10.1016/j.electacta.2015.07.042.

Khan, M.J., Bawra, N., Verma, A. et al. (2021). Cultivation of diatom *Pinnularia saprophila* for lipid production: a comparison of methods for harvesting the lipid from the cells. *Bioresource Technology 319*: 124129. https://doi.org/https://doi.org/10.1016/j.biortech.2020.124129.

Khan, M.J., Singh, N., Mishra, S. et al. (2022). Impact of light on microalgal photosynthetic microbial fuel cells and removal of pollutants by nanoadsorbent biopolymers: updates, challenges and innovations. *Chemosphere 288*: 132589. https://doi.org/https://doi.org/10.1016/j.chemosphere.2021.132589.

Khandelwal, A., Chhabra, M., and Yadav, P. (2020). Performance evaluation of algae assisted microbial fuel cell under outdoor conditions. *Bioresource Technology 310*: 123418. https://doi.org/https://doi.org/10.1016/j.biortech.2020.123418.

Khazraee Zamanpour, M., Kariminia, H.-R., and Vosoughi, M. (2017). Electricity generation, desalination and microalgae cultivation in a biocathode-microbial desalination cell. *Journal of Environmental Chemical Engineering 5* (1): 843–848. https://doi.org/https://doi.org/10.1016/j.jece.2016.12.045.

Kiran Kumar, V., Man Mohan, K., Manangath, S.P., and Gajalakshmi, S. (2018). Terracotta separator based plant microbial fuel cell for bioelectricity and catholyte production. *International Journal of Applied Engineering Research* 13: 14948–14955.

Kuo, F. S., Chien, Y. H., and Chen, C. J. (2012). Effects of light sources on growth and carotenoid content of photosynthetic bacteria Rhodopseudomonas palustris. *Bioresource Technology 113*: 315–318. https://doi.org/10.1016/j.biortech.2012.01.087.

Lai, Y.C., Liang, C.M., Hsu, S.C. et al. (2017). Polyphosphate metabolism by purple non-sulfur bacteria and its possible application on photo-microbial fuel cell. *Journal of Bioscience and Bioengineering 123* (6): 722–730. https://doi.org/10.1016/j.jbiosc.2017.01.012.

Lakshmidevi, R., Gandhi, N.N., and Muthukumar, K. (2020). Carbon neutral electricity production from municipal solid waste landfill leachate using algal-assisted microbial fuel cell. *Applied Biochemistry and Biotechnology 191* (2): 852–866. https://doi.org/10.1007/s12010-019-03160-5.

Lan, J.C.W., Raman, K., Huang, C.M., and Chang, C.M. (2013). The impact of monochromatic blue and red LED light upon performance of photo microbial fuel cells (PMFCs) using *Chlamydomonas reinhardtii* transformation F5 as biocatalyst. *Biochemical Engineering Journal 78*: 39–43. https://doi.org/10.1016/j.bej.2013.02.007.

Lee, S., Boo, C., Elimelech, M., and Hong, S. (2010). Comparison of fouling behavior in forward osmosis (FO) and reverse osmosis (RO). *Journal of Membrane Science*. https://doi.org/10.1016/j.memsci.2010.08.036.

Lin, C.C., Wei, C.H., Chen, C.I. et al. (2013). Characteristics of the photosynthesis microbial fuel cell with a *Spirulina platensis* biofilm. *Bioresource Technology 135*: 640–643. https://doi.org/10.1016/j.biortech.2012.09.138.

Logan, B., Cheng, S., Watson, V., and Estadt, G. (2007). Graphite fiber brush anodes for increased power production in air-cathode microbial fuel cells. *Environmental Science and Technology 41* (9): 3341–3346. https://doi.org/10.1021/es062644y.

McCormick, A.J., Bombelli, P., Bradley, R.W. et al. (2015). Biophotovoltaics: oxygenic photosynthetic organisms in the world of bioelectrochemical systems. *Energy and Environmental Science 8* (4): 1092–1109. https://doi.org/10.1039/c4ee03875d.

Mekuto, L., Olowolafe, A.V.A., Pandit, S. et al. (2020). Microalgae as a biocathode and feedstock in anode chamber for a self-sustainable microbial fuel cell technology: a review. *South African Journal of Chemical Engineering 31*: 7–16. https://doi.org/https://doi.org/10.1016/j.sajce.2019.10.002.

Melis, A. (2009). Solar energy conversion efficiencies in photosynthesis: minimizing the chlorophyll antennae to maximize efficiency. *Plant Science 177* (4): 272–280. https://doi.org/https://doi.org/10.1016/j.plantsci.2009.06.005.

Muzziotti, D., Adessi, A., Faraloni, C. et al. (2017). Acclimation strategy of *Rhodopseudomonas palustris* to high light irradiance. *Microbiological Research 197*: 49–55. https://doi.org/https://doi.org/10.1016/j.micres.2017.01.007.

Nadzri, N.A.A., Mohd Yasin, N.H., Abu Bakar, M.H. et al. (2023). Photosynthetic microbial desalination cell (PhMDC) using *Chlamydomonas* sp. (UKM6) and *Scenedesmus* sp. (UKM9) as biocatalysts for electricity production and water treatment. *International Journal of Hydrogen Energy 48* (31): 11860–11873. https://doi.org/https://doi.org/10.1016/j.ijhydene.2023.01.142.

Navaneeth, B., Hari Prasad, R., Chiranjeevi, P. et al. (2015). Implication of composite electrode on the functioning of photo-bioelectrocatalytic fuel cell operated with heterotrophic-anoxygenic condition. *Bioresource Technology 185*: 331–340. https://doi.org/10.1016/j.biortech.2015.02.065.

Pant, D., Van Bogaert, G., Diels, L., and Vanbroekhoven, K. (2010). A review of the substrates used in microbial fuel cells (MFCs) for sustainable energy production. *Bioresource Technology 101* (6): 1533–1543. https://doi.org/10.1016/j.biortech.2009.10.017.

Pisciotta, J. M., Zou, Y., and Baskakov, I. V. (2010). Light-dependent electrogenic activity of cyanobacteria. *PLoS ONE 5* (5). https://doi.org/10.1371/journal.pone.0010821.

Qi, X., Ren, Y., Liang, P., and Wang, X. (2018). New insights in photosynthetic microbial fuel cell using anoxygenic phototrophic bacteria. *Bioresource Technology 258* (266): 310–317. https://doi.org/10.1016/j.biortech.2018.03.058.

Rosenbaum, M., He, Z., and Angenent, L.T. (2010). Light energy to bioelectricity: photosynthetic microbial fuel cells. *Current Opinion in Biotechnology 21* (3): 259–264. https://doi.org/10.1016/j.copbio.2010.03.010.

Sarma, M.K., Kaushik, S., and Goswami, P. (2016). Cyanobacteria: a metabolic power house for harvesting solar energy to produce bio-electricity and biofuels. *Biomass and Bioenergy 90*: 187–201. https://doi.org/10.1016/j.biombioe.2016.03.043.

Sasaki, K., Watanabe, M., Suda, Y. et al. (2005). Applications of photosynthetic bacteria for medical fields. *Journal of Bioscience and Bioengineering 100* (5): 481–488. https://doi.org/10.1263/jbb.100.481.

Sathe, S.M., Chakraborty, I., Dubey, B.K., and Ghangrekar, M.M. (2022). Microbial fuel cell coupled Fenton oxidation for the cathodic degradation of emerging contaminants from wastewater: applications and challenges. *Environmental Research 204*: 112135. https://doi.org/https://doi.org/10.1016/j.envres.2021.112135.

Sathish-Kumar, K. (2018). A preliminary study on comparing the efficacy of anaerobic inocula and facultative inocula for application in microbial fuel cells. *International Journal for Research in Engineering Application & Management (IJREAM) 04* (7): https://doi.org/10.18231/2454-9150.2018.0993.

Shukla, M. and Kumar, S. (2018). Algal growth in photosynthetic algal microbial fuel cell and its subsequent utilization for biofuels. *Renewable and Sustainable Energy Reviews 82* (October 2016): 402–414. https://doi.org/10.1016/j.rser.2017.09.067.

Strik, D.P.B.T.B., Terlouw, H., Hamelers, H.V.M., and Buisman, C.J.N. (2008). Renewable sustainable biocatalyzed electricity production in a photosynthetic algal microbial fuel cell (PAMFC). *Applied Microbiology and Biotechnology 81* (4): 659–668. https://doi.org/10.1007/s00253-008-1679-8.

Strik, D.P.B.T.B., Timmers, R.A., Helder, M. et al. (2011). Microbial solar cells: applying photosynthetic and electrochemically active organisms. *Trends in Biotechnology 29* (1): 41–49. https://doi.org/10.1016/j.tibtech.2010.10.001.

Suwan, D., Chitapornpan, S., Honda, R. et al. (2014). Conversion of organic carbon in food processing wastewater to photosynthetic biomass in photo-bioreactors using different light sources. *Environmental Engineering Research 19* (3): 293–298. https://doi.org/10.4491/eer.2014.S1.009.

Taşkan, B. and Taşkan, E. (2022). Sustainable bioelectricity generation using *Cladophora* sp. as a biocathode in membrane-less microbial fuel cell. *Bioresource Technology 347*: 126704. https://doi.org/https://doi.org/10.1016/j.biortech.2022.126704.

de Torquato, L.D. and Grattieri, M. (2022). Photobioelectrochemistry of intact photosynthetic bacteria: advances and future outlook. *Current Opinion in Electrochemistry 34* (Scheme 1): 101018. https://doi.org/10.1016/j.coelec.2022.101018.

Venkata Mohan, S., Nikhil, G.N., Chiranjeevi, P. et al. (2016). Waste biorefinery models towards sustainable circular bioeconomy: critical review and future perspectives. *Bioresource Technology 215*: 2–12. https://doi.org/https://doi.org/10.1016/j.biortech.2016.03.130.

Venkata Subhash, G., Chandra, R., and Venkata Mohan, S. (2013). Microalgae mediated bio-electrocatalytic fuel cell facilitates bioelectricity generation through oxygenic photomixotrophic mechanism. *Bioresource Technology 136*: 644–653. https://doi.org/10.1016/j.biortech.2013.02.035.

Wu, X.Y., Song, T.S., Zhu, X.J. et al. (2013). Construction and operation of microbial fuel cell with *Chlorella vulgaris* biocathode for electricity generation. *Applied Biochemistry and Biotechnology 171* (8): 2082–2092. https://doi.org/10.1007/s12010-013-0476-8.

Wu, Y., Wang, Z., Zheng, Y. et al. (2014). Light intensity affects the performance of photo microbial fuel cells with *Desmodesmus* sp. A8 as cathodic microorganism. *Applied Energy 116*: 86–90. https://doi.org/https://doi.org/10.1016/j.apenergy.2013.11.066.

Wu, J.Y., Lay, C.H., Chia, S.R. et al. (2021). Economic potential of bioremediation using immobilized microalgae-based microbial fuel cells. *Clean Technologies and Environmental Policy 23* (8): 2251–2264. https://doi.org/10.1007/s10098-021-02131-x.

Xiao, L. and He, Z. (2014). Applications and perspectives of phototrophic microorganisms for electricity generation from organic compounds in microbial fuel cells. *Renewable and Sustainable Energy Reviews 37*: 550–559. https://doi.org/10.1016/j.rser.2014.05.066.

Xiao, L., Young, E.B., Berges, J.A., and He, Z. (2012). Integrated photo-bioelectrochemical system for contaminants removal and bioenergy production. *Environmental Science and Technology 46* (20): 11459–11466. https://doi.org/10.1021/es303144n.

Xiao, L., Ge, Z., Kelly, P. et al. (2014). Evaluation of normalized energy recovery (NER) in microbial fuel cells affected by reactor dimensions and substrates. *Bioresource Technology 157*: 77–83. https://doi.org/10.1016/j.biortech.2014.01.086.

Xing, D., Zuo, Y., Cheng, S. et al. (2008). Electricity generation by *Rhodopseudomonas palustris* DX-1. *Environmental Science and Technology 42* (11): 4146–4151. https://doi.org/10.1021/es800312v.

Xu, Z., Chen, S., Guo, S. et al. (2021). New insights in light-assisted microbial fuel cells for wastewater treatment and power generation: a win-win cooperation. *Journal of Power Sources 501* (February): 230000. https://doi.org/10.1016/j.jpowsour.2021.230000.

Zafar, S.U., Mehra, A., Nesamma, A.A., and Jutur, P.P. (2023). Innovations in algal biorefineries for production of sustainable value chain biochemicals from the photosynthetic cell factories. *Algal Research 69*: 102949. https://doi.org/https://doi.org/10.1016/j.algal.2022.102949.

Zheng, W., Cai, T., Huang, M., and Chen, D. (2017). Comparison of electrochemical performances and microbial community structures of two photosynthetic microbial fuel cells. *Journal of Bioscience and Bioengineering 124* (5): 551–558. https://doi.org/10.1016/j.jbiosc.2017.05.013.

Zhou, Q., Zhang, P., and Zhang, G. (2014). Biomass and carotenoid production in photosynthetic bacteria wastewater treatment: effects of light intensity. *Bioresource Technology 171*: 330–335. https://doi.org/10.1016/j.biortech.2014.08.088.

Zhou, Q., Zhang, P., Zhang, G., and Peng, M. (2015). Biomass and pigments production in photosynthetic bacteria wastewater treatment: Effects of photoperiod. *Bioresource Technology 190*: 196–200. https://doi.org/10.1016/j.biortech.2015.04.092.

8

The Development of Bryophyte Microbial Fuel Cell Systems

Iryna Rusyn[1], *Wilgince Apollon*[2], *and Soumya Ghosh*[3,4]

[1] *Department of Ecology and Sustainable Environmental Management, Viacheslav Chornovil, Institute of Sustainable Development, Lviv Polytechnic National University, Lviv, Ukraine*
[2] *Department of Agricultural and Food Engineering, Faculty of Agronomy, Autonomous University of Nuevo León, General Escobedo, Nuevo León, México*
[3] *Department of Genetics, Faculty of Natural and Agricultural Sciences, University of the Free State, Bloemfontein, South Africa*
[4] *Natural & Medical Science Research Center, University of Nizwa, Nizwa, Oman*

8.1 Introduction

Photosynthesis-assisted power generation presents a hybrid bio-electrotechnology, the successful functioning of which depends on the biological (photosynthetic and electro-active organisms) and technological components (electrodes and electrical circuit, substrate and container), which are equally important. However, the biological components are crucial and serve as a driving force of bioelectricity generation and determine the place of its application. Thereby, the technology of photosynthesis-assisted power generation besides vascular plants tries to explore the possibility of exploitation of different classes of photosynthetic organisms such as anoxygenic and oxygenic photosynthetic bacteria, cyanobacteria, and non-vascular plants, microalgae as well as bryophytes.

Traditional plant-microbial fuel cells (PMFCs) have been developed on typical plants, i.e. vascular plants, which excrete photosynthates and secondary metabolite compounds through phloem tissue and root exudation into the soil rhizosphere. In this way, rhizodeposit and dead root cells serve as electron sources for electro-active microorganisms (Strik et al. 2008; Timmers et al. 2012a). However, bryophytes have hair-like rhizoids instead of roots (Ligrone et al. 2000), and it was

Photosynthesis-Assisted Energy Generation: From Fundamentals to Lab Scale and In-Field Applications, First Edition. Edited by Sathish-Kumar Kamaraj and Iryna Rusyn.

unclear how electricity generation occurs in other photosynthetic plants that lack vascular tissue. The first bryophyte MFC, based on the forest moss *Dicranum montanum*, was reported by Hubenova and Mitov in 2011 and demonstrated successful operation. Recent studies of bryophytes determine them as a promising object of photosynthesis-assisted power generation technology due to high-stress resistance and survival in a wide range of conditions, with the absence of requirements particular conditions of humidity, temperature and soil nutrients. Therefore, developments of mosses-based MFCs represent a significant interest.

Thereby, we have collected all existing information on this topic about and detailly reviewed (i) features of mosses that give them advantages of use in photosynthesis-assisted power generation, (ii) diversity of moss-driven MFCs and their application in powering of electrical devices, (iii) achievements and challenges of application bryophytes-based MFCs on green roofs, and also (iv) perspectives and limitations of lab-scale and in-field application of bryophyte MFC systems.

8.1.1 Physiological Peculiarities of Mosses as an Object for Electro-biotechnology

Mosses, as a constituent part of electro-biotechnology can be applied both indoors and outdoors on green roofs. The plant selection for extensive green roofing is considered to be an important criterion as they need to survive at intense solar radiation, wind exposure, low nutrient supply, limited root area, and essentially to be drought and heat/frost resistant (Konášová 2014). Although a majority of the green roofs are planted with drought-resistant, higher plant species, grasses, herbs, succulents (Oberndorfer et al. 2007; Emilsson 2008), and other plant varieties such as mosses have gained much value and have been extensively used in the recent times for extensive greening roofs in the urban areas because of their stress resistance, low nutrient and substrate demands, and excellent water retention (Grant 2006; Studlar and Peck 2009; Heim et al. 2014). Mosses display stunning physiological range and occupy some of Earth's most stressful habitats, including the cold deserts of Antarctica and hot deserts of North America (Wolf et al. 2010). Desiccation-tolerant mosses have the ability to absorb up to 1400% of their dry mass in water (Michel et al. 2013). For example, *Racomitrium canescens* moss holds 10 times its weight in water, providing the most significant benefits in water retention (Anderson et al. 2010). The absence of a request for a large amount of substrate matters because many roofs cannot support the large weight of the substrate necessary to grow other plant species.

Observation of the urban areas shows that they can grow almost on any substrate, with some particular preferences (e.g. calcicole or calcifuge) (de Carvalho et al. 2019). Notably, the moss can also be transported via wind and bird dispersal.

For instance, in the United Kingdom, through this unique transportation system, the flat asphalt roofs are usually dominated by *Ceratodon purpureus* and *Syntrichia (Tortula) ruralis*. These mosses are also associated with slates and tiles along with *Bryum capillare* and *Hedwigia ciliate* (Porley and Hodgetts 2005). Likewise, in the Northwest of America, it has been observed that both *Racomitrium* and *Dicranoweisia* have colonized the rooftops following a similar approach (Schenk 1997). To be mentioned for an effective roof moss cultivation, it is always necessary to select the right moss species in accordance with the area/climate. This can be determined with prior experimentation with the root moss. The less prominent moss can also be chosen, provided their natural habitat is similar to the rooftops. Another valuable characteristic of mosses is the ability to be grown vegetatively from tiny fragments of any portion of the plant, making them amenable to culture (Stark et al. 2004). Several companies are involved in making these moss mats. German and French companies are engaged in developing moss mats/panels from *Polytrichum* (Glime 2007) and *C. purpureus* (Chiaffredo 2004), respectively. On the other hand in Japan, composite mats comprised of *R. canescens* rooted in cotton and polyvinyl chloride netting are largely used (Ondimu and Murase 2007).

Many mosses are well suited to harsh rooftop conditions (Anderson et al. 2010). Poikilohydric mosses can spend long periods in a desiccated state without suffering damage. They are also capable of rehydrating in short periods; that is often lasted for 20 minutes (Kimmerer 2003). Mosses have more efficient prolonged drought tolerance mechanisms than vascular plants, which can easily become water-stressed under the sunny and low moisture conditions commonly found on roofs (Anderson et al. 2010). Unlike many vascular plants that require regular watering in climates with periodic drought, mosses can survive long periods of drought-induced dormancy (Nagase and Thuring 2006; Studlar and Peck 2009).

The integral properties of these mosses chosen for green roofs are their key features. For instance, the intensity of desiccation with varying tolerance abilities of some of these mosses to withstand a higher rate of water loss is one of the factors that play major roles in green roofing. These species possess constitutive desiccation tolerance mechanisms (Wang et al. 2009; de Carvalho et al. 2014) in comparison to the less desiccation-tolerant ones that depend on the alteration of dehydration-rehydration cycles (Stark 2017; de Carvalho et al. 2019). However, in the Mediterranean-type climate, the presence of these mosses contributes to sustainable irrigation of green roofs (Brandao et al. 2017; Paco et al. 2019). Most desiccation-resistant species exhibit high values of photosynthetic fitness following desiccation/rehydration events. Notably, the high degree of rehydration of these most tolerant species demonstrates a decrease in photosynthetic activity with increased photo-inhibition and non-photochemical quenching process. Especially during the hot Mediterranean summers, these mosses lose their bright green color to dark-green brown (Paco et al. 2019) due to the loss of cellular water and subsequent shutdown of their metabolic process (Oliver et al. 2005; Wang et al. 2009; de Carvalho et al. 2011; de Carvalho et al. 2014; Stark 2017; de Carvalho et al. 2018), allows them to tolerate and absorb high solar radiation, making them suitable for green roof application (de Carvalho et al. 2019). Conversely, during the Mediterranean cold, wet winter, these species could be prevented from hydration-desiccation cycles to restore their photochemical performance (de Carvalho et al. 2020).

Despite mosses are generally adapted to photosynthetic optima at middle temperatures (10–28 °C), many of the habitats of moss biocrusts do endure higher temperatures (38–40 °C), (Lange, 2003; Bu et al. 2017). The artificial moss biocrusts can also resist short-time heat stress and possess CO_2-induced photosynthetic thermotolerance in markedly different ways than vascular plants (Coe et al. 2012; Bu et al. 2017). In addition, mosses are known for their cold/freezing tolerance. The Arctic polar region, characterized by a climate with freezing winters and cold summers, is reached by numerous moss species. More than 380 bryophyte species have been recorded from the Svalbard Archipelago (Byun et al. 2020). Phenotypic plasticity is one of the most prominent characteristics enabling bryophytes to fit into and colonize polar environments through considerable tolerance to dehydration and frost stress (La Farge et al. 2013; Stanton and Reeb 2016)—Arctic moss *Bryum* sp. KMR5045, due to inherent low-temperature tolerance, permitted it to maintain slow growth even at 2 °C (Byun et al. 2020).

Mosses are also UV-resistant plants and successfully survive in desert ecosystems where they are exposed to numerous adverse environmental conditions, such as strong light, extreme temperature, salinity, and water deficit. Bryophytes can recover from UV-B stress and still flourish in desert regions (Hui et al. 2013; Hui et al. 2016; Rong et al. 2016). Moss crust has great ecological significance in desert areas, where it is the dominant component of biological soil crusts and a major pioneer species in community succession processes (Rong et al. 2016). UV-B protection is determined via pigmentation as well as anthocyanin-like compounds and other flavonoid UVR absorbing compounds (Ekwealor et al. 2022). The resistance depends on the species of moss. *S. ruralis* moss has a more plastic, acclimatable UVR response than *Syntrichia caninervis*, and through their unique interaction with these stressors as poikilohydric organisms (Ekwealor et al. 2022). Obviously, due to the set of valuable characteristics mentioned above, mosses are a promising candidate for use in the PMFC.

8.2 Moss-Driven Microbial Fuel Cells

Plants from the moss category have only been started to drive bioelectrochemical technology such as bioelectrochemical systems (BESs), i.e. plant-MFC (PMFC). The first mosses were successfully applied in PMFC in 2011 (Hubenova and

Mitov 2011). A forest moss *D. montanum* from the tree roots in Rodopi mountain (Bulgaria) was placed in a single-chamber MFC exploiting forest soil from the natural habitat. The experiment lasted 70 days in natural conditions during two seasons of summer and autumn. The exponential growth of bioelectricity parameters of bryophyte-MFC correlated with the microorganism's development on the electrode surface as well as in the soil of B-MFC. The symbiotic connections between bacteria and mosses and their benefits for current outputs were shown. B-MFC was reached voltage values of 550–790 mV and a current density of 20–40 mA/m^2 after the microorganisms' layer was formed (Hubenova and Mitov 2011). This work served as a starting point that brought attention to the possibility of using bryophytes in the PMFC.

With the main idea to improve bioelectricity production, Castresana and coworkers built different PMFCs that were operated with moss (Castresana et al. 2018). The anode compartment was constructed using 4 cm^2 of carbon felt (2×2 cm), which consisted of a 7 mm thick rectangular piece. Prior to use, the anodic electrode material was pretreated to improve the quality of the nanofibers in terms of roughness and hydrophilicity. These parameters are crucial for biofilm formation, which facilitates the transport of electrons from the anodic to the cathodic compartment. This would lead to a higher performance of the PMFC system. The distance between the anode electrode and the cathode was 0.5 cm. Both anode and cathode electrodes were installed in parallel positions. The anode was placed on the ground (i.e. was submerged), and the cathode was placed at the top of the ground (air cathode). Therefore, various types of materials have been used as the cathode electrode. Hence, four types of PMFCs were built, each with a different cathode material. The first one was built with 2×2×0.70 cm of carbon felt; the second made of 3×1 cm of bare fluorine-doped tin oxide-coated glass (glass/FTO, 3×1 cm, Sigma-Aldrich) as the cathode; the third with the same glass/FTO cathode but treated with CuO (copper oxide); and the fourth also with glass/FTO treated with CuO and Nafion as cathode electrode. The PMFCs were operated in long-term operation by connecting to an external 510 Ω resistor. Furthermore, the PMFCs were installed and operated at room temperature, that is, between the ranges of 17 and 20 °C during the experiment. The results indicated that the glass cathode PMFC/FTO treated with CuO and Nafion (NF), achieved higher power generation performance with a power density fly of 2.5 mW/m^2 compared to the other PMFCs (Castresana et al. 2018). However, this value was lower than the recently reported power density of a moss-enriched MFC (Bombelli et al. 2016; Rusyn and Hamkalo 2020; Çek et al. 2022). Table 8.1 summarizes the performance of different types of MFCs previously embedded with moss.

Another moss-driven MFC exploiting both metal electrodes was designed recently by Çek et al. (2022). Such an unusual approach in electrode selection was

Table 8.1 Diversity of Moss-driven microbial fuel cells.

Moss species	Application	Power output	Electrode system	Module volume	Substrate	References
Amblystegium serpens, Brachythecium rutabulum Bryum capillare, Bryum dichotomum, Ceratodon purpureus, Syntrichia ruralis	Power a radio or an environmental sensor	2.6 mW/m²	Anode: mixture of tissue paper and carbon fiber Cathode: carbon paper, loaded with Pt	8 pots: 110 mm (W) × 75 mm (L) × 50 mm (H) 10 pots: 70 mm (W) × 70 mm (L) × 60 mm (H)	BCD medium	Bombelli et al. 2016
Dicranum montanum	First sample of B-MFC	20–40 mA/m²	Not indicated	Not indicated	Forest soil	Hubenova and Mitov 2011
Hypnum cupressiforme, Leucobryum glaucum, Polytrichum commune	Green roof	0. 50–0.56 W/m² PGA	Zn-steel plate anode Graphite rods cathode	5l, 0.0525 m²	Soil	Rusyn and Hamkalo 2020
Physcomitrella patens	Power a radio or an environmental sensor	6.7 mW/m²	Anode: mixture of tissue paper and carbon fiber Cathode: carbon paper, loaded with Pt	8 pots: 10 mm (W) × 75 mm (L) × 50 mm (H) 10 pots: 70 mm (W) × 70 mm (L) × 60 mm (H) BCD medium	BCD medium	Bombelli et al. 2016
Pleurocapous moss	Energy harvesting from moss	2.5 mW/m²	Carbon felt anode Cathode:FTO/CuxO/Nf	1 pot: 60 mm (W) × 90 mm (L) × 30 mm (H)	Soil	Castresana et al. 2018
Not indicated	Energy harvesting from moss	20 mW/m²	Anode: aluminum Cathode: silver electrodes	Not indicated	Soil	Çek et al. 2022

used due the lower electrical resistivity of metals in comparison to the traditional carbon electrodes which possess inert and high bacterial adhesion properties. Moreover, aluminum was preferred as the anode electrode because of its amphoteric easy supply and good electrochemical capacity. Silver was preferred as the cathode electrode because it has a good reduction potential (0.80 V at Ag^+/Ag). Two types of MFCs were designed with moss (20 g), water (50 ml) (MFC 1), and moss (20 g), water (50 ml) (MFC 2), and natural soil (60 g) using aluminum and silver as electrodes. Experiments were performed at 20–30 °C temperatures. Biofilm formed on the electrode surfaces with *Bacillus thuringiensis* strain, *Bacillus cereus* strain, and *Enterococcus faecium* strain bacteria acting as biocatalysts of bryophyte-MFC were revealed in this study. The highest power densities for MFC 1 and MFC 2 were calculated as 20 and 16.13 mW/m, respectively (Çek et al., 2022).

On the other hand, perennial plants such as *Stevia* (*Stevia rebaudiana*), which is not a moss plant, were also used in a single-chamber MFC for simultaneous nutrient recovery and bioelectricity generation (Apollon et al. 2023). The anodic chamber was made with graphite felt (648 cm^2) and 270 cm^2 stainless steel mesh (9.5 mm 610 × 610 mm) as cathode. The PMFCs were operated in the open field for 43 days where the environmental factors were not considered or controlled. A total of four PMFCs were evaluated: PMFC1 uninoculated; PMFC2 inoculated with 643 ml of cow urine; PMFC3 inoculated with 643 ml of goat urine; and PMFC4 inoculated with 643 ml sheep urine. The anode and cathode of each PMFC were connected with a 1000 Ω resistor. Stevia was first used in PMFC to improve its power output performance. Hence, PMFC inoculated with sheep urine showed a higher power density of 46.97 ± 0.67 mW/m^2 compared to the other PMFCs. The results indicated a 98% increase in terms of power generation using this kind of reactor. Furthermore, in a previous study by Apollon et al. (2022a), the same perennial plant (*Stevia*) was used in MFC. The anode and cathode were both separated with a clay cup membrane (20 cm tall with a wall thickness of 3 mm and an inner diameter of 9 cm). The maximum power density of 43.68 ± 3.05 mW/m^2 was achieved in PMFC inoculated with cow urine (for 30 days of operation). Close values of 55.05 and 61.72 mW/m^2 were received with PMFC based on waterlogger plants *Ipomoea aquatica* and *Oryza sativa* ssp. Indica (Arends et al. 2013, Liu et al. 2014). The polarization curve efficiency of the PMFC based on other perennial plants such as *Amaranthus viridis* and annual *Triticum aestivum* ranges from 7.62 to 194.45 mW/m^2 (Arulmani et al. 2021), and PMFC with perennial ornamental plant *Helcina soleirolii* was reached 188.33 mW/m^2. These yields were higher than those reported by Castresana et al. (2018) and Çek et al. (2022) using MFC embedded with moss plants. According to the above, it can be argued that moss-operated MFCs proved to be an excellent option to increase the performance of MFCs by improving their operation properties, electrode materials, and design. However, more studies are required on the use of moss in MFC to bring this technology to the field of application and commercialization.

8.3 Indoor Application of Moss-PMFC

In the study by Bombelli et al. (2016), a prototype bryophyte-based MFC (bryoMFC) was built to power a radio or environmental sensor. Hence, a total of eight identical bryoMFCs were constructed using a plastic chamber (110 mm × 75 mm × 50 mm) as an electrode. The anode chamber was made of 10 g tissue paper and 1 g carbon fiber. Prior to the system setup, the carbon cloth strands were separated into individual fibers (cutting at 4 cm ± 2 cm in length), creating a paper pulp/carbon fiber matrix. This paper/carbon fiber mixture was formed using a blender with the addition of approximately 300 ml of deionized water (DIW). This allowed the moss to be in direct contact with the anode electrode surface. For the cathode compartment, Pt-loaded carbon paper (3 mg loading per cm^2) was used. After system setup, the bryoMFCs were subsequently embedded with the *Physcomitrella patens* filamentosa culture, which was maintained at a temperature of 20 ± 2 °C under moderate light (for 24 hours (12 hours light/dark) in near-sterile conditions. An MFC without a plant was used as a control in this experiment. The anodic and cathodic electrodes were connected using an external resistance of 1 MΩ for 7 days. This connection allowed the stabilization of the system and greater bacterial growth on the anode surface. Subsequently, the system was operated in a long-term experiment for 70 days. This experiment was created and characterized with the objective of evaluating the impact of mosses on the MFCs in terms of electricity generation. Sterile and near-sterile bryoMFCs were tested during this study. The results showed that non-sterile bryoMFCs (operated with environmental moss samples) achieved the highest power density of $6.7 \pm 0.6\,mW/m^2$ compared to the other bryoMFCs operating in near-sterile conditions evaluated in this study (Bombelli et al. 2016). These results confirm the importance of the microbial populations and anodophilic microorganism for delivering electrons to the anode in a bryoMFC. The study's findings indicated that the novel 3D bryoMFC anode matrix based on environmental moss samples had the ability to power the radio receiver. This means that mosses, like the other plant species used previously in MFC, carry out the process of photosynthesis, releasing enough exudates to generate electricity through the activity of microorganisms as catalysts.

On the other hand, in the study by Bombelli and co-workers, bryophytes were chosen because they are poikilohydric plants with greater tolerance to dehydration compared to vascular plants. This was a good strategy to use this category of plants in MFC because they have a particular physiology; that is, they allow the accumulation of both nutrients and water that the plant needs for its optimal development. In addition, bryophytes, like cactus (e.g. *Opuntia* spp.), can survive in adverse conditions (Apollon et al. 2020). Therefore, bryophytes provide adequate food sources, such as exudates, to increase the activity of microbes present on the anode surface. In contrast, *Opuntia* spp. are Crassulacean acid metabolism (CAM) plants that gradually release organic compounds (exudates) to feed microbial consortia. In the study by Apollon et al. (2022b), four species of *Opuntia* (*O. ficus-indica*, *O. joconostle*, *O. robusta*, and *O. albicarpa*) were evaluated using an MFC built with a clay stick as a proton exchange membrane (experiment carried out under instauration conditions). The anode was made of graphite felt and the cathode was made of zinc sheet. Highest maximum power density of $6.07 \pm 2.41\,mW/m^2$ was obtained in *Opuntia joconostle*. This performance was equal to the maximum power density achieved in *P. patens* (Bombelli et al., 2016). This means that *O. joconostle*, although it is a CAM plant, has the same potential as moss *P. patens* to drive the MFC system.

8.4 Bryophyte PMFC as a Source of Photosynthesis-Associated Energy Generation on Green Roofs

8.4.1 Eco-environmental Value of Green Roofs

Green roofs represent an effective nature-based solution to environmental problems arising from the exploitation of nonrenewable energy sources and rapid urbanization resulting in global climate change. The arrangement of green roofs is an effective tool for sustainable development, climate regulation and water management, energy saving, and carbon sequestration (Liu et al. 2021) (Figure 8.1). The advantages of additional greening of the city on the roofs include mitigation of the urban microclimate (Sahnoune et al. 2021; Manso et al. 2021; Park et al. 2022), reduction of energy demand for the cooling of buildings in

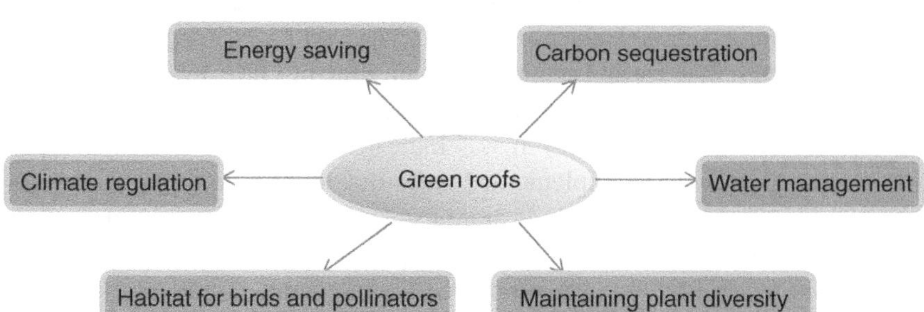

Figure 8.1 Eco-environmental value of green roofs.

summer, and insulation buildings from summer heat (Evangelisti et al. 2020; Bhardwaj et al. 2021), thereby reducing CO_2-dependent energy costs for heating and air-conditioning. Green roofs decrease urban heat island temperatures by 2.5–6 °C (Convertino et al. 2022; Quaranta et al. 2021). Absorbing water and gases on green roof surfaces solving the problem of rainwater runoff turning water for vegetation purposes, and helps to reduce urban flooding (Rahman et al. 2022; Shahmohammad et al. 2022) and making possible to reduce greenhouse gas concentrations (Dimitrijevic et al. 2018; Kuronuma et al. 2018). Research conducted in 2021 by Quaranta et al. shows that greening of 35% of the total EU's urban surface would avoid up to 55.8 Mtons/year CO_2 equivalent of greenhouse gas emissions, reducing energy demand for the cooling of buildings in summer by up to 92 TWh per year, with a net present value of more than 364 billion Euro (Quaranta et al. 2021). Moreover, green roofs provide additional benefits by making habitat spaces for birds and pollinators and maintaining plant diversity (Baldock et al. 2019). Because of these multiple benefits, green roofs can be an important climate and environmental regulator, meeting the concepts of sustainable urban development and the European strategy of Carbon neutrality.

8.4.2 Comparative Analysis of Bryophyte-PMFC with other ones using on Green Roofs

With the development of biotechnology of photosynthesis-associated energy generation from 2008 (PMFC) (De Schamphelaire et al. 2008; Kaku et al. 2008; Strik et al. 2008), green roofs were considered as a promising source of bioelectricity following the study conducted by Helder et al. (2013). However, only a few investigations explore plant-microbial power generation on green roofs. PMFC based on coastal grass *Spartina anglica* were placed on a roof-top in Wageningen, the Netherlands during 8 months from August 2009 to March 2010 (Helder et al. 2013). Current and power densities were not stable under outdoor extreme conditions and faced freezing of plants during winter of temperate oceanic climatic zones. Nonetheless, in total, 12 PMFCs with a combined surface area of 0.147 m^2 generated 9.096 kJ over 221 days (Helder et al. 2013). The next PMFC for green roofs was developed in 2017, in a dry and semiarid climate zone in Chile. Evaluation of succulent plants *Sedum* as a driver for PMFCs was conducted over 360 days on green roof ecosystem (Tapia et al. 2017). It was indicated that *Sedum hybridum*-based PMFC technology could be implemented and suitable for monitoring water content. It could serve as a potential indicator of soil water content in semiarid conditions on green roofs (Tapia et al. 2018). Sedge *Carex hirta* and fountain grass *Chinese pennisetum* PMFCs were developed for temperate continental and subtropical climate zone (Rusyn and Hamkalo 2020, Guan and Yu 2021). Sedge-based PMFCs were characterized by the highest power output values, as shown in Table 8.2.

Table 8.2 Comparison of mosses-PMFC efficiency with other ones using on green roofs.

Plant species	Power generation	Region and climate[a]	References
Coastal grass *Spartina anglica*	88 mW/m^2	Netherlands, temperate oceanic	(Helder et al. 2013)
Succulent *Sedum hybridum*	92 mW/m^2	Chile, dry semiarid	(Tapia et al. 2017)
Sedge *Carex hirta*	970 mW/m^2 PGA	Ukraine, temperate continental	(Rusyn et al., 2022)
Moss *Hypnum cupressiforme*	560 mW/m^2 PGA	Ukraine, temperate continental	(Rusyn and Hamkalo 2020)
Moss *Polytrichum commune*	541 mW/m^2 PGA	Ukraine, temperate continental	(Rusyn and Hamkalo 2020)
Moss *Leucobryum glaucum*	502 mW/m^2 PGA	Ukraine, temperate continental	(Rusyn and Hamkalo 2020)
Grass *Chinese pennisetum*	2.86 mW/m^2	Taiwan, subtropical	(Guan and Yu 2021)

[a] European climate zones based on information provided by EUCA15000.

Perennial herbs and succulent plants, particularly *Crassulaceae*, especially heat and drought-tolerant, are widely used and suitable for green roof technology (Shahmohammad et al. 2022). Low bioelectricity output obtained on green roofs (Tapia et al. 2017, Guan and Yu 2021) in comparison with laboratory conditions and the instability of plants for exploitation on green roofs (Helder et al. 2013) (Table 8.2) indicate the importance of plant selection and conditions for its existence for stable bioelectricity generation on green roofs. Grass plant *S. anglica* has been shown to be cold tolerant down to 5 °C; hence, plants suffered from severe cold stress and died consequently (Helder et al. 2013). Therefore, the performance of bryophyte PMFCs on green roofs based on mosses, which are unpretious to adverse growing conditions, frost-resistant, and know their ability to generate bioelectricity (Hubenova and Mitov 2011; Bombelli et al. 2016; Castresana et al. 2018; Rusyn and Hamkalo 2020; Çek et al. 2022) presents great interest.

Three types of PMFC based on widespread species of mosses *Hypnum cupressiforme* Hedw., *Polytrichum commune* Hedw. and *Leucobryum glaucum* Hedw. were placed on roof-top of building of the National University Lviv Polytechnic, Lviv, Ukraine from September 2014 to June 2015 (Rusyn and Hamkalo 2020). The maximal power output in polarization analysis was 560 mW/m^2 PGA (Figure 8.2). Together temperature and rainfall had a significant impact on the functioning of rooftop moss-PMFCs and the generation of bioelectricity. The impact of long-term rainfall on the bioelectricity generation of moss-PMFCs was favorable, but

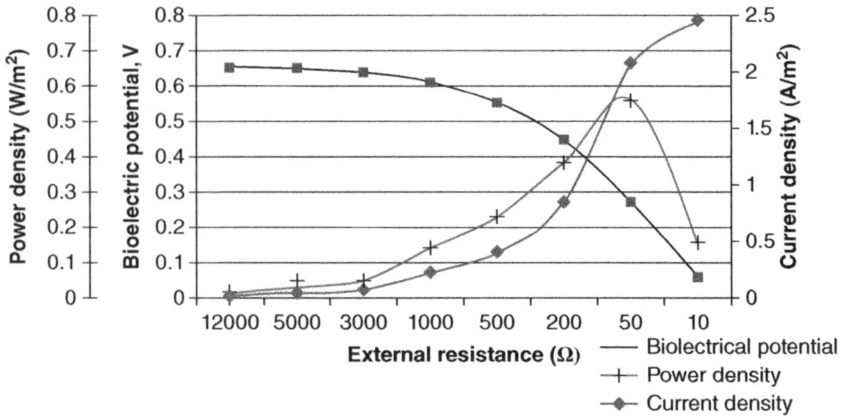

Figure 8.2 Effect of short-term action of external resistance on bioelectric potential, power density, and current density of moss-PMFC during the autumn period, normalized to $1\,m^2$ plant growth area (PGA), covered by plants and electrodes (Rusyn and Hamkalo 2020).

only at optimum temperature. Electro-biosystems operated at full capacity at temperatures above +10 °C and without prolonged droughts. The temperature below +10 °C or/and a long-term decrease of the precipitation below 5 mm/day caused a decrease in current output (Rusyn and Hamkalo 2020). During frosts and extended droughts, bioelectricity generation significantly decreased and recovered when optimal conditions were restored.

During the winter, PMFCs on the roof were in difficult weather conditions, with sharp temperature fluctuations up to −19 °C and thaws; hence, connections between the electrodes of part of PMFCs were damaged due to the effect of frozen water. The PMFC configuration, where several cathodes were connected in one cathode system, provided a larger area of contact with the substrate and allowed it to reach slightly higher values of bioelectricity. However, numerous contacts between the electrodes were more susceptible to damage and were more damaged after the winter period. The most optimal configuration of the electrode system for generating bioelectricity and resistance to weather damage factors was a system consisting of one monolithic graphite cathode located in the depth of the soil under the rhyzoides. This PMFC was stable and operated at the initial level after the winter period (Rusyn and Hamkalo 2020).

Moss driven PMFCs, functioning at the initial level after the winter period on green roofs, reveal the prospects of further research of moss-based electro-biosystems at a temperate continental climate zone after optimization in such directions as installing PMFC containers in the soil of an extensive and intensive green roofs or thermal insulation of PMFC containers and use of warm from waste heating gases of building for warming of PMFC systems.

8.5 Perspectives of Bryophyte PMFC

The first PMFCs were developed based on wetland plants (Helder et al. 2010; Takanezawa et al. 2010; Timmers et al. 2012b). Nowadays dozens of different types of PMFC based on both swamp (Moqsud et al. 2017; Regmi et al. 2018; Oodally et al. 2019) and drought-resistant plants (Sophia and Sreeja 2017; Rusyn and Hamkalo 2019; Rusyn and Valko 2019; Cheng et al. 2020; Jawre and Sandhu 2021) are now known. The study of moss-based PMFCs has only just received its development. However, the results of their applications for powering radio and environmental sensors, as well as for obtaining energy on green roofs (Bombelli et al. 2016; Rusyn and Hamkalo 2020), indicate this as a promising direction. Requirements of plants suitable to sustainable PMFC, especially for green roof applications, include drought resistance and frost resilience, stress tolerance, and endurance under limited substrate surface for root or rhizoids development and at low nutrient reserves as well as a high bioelectricity performance. Mosses possess all these characteristics listed above, as known mosses are widespreading roof weeds. Their stress-resistance and unpretentious, the formation a green roof carpet, and the ability to survive in a wide temperature range give advantages to mosses as candidates for PMFC systems for indoor use as well as for green roofs.

In recently developed bryophyte PMFCs despite the obvious progress and achievements, still remain problems such as, at outdoor: (i) freezing of plants in winter, (ii) damage to electrode systems by ice, (iii) dropping of bioelectricity generation at temperature below 0 °C, and also common (iv) the problem of increasing capacity, which has value for expanding and application scaling of moss-driven PMFCs. Development of moss-FM is considered a successful important step; however, a 10-hour battery charge by the energy of moss-PMFC powers the radio only for 80 seconds (Bombelli et al. 2016). A possible way to increase power could be realized by a combination of parallel-serial connections of PMFC modules (Pamintuan et al 2018), successfully implemented in works (Apollon et al. 2020; Apollon et al. 2022b), demonstrating the real-time powering of different low-power consuming devices, and as well as the use of multi-electrode systems in one PMFC module (Rusyn et al. 2021). The use of PMFC modules inside the soil, versus non-thermally insulated open separate containers exposed much more to the cold and wind, and the use of heat from waste gases from building's heating systems can be used to maintain the temperature above 0 °C, which is necessary for the stable functioning of the PMFC whole conditions temperate continental climate. Optimizing of the technological components of PMFC, providing optimal conditions for photosynthesis-dependent rhizodeposition of plants, and the development of an electro-active microbiome are the steps that bring the time of widespread practical application of bryophyte-based PMFC systems.

8.6 Conclusions

A wide range of physiological peculiarities of mosses contribute to the considering them as a prominent object for electro-biotechnology of plant-microbial bioelectricity. Complexes of the valuable characteristics of mosses are mentioned as follows: (i) Ability to grow almost on any substrate including ones with limited nutrient composition and substrate volume, (ii) Easy reproduction by vegetatively from small fragments of any portion of the plant, (iii) The surviving in long periods of drought and excellent water retention, (iv) UV resistance, (v) Heat and photosynthetic thermotolerance, and (vi) Cold/freezing tolerance, determine mosses as a promising candidate to photosynthesis assisted technology. Only a few moss-driven MFCs have been developed at the moment, reaching 2.5–20 mW/m^2 demonstrating the same potential as CAM plant MFC systems and the ability to power commercial radio and environmental sensors. Operation of Moss-MFC on a green roof for nine months at a temperate continental climate zone in a pilot study that reached 0.50–0.56 W/m^2 PGA opens the prospects using moss-based electro-biosystems at these conditions after optimization to prevent energy loss during cold and drought periods. Bryophyte-MFCs proved to be a good fundament to enhance their performance through improving operational properties, electrode materials, and design, as well as parallel-serial stacking of PMFC modules. However, more studies are required on the application of moss in MFC to bring this technology to the field of application and commercialization.

References

Anderson, M., Lambrinos, J.G., and Schroll, E. (2010). The potential value of mosses for stormwater management in urban environments. *Urban Ecosystem* 13: 319–332. https://doi.org/10.1007/s11252-010-0121-z.

Apollon, W., Kamaraj, S.K., Silos-Espino, H. et al. (2020). Impact of *Opuntia* species plant bio-battery in a semi-arid environment: demonstration of their applications. *Applied Energy* 279: 115788. https://doi.org/10.1016/j.apenergy.2020.115788.

Apollon, W., Vidales-Contreras, J., Rodríguez-Fuentes, H. et al. (2022a). Livestock's urine-based plant microbial fuel cells improve plant growth and power generation. *Energies* 15 (19): 6985. https://doi.org/10.3390/en15196985.

Apollon, W., Valera-Montero, L.L., Perales-Segovia, C. et al. (2022b). Effect of ammonium nitrate on novel cactus pear genotypes aided by biobattery in a semi-arid ecosystem. *Sustainable Energy Technologies and Assessments* 49: 101730. https://doi.org/10.1016/j.seta.2021.101730.

Apollon, W., Isabel Luna-Maldonado, A., Kamaraj, S.K. et al. (2023). Self-sustainable nutrient recovery associated to power generation from livestock's urine using plant-based bio-batteries. *Fuel* 332: 126252. https://doi.org/10.1016/j.fuel.2022.126252.

Arends, J.B., Speeckaert, J., Blondeel, E., Vrieze, J.D., Boeckx, P., Verstraete, W.H., Rabaey, K., & Boon, N. (2013). Greenhouse gas emissions from rice microcosms amended with a plant microbial fuel cell. *Applied Microbiology and Biotechnology* 98: 3205-3217. https://doi.org/10.1007/s00253-013-5328-5.

Arulmani, S.R., Gnanamuthu, H.L., Kandasamy, S. et al. (2021). Sustainable bioelectricity production from *Amaranthus viridis* and *Triticum aestivum* mediated plant microbial fuel cells with efficient electrogenic bacteria selections. *Process Biochemistry* 107: 27–37. https://doi.org/10.1016/j.procbio.2021.04.015.

Baldock, K.C., Goddard, M.A., Hicks, D.M. et al. (2019). A systems approach reveals urban pollinator hotspots and conservation opportunities. *Nature Ecology & Evolution* 3: 363–373.

Bhardwaj, A.A., Chaudhary, D.S., and Varandani, A. (2021). Impact of green roof on heating and cooling in context of different climatic zones in India. *EPRA International Journal of Research & Development (IJRD).* https://doi.org/10.36713/epra7466.

Bombelli, P., Dennis, R.J., Felder, F. et al. (2016). Electrical output of bryophyte microbial fuel cell systems is sufficient to power a radio or an environmental sensor. *Royal Society Open Science* 3 (10): 160249. https://doi.org/10.1098/rsos.160249.

Brandao, C., Cameira, M.D., Valente, F. et al. (2017). Wet season hydrological performance of green roofs using native species under Mediterranean climate. *Ecological Engineering* 102: 596–611. https://doi.org/10.1016/j.ecoleng.2017.02.025.

Bu, C., Wang, C., Yang, Y. et al. (2017). Physiological responses of artificial moss biocrusts to dehydration-rehydration process and heat stress on the Loess Plateau, China. *Journal of Arid Land* 9: 419–431. https://doi.org/10.1007/s40333-017-0057-8.

Byun, M.Y., Seo, S., Lee, J. et al. (2020). Transfection of *Arctic Bryum* sp. KMR5045 as a model for genetic engineering of cold-tolerant mosses. *Frontiers in Plant Science* 11: https://doi.org/10.3389/fpls.2020.609847.

Castresana, P.A., Martinez, S.M., Freeman, E. et al. (2018). Electricity generation from moss with light-driven microbial fuel cells. *Electrochimica Acta* 298: 934–942. https://doi.org/10.1016/j.electacta.2018.12.108.

Çek, N., Erensoy, A., Ak, N. et al. (2022). High-efficiency, environment-friendly moss-enriched microbial fuel cell. *International Journal of Chemical Reactor Engineering* 20: 1131–1140. https://doi.org/10.1515/ijcre-2021-0149.

Cheng, T.H., Ching, K.B., Uttraphan, C., and Heong, Y.M. (2020). Electrical energy production from plant biomass: an analysis model development for pandanus amaryllifolius plant microbial fuel cell. *Indonesian Journal of Electrical Engineering and Computer Science* 18 (3): 1163–1171. https://doi.org/10.11591/ijeecs.v18.i3.

Chiaffredo, M.R. (2004). When mosses recreate the landscape on the roof. *MCK Environnement and Bryotec Technology*.

Coe, K.K., Belnap, J., Grote, E.E., and Sparks, J.P. (2012). Physiological ecology of desert biocrust moss following 10 years exposure to elevated CO_2: evidence for enhanced photosynthetic thermotolerance. *Physiologia Plantarum* 144 (4): 346–356. https://doi.org/10.1111/j.1399-3054.2012.01566.x.

Convertino, F., Schettini, E., Blanco, I. et al. (2022). Effect of leaf area index on green facade thermal performance in buildings. *Sustainability* https://doi.org/10.3390/su14052966.

De Carvalho, R.C., Branquinho, C., and Da Silva, J.M. (2011). Physiological consequences of desiccation in the aquatic bryophyte *Fontinalis antipyretica*. *Planta* 234: 195–205. https://doi.org/10.1007/s00425-011-1388-x.

De Carvalho, R.C., Da Silva, A.B., Soares, R. et al. (2014). Differential proteomics of dehydration and rehydration in bryophytes: evidence towards a common desiccation tolerance mechanism. *Plant, Cell and Environment* 37: 1499–1515. https://doi.org/10.1111/pce.12266.

De Carvalho, R.C., Dos Santos, P., and Branquinho, C. (2018). Production of moss-dominated biocrusts to enhance the stability and function of the margins of artificial water bodies. *Restoration Ecology* 26: 419–421. https://doi.org/10.1111/rec.12688.

De Carvalho, R.C., Mauricio, A., Pereira, M.F. et al. (2019). All for one: the role of colony morphology in bryophyte desiccation tolerance. *Frontiers in Plant Science* 10: https://doi.org/10.3389/fpls.2019.01360.

De Carvalho, R.C., Do Paco, T.A., Branquinho, C., and Da Silva, J.M. (2020). Using chlorophyll a fluorescence imaging to select desiccation-tolerant native moss species for water-sustainable green roofs. *Watermark* 12: https://doi.org/10.3390/w12061748.

De Schamphelaire, L., Van Den Bossche, L., Hai, S.D. et al. (2008). Microbial fuel cells generating electricity from rhizodeposits of rice plants. *Environmental Science & Technology* 42 (8): 3053–3058. https://doi.org/110.1021/es071938w.

Dimitrijevic, D., Zivcovic, P., Brancovic, J. et al. (2018). Air pollution removal and control by green living roof systems. *Acta Technica Corviniensis – Bulletin of Engineering* 11: 47–50.

Ekwealor, J.T., Kosina, S.M., Bowen, B.P. et al. (2022). Ultraviolet radiation and dehydration stress induce overlapping transcriptional and metabolic responses in *Syntrichia mosses*. *BioRxiv* https://doi.org/10.1101/2022.09.14.508022.

Emilsson, T. (2008). Vegetation development on extensive vegetated green roofs: influence of substrate composition, establishment method and species mix. *Ecological Engineering* 33: 265–277. https://doi.org/10.1016/j.ecoleng.2008.05.005.

Evangelisti, L., Guattari, C., Grazieschi, G. et al. (2020). On the energy performance of an innovative green roof in the mediterranean climate. *Energies* https://doi.org/10.3390/en13195163.

Glime, J. (2007). Construction. In: *Bryophyte Ecology: Physiological Ecology*. Houghton: Michigan Technological University and the International Association of Bryologists.

Grant, G. (2006). Extensive green roofs in London. *Urban Habitats* 4: 51–65.

Guan, C.Y. and Yu, C.P. (2021). Evaluation of plant microbial fuel cells for urban green roofs in a subtropical metropolis. *Science of The Total Environment* 765: 142786.

Heim, A., Lundholm, J.T., and Philip, L.J. (2014). The impact of mosses on the growth of neighbouring vascular plants, substrate temperature and evapotranspiration on an extensive green roof. *Urban Ecosystem* 17: 1119–1133. https://doi.org/10.1007/s11252-014-0367-y.

Helder, M., Strik, D.P.B.T.B., Hamelers, H.V.M. et al. (2010). Concurrent bioelectricity and biomass production in three plant-microbial fuel cells using *Spartina anglica*, *Arundinella anomala* and *Arundo donax*. *Bioresource Technology* 101 (10): 3541–3547. https://doi.org/10.1016/j.biortech.2009.12.124.

Helder, M., Strik, D.P.B.T.B., Timmers, R.A. et al. (2013). Resilience of roof-top plant-microbial fuel cells during Dutch winter. *Biomass & Bioenergy* 51: 1–7. https://doi.org/10.1016/j.biombioe.2012.10.011.

Hubenova, Y. and Mitov, M. (2011). Bacterial mutualism in the mosses roots applicable in Bryophyta-microbial fuel cell. *Communications in Agricultural and Applied Biological Sciences* 76: 63–65.

Hui, R., Li, X., Chen, C. et al. (2013). Responses of photosynthetic properties and chloroplast ultrastructure of *Bryum argenteum* from a desert biological soil crust to elevated ultraviolet-B radiation. *Physiologia Plantarum* 147 (4): 489–501. https://doi.org/10.1111/j.1399-3054.2012.01679.x.

Hui, R., Zhao, R., Liu, L.C. et al. (2016). Effects of UV-B, water deficit and their combination on *Bryum argenteum* plants. *Russian Journal of Plant Physiology* 63: 216–223. https://doi.org/10.1134/S1021443716020084.

Jawre, A.K. and Sandhu, S.S. (2021). Production of green electricity from *Cynodon dactylon* in plant-bio-photovoltaic device. *International Journal of Environmental Science and Technology* 19: 5623–5630. https://doi.org/10.1007/s13762-021-03518-5.

Kaku, N., Yonezawa, N., Kodama, Y., and Watanabe, K. (2008). Plant/microbe cooperation for electricity generation in a rice paddy field. *Applied Microbiology and Biotechnology* 79 (1): 43–49. https://doi.org/10.1007/s00253-008-1410-9.

Kimmerer, R. (2003). *Gathering Moss: A Natural and Cultural History of Mosses*. Corvallis: Oregon State University.

Konásova, S. (2014). The role of green roofs in sustainable construction. In: *Proceedings of Creative Construction Conference*, Prague, Czech Republic (21–24 June 2014), 470–478. Elsevier Procedia.

Kuronuma, T., Watanabe, H., Ishihara, T. et al. (2018). CO_2 payoff of extensive green roofs with different. *Vegetation Species Sustainability* 10 (2256): 1–12. https://doi.org/10.3390/su10072256.

La Farge, C., Williams, K.H., and England, J.H. (2013). Regeneration of little ice age bryophytes emerging from a polar glacier with implications of totipotency in extreme environments. *Proceedings of the National Academy of Sciences of the USA*. 110: 9839–9844. https://doi.org/10.1073/pnas.1304199110.

Lange, O.L. (2003). Photosynthesis of soil-crust biota as dependent on environmental factors. In: *Biological Soil Crusts: Structure, Function, and Management* (ed. J. Belnap and O.L. Lange), 217–240. Berlin Heidelberg: Springer-Verlag.

Ligrone, R., Ducket, J.G., and Renzaglia, K.S. (2000). Conducting tissues and phyletic relationships of bryophytes. *Philosophical Transactions of the Royal Society of London. Series B, Biological Sciences 355* (1398): 795–813. https://doi.org/10.1098/RSTB.2000.0616.

Liu, S., Song, H., Wei, S. et al. (2014). Bio-cathode materials evaluation and configuration optimization for power output of vertical subsurface flow constructed wetland - microbial fuel cell systems. *Bioresource Technology* 166: 575–583. https://doi.org/10.1016/j.biortech.2014.05.104.

Liu, H., Kong, F., Yin, H. et al. (2021). Impacts of green roofs on water, temperature, and air quality: a bibliometric review. *Building and Environment* 196: 107794. https://doi.org/10.1016/j.buildenv.2.

Manso, M., Teotónio, I., Silva, C.M., and Cruz, C.O. (2021). Green roof and green wall benefits and costs: a review of the quantitative evidence. *Renewable & Sustainable Energy Reviews* 135: 110111. https://doi.org/10.3390/w14193188.

Michel, P., Payton, I.J., Lee, W.G., and During, H.J. (2013). Impact of disturbance on above-groundwater storage capacity of bryophytes in New Zealand indigenous tussock grassland ecosystems. *New Zealand Journal of Ecology* 37: 114–136.

Moqsud, A.M., Gazalia, T.A., Omineb, K., and Nakataa, Y. (2017). Green electricity by water plants in organic soil and marine sediment through microbial fuel cell. *Energy Sources, Part A: Recovery, Utilization, and Environmental Effects* 39 (2): 160–165. https://doi.org/10.1080/15567036.2016.1159263.

Nagase, A. and Thuring, C. (2006). Plant response to drought on extensive green roofs: the effect of temperature substrate type and substrate depth. In: *Proceedings Greening Rooftops for Sustainable Communities Conference*. Boston, Massachusetts, May 11–12 2006.

Oberndorfer, E., Lundholm, J., Bass, B. et al. (2007). Green roofs as urban ecosystems: ecological structures, functions, and services. *Bioscience* 57: 823–833. https://doi.org/10.1641/B571005.

Oliver, M.J., Velten, J., and Mishler, B.D. (2005). Desiccation tolerance in bryophytes: a reflection of the primitive strategy for plant survival in dehydrating habitats? *Integrative and Comparative Biology* 45: 788–799. https://doi.org/10.1093/icb/45.5.788.

Ondimu, S.N. and Murase, H. (2007). Combining Galerkin methods and neural network analysis to inversely determine thermal conductivity of living green roof materials. *Biosystems Engineering* 96: 541–550. https://doi.org/10.1016/j.biosystemseng.2006.12.007.

Oodally, A., Gulamhussein, M., and Randall, D.G. (2019). Investigating the performance of constructed wetland microbial fuel cells using three indigenous South African wetland plants. *Journal of Water Process Engineering* 32: 100930. https://doi.org/10.1016/j.jwpe.2019.100930.

Paco, T., De Carvalho, R.C., Arsenio, P., and Martins, D. (2019). Green roof design techniques to improve water use under mediterranean conditions. *Urban Science* 3: https://doi.org/10.3390/urbansci3010014.

Pamintuan, K.R., Clomera, J.A., Garcia, K.G. et al. (2018). Stacking of aquatic plant-microbial fuel cells growing water spinach (*Ipomoea aquatica*) and water lettuce (*Pistia stratiotes*). In: *IOP Conference Series: Earth and Environmental Science. The 4th International Conference on Water Resource and Environment (WRE 2018)* (17–21 July 2018), vol. 191, 012054. Kaohsiung City, Taiwan: IOP Publishing https://doi.org/10.1088/1755-1315/191/1/012054.

Park, J., Shin, Y., Kim, S. et al. (2022). Efficient plant types and coverage rates for optimal green roof to reduce urban heat island effect. *Sustainability* https://doi.org/10.3390/su14042146.

Porley, R. and Hodgetts, N. (2005). *Mosses and Liverworts*. London: Harper-Collins.

Quaranta, E., Dorati, C., and Pistocchi, A. (2021). Water, energy and climate benefits of urban greening throughout Europe under different climatic scenarios. *Scientific Reports* 11: https://doi.org/10.1038/s41598-021-88141-7.

Rahman, M.A., Alim, M.A., Jahan, S., and Rahman, A. (2022). Vegetated roofs as a means of sustainable urban development: a scoping review. *Watermark* 14 (19): https://doi.org/10.3390/w14193188.

Regmi, R., Nitisoravut, R., Charoenroongtavee, S. et al. (2018). Earthen pot-plant microbial fuel cell powered by *Vetiver* for bioelectricity production and wastewater treatment. *Clean* 46 (3): 1700193. https://doi.org/10.1002/clen.201700193.

Rong, H., Ruiming, Z., Gang, L. et al. (2016). Effects of UV-B radiation and light recovery on physiological property and cell ultrastructure of *Bryum argenteum*. *Environmental Sciences* 31 (1): 38–43. https://doi.org/10.5846/stxb201504180796.

Rusyn, I.B. and Hamkalo, K.R. (2019). Use of *Carex hirta* in electro-biotechnological systems on green roofs. *Regulatory Mechanisms in Biosystems* 10 (1): 39–44. https://doi.org/10.15421/021906.

Rusyn, I.B. and Hamkalo, K.R. (2020). Electro-biosystems with mosses on green roofs. *Environmental Research, Engineering and Management* 76 (1): 20–31. https://doi.org/10.5755/j01.erem.76.1.22212.

Rusyn, I.B. and Valko, B.T. (2019). Container landscaping with *Festuca arundinaceae* as bioelectrical minisystems in modern buildings. *International Journal of Energy for a Clean Environment* 20 (3): 211–229. https://doi.org/10.1615/InterJEnerCleanEnv.2019026674.

Rusyn, I.B., Medvediev, O.V., and Valko, B.T. (2021). Enhancement of bioelectric parameters of multi-electrode plant–microbial fuel cells by combining of serial and parallel connection. *International Journal of Environmental Science and Technology* 18 (6): 1323–1334. https://doi.org/10.1007/s13762-020-02934-3.

Rusyn, I., Fihurka, O., and Dyachok, V. (2022). Effect of plants morphological parameters on plant-microbial feel cell efficiency. *Innovative Biosystems and Bioengineering* 6 (3–4): 161–168. https://doi.org/10.20535/ibb.2022.6.3-4.273108.

Sahnoune, S., Benhassine, N., Bourbia, F., and Hadbaoui, H. (2021). Quantifying the effect of green roof and urban green infrastructure Rario on urban heat island mitigation – semi Ari climate. *Journal of Fundamental and Applied Sciences* 13: 199–224. https://doi.org/10.4314/JFAS.V13I1.12.

Schenk, G. (1997). *Moss Gardening: Including Lichens, Liverworts, and Other Miniatures*. Portland, OR, USA: Timber Press.

Shahmohammad, M., Hosseinzadeh, M., Dvorak, B. et al. (2022). Sustainable green roofs: a comprehensive review of influential factors. *Environmental Science and Pollution Research International* 29: 78228–78254. https://doi.org/10.1007/s11356-022-23405-x.

Sophia, A.C. and Sreeja, S. (2017). Green energy generation from plant microbial fuel cells (PMFC) using compost and a novel clay separator. *Sustainable Energy Technology* 21: 59–66. https://doi.org/10.1016/j.seta.2017.05.001.

Stanton, D.E. and Reeb, C. (2016). Morphogeometric approaches to non-vascular plants. *Frontiers in Plant Science* 7: 916. https://doi.org/10.3389/fpls.2016.00916.

Stark, L.R. (2017). Ecology of desiccation tolerance in bryophytes: a conceptual framework and methodology. *Bryologist* 120: 130–165. https://doi.org/10.1639/0007-2745-120.2.130.

Stark, L.R., Nichols, L., McLetchie, D.N. et al. (2004). Age and sex-specific rates of leaf regeneration in the Mojave Desert moss Syntrichia caninervis. *American Journal of Botany* 91 (1): 1–9.

Strik, D.P.B.T.B., Hamelers, H.V.M., Snel, J.F.H., and Buisman, C.J. (2008). Green electricity production with living plants and bacteria in a fuel cell. *International Journal of Energy Research* 32 (9): 870–876. https://doi.org/10.1002/er.1397.

Studlar, S.M. and Peck, J.E. (2009). Extensive green roofs and mosses: reflections from a pilot study in terra Alta, West Virginia. *BioOne* 26: https://doi.org/0.1639/0747-9859-26.2.52.

Takanezawa, K., Nishio, K., Kato, S. et al. (2010). Factors affecting electric output from rice-paddy microbial fuel cells. *Bioscience, Biotechnology, and Biochemistry* 74: 1271–1273. https://doi.org/10.1271/bbb.90852.

Tapia, N.F., Rojas, C., Bonilla, C.A., and Vargas, I.T. (2017). Evaluation of *Sedum* as driver for plant microbial fuel cells in a semi-arid green roof ecosystem. *Ecological Engineering* 108 (A): 203–210. https://doi.org/10.1016/j.ecoleng.2017.08.017.

Tapia, N.F., Rojas, C., Bonilla, C.A., and Vargas, I.T. (2018). A new method for sensing soil water content in green roofs using plant microbial fuel cells. *Sensors (Basel, Switzerland)* 18: https://doi.org/10.3390/s18010071.

Timmers, R.A., Strik, D.P.B.T.B., Arampatzoglou, C. et al. (2012a). Rhizosphere anode model explains high oxygen levels during operation of a *Glyceria maxima* PMFC. *Bioresource Technology* 108: 60–67. https://doi.org/10.1016/j.biortech.2011.10.088.

Timmers, R.A., Rothballer, M., Strik, D.P.B.T.B. et al. (2012b). Microbial community structure elucidates performance of *Glyceria maxima* plant microbial fuel cell. *Applied Microbiology and Biotechnology* 94 (2): 537–548. https://doi.org/10.1007/s00253-012-3894-6.

Wang, X.Q., Yang, P.F., Liu, Z. et al. (2009). Exploring the mechanism of *Physcomitrella patens* desiccation tolerance through a proteomic strategy. *Plant Physiology* 149: 1739–1750. https://doi.org/10.1104/pp.108.131714.

Wolf, L., Rizzini, L., Stracke, R. et al. (2010). The molecular and physiological responses of *Physcomitrella patens* to ultraviolet-B radiation. *Plant Physiology* 153: 1123–1134. https://doi.org/10.1104/pp.110.154658.

9

Duckweeds as Biocatalysts in Plant-based Biofuel Cell

Yolina Hubenova[1, 2] *and Mario Mitov*[3]

[1] *Department of Biochemistry and Microbiology, Plovdiv University "Paisii Hilendarski", Plovdiv, Bulgaria*
[2] *Department of Electrocatalysis and Electrocrystallization, Institute of Electrochemistry and Energy Systems "Acad. E. Budevski" – Bulgarian Academy of Sciences, Sofia, Bulgaria*
[3] *Innovative Center for Eco Energy Technologies, South-West University "Neofit Rilski", Blagoevgrad, Bulgaria*

9.1 Introduction to Plant-based Microbial Fuel Cells

Microbial fuel cells (MFCs) are devices for simultaneous electricity generation, wastewater purification, and/or synthesis of valuable raw materials. Many bacterial and yeast species have been proven to be exoelectrogenic, i.e. they can transfer electrons to the anode and use it as an alternative final electron acceptor to implement their respiration processes, which is why they are called anode-respiring bacteria. In general, the extracellular electron transfer (EET) from the microbes to the anode has been classified into direct and indirect EET mechanisms. The first group includes the EET implemented by microbial biofilms (Santoro et al. 2017; Hubenova and Mitov 2015), while the second is based on the communication of the bacteria with the anode via electron shuttles, or the so-called electron transfer mediators (EMs). Two types of EM are distinguished: (i) endogenous mediators (EnMs) produced and secreted by the (micro)organisms, redox-active substances, and (ii) artificially added redox molecules or exogenous mediators (ExMs). The establishment of new exoelectrogenic species continues (Hubenova et al. 2022; Hubenova et al. 2022; Hubenova et al. 2023) and the MFC technology is further developed.

Photosynthesis-Assisted Energy Generation: From Fundamentals to Lab Scale and In-Field Applications, First Edition. Edited by Sathish-Kumar Kamaraj and Iryna Rusyn.

Fifteen years ago, Verstraete et al. examined the possibility of the bacteria in the rhizodeposits of rice plants being used as anodic biocatalysts (De Schamphelaire et al. 2008). During the next four years, the group of Buisman including Strick, Helder, Timmers, and Hamelers characterized and explained the principles of plant microbial fuel cells (P-MFCs) (Strik et al. 2008; Helder et al. 2010; Timmers et al. 2010; Strik et al. 2011). *Spartina anglica, Arundinella anomala, Glyceria maxima,* and *Arundo donax* were the plants (Helder et al. 2010), which have been used in the development of biofuel cell systems. The principles were officially recognized at the First PowerPlant Symposium, held in 2011 in Gent, Belgium. On the one hand, they are based on the specific relations between the exoelectrogenic bacteria in the rhizodeposits and the plant's roots, and on the other hand, the bacteria contact with the anode buried in the soil. The plants produce organic matter from the CO_2, water, and the energy of sunlight. A part of these organics are assimilated by the bacteria during symbiosis and converted again into energy necessary for bacterial development. When an electrode is present, electrons from the bacteria are transferred extracellularly to it, and if the electric circuit is closed, a current flows.

Bombelli et al. (2013) compared the performance of *Oryza sativa* and the associated weed *Echinochloa glabrescens* and established that the total charge accumulation of 6.75 ± 0.87 C was six times higher when *O. sativa* was used, expressed in an average power output of 0.980 ± 0.059 GJ/ha/yr. McCormick et al. (2015) contextualized the specific use of light-harvesting bioelectrochemical systems (BESs) and called them biophotovoltaic systems (BPVs). BPVs utilize photosynthetic organisms, such as algae and cyanobacteria, which can produce organics harvesting light energy and converting part of the biological energy into electrical current. A new horizon for the application of water plants as biocatalysts in fuel cells without added fuel (substrate) has been proposed. Two independent groups verify bryophytes (mosses), having roots in soils and growing in areas rich in water, in the so-called bryophyte microbial fuel cell (Hubenova and Mitov 2011; Bombelli et al. 2016) and prove that mosses are appropriate to be grown in P-MFC even to supply small power consumers (Bombelli et al. 2016).

9.2 Biofuel Cells Using Aquatic Higher Plants as Anodic Biocatalysts

As next, aquatic plants, floating on the water's surface, have been investigated for the possibility to be utilized as biocatalysts in P-MFC generating electrical energy. Because of the lack of soil and unidentified mechanisms, they have been called plant biofuel cells (P-BFCs).

9.2.1 Structure of the Duckweeds

The duckweed, known also as water lens, is a flowering aquatic plant belonging to the Lemnaceae family. Duckweed grows in freshwater environments such as ponds, swamps, slow-moving streams and rivers, and wetlands and is distributed in five genera *Lemna*, *Spirodela*, *Landoltia*, *Wolffia*, and *Wolffiella*. *Lemna* species are free-floating thalli on the water surface with a size of about 5 mm in length, while *Spirodela polyrhiza* grows in colonies forming a dense covering on the water surface. The plants of *Spirodela* are bigger, round, flat discs 0.5–1.0 cm wide. The main difference between the representatives of *Lemna* and *Wolffia* is that the latter are rootless. Some representatives of the genus *Lemna* have a single root, while others like *Spirodela* and *Landoltia* have multiple roots. Duckweeds are flowering plants producing seeds under appropriate conditions (sexual reproduction). However, the rapid colonization of new water is thought to be due to their vegetative (asexual) reproduction by the formation of daughter plants from the vegetative buds of the adult plants (Figure 9.1). The duckweed plants are often observed as clusters of fronds derived from a mother frond. The duckweed possesses air

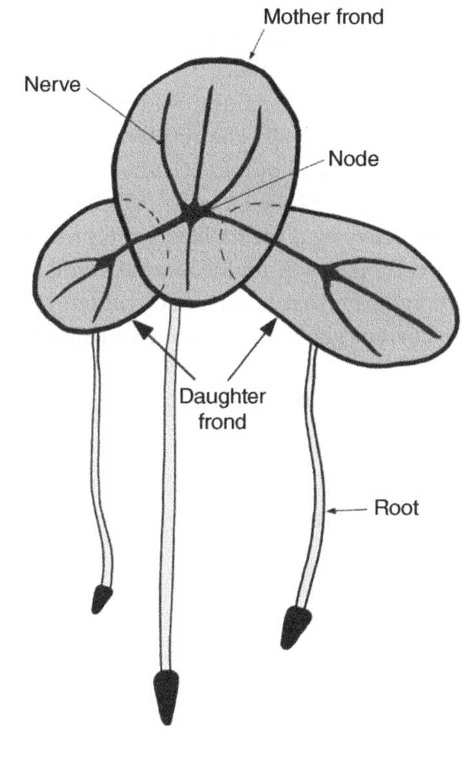

Figure 9.1 Scheme of the common structure of the duckweed *Lemna minor*. Each leaf has a single root. Source: More information: https://plants.usda.gov/home/plantProfile?symbol=LEMI3.

pockets called aerenchyma, which is responsible for buoyancy and a meristem zone is a site where an organized structure of dividing cells is located. The duckweeds have vascular tissue (nerves) consisting of elongated cells that branch off from the node and run nearly to the tips of the fronds. To survive the winter, duckweed produces specialized buds called turions. They detach from the parent stem and sink to the bottom of the pond. In the spring, the turions begin to break down the reserve carbohydrate (starch) and float back to the surface to form the typical duckweed stem.

9.2.2 The Electrical Parameters Achieved by Plant Biofuel Cells (P-BFC)

The first species investigated as an anode biocatalyst was the duckweed *Lemna minuta*, which was grown autotrophically in a specially designed fuel cell called direct photosynthetic plant fuel cell (DPPFC) (Hubenova and Mitov 2012). Each fuel cell consisted of a glass body, in which a carbon cloth located near the water surface was used as an anode, while graphite granules, placed into a water filter playing the role of a separator, served as a cathode (Figure 9.2). The requirements for the growth and development of the duckweed have been met and at the same time, the plants were in a contact with the anode. The electrical circuit flows when the circuit has been closed through an external resistance.

The experiments have been conducted under both natural sunlight (over 10 000°lx) and artificial light (800°lx) illuminance as well. The results show the significant contribution of solar energy to the electrical outputs (Hubenova and Mitov 2012). The very low total microbial number of $(7.2 \pm 1.1) \times 10^3$ cells/ml and the absence of fungal proliferation on the selective agar media have been proven for the anolyte after six weeks of the fuel cell operation under natural sunlight (Hubenova and Mitov 2012). These results indicate that not the bacteria but the plants are the leading participants in the current generation. By measuring the

Figure 9.2 An image of the P-BFC experimental setup using duckweeds as anodic biocatalysts.

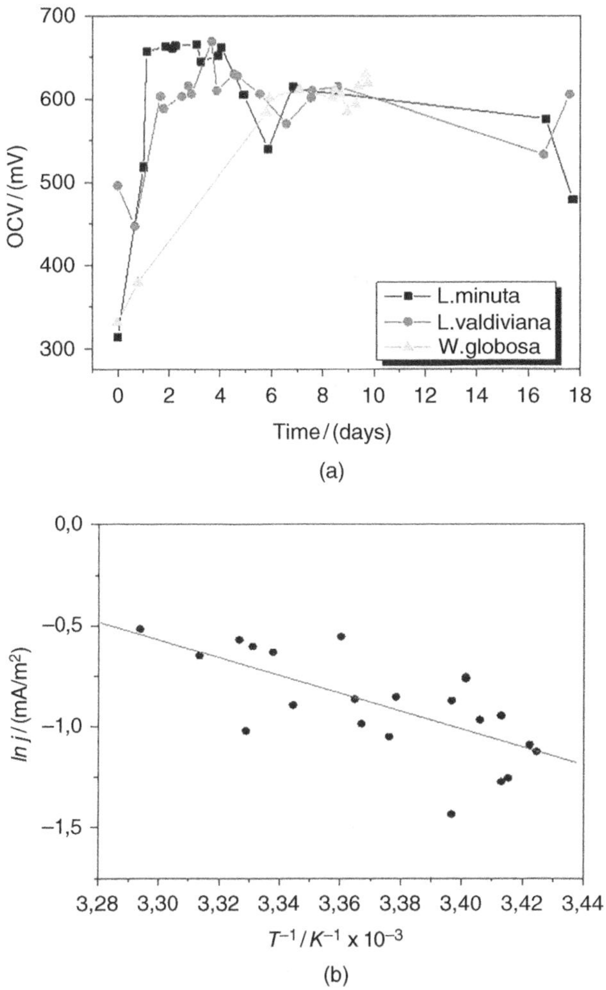

Figure 9.3 (a) OCV over time registered for three different used duckweed species grown in a fuel cell; (b) Dependance of the current density generated by *L. minuta*—P-BFC on the temperature presented as an Arrhenius plot.

anodic and cathodic potentials over time, it has been shown that after four days of polarization caused by a temporary switching resistance, the anodic potential dropped to values of -200 mV (versus SHE), while the cathodic potentials were relatively constant (+500 mV) (Hubenova and Mitov 2012). The summary results for three species of duckweed *L. minuta*, *Lemna valdiviana*, and *W. globose,* grown in the fuel cell, presented in Figure 9.3a, show similar open circuit voltage (OCV). The utilization of the identical cathode and its constant high positive value potential suggests that the OCV is determined exclusively by the anode potential, which in turn depends on the plants grown in the anodic compartment.

The current generated by *L. minuta*-DPPFC over time varied from 0.4 Am^{-2} to 1.6 Am^{-2} on the eleventh day (Hubenova and Mitov 2012). It has been suggested that the fluctuations are related to the temperature and the illuminance. The measured current density follows that of the temperature during the natural day/night cycle. The low correlation coefficient of 0.683 of the relationship between the current density and the temperature (Figure 9.3b) indicates that the temperature is not a dominating factor for the current generation in the explored system. The comparison of the *L. minuta*-based biofuel cell under different illumination conditions and the establishment that under lab conditions the generated current was one order of magnitude lower than that under natural sunlight shows the greater importance of the light intensity. Higher electric parameters were observed during the light part of the periodic alteration of day/night and the current values strictly followed the day/night cycle (Hubenova and Mitov 2012). After dawn, the current gradually raised and reached the highest values at the maximal light intensity and then decreased. The dependence of the current measured during the first half of the day on the light intensity fits an equation very similar to those describing photosynthesis light response (*P–I*) curves (Jassby and Platt 1976), suggesting that the current generation is related to the natural physiological response of the duckweed to the photoperiod.

Using *L. minuta* as an anodic biocatalyst in a DPPFC under natural sunlight leads to achieving a high power density of $380 \pm 19\,mW/m^2$, corresponding to $120 \pm 6\,GJ/ha/year$ (Bombelli et al. 2013). The first examination of the putative mechanism of EET between the duckweeds and the anode has been carried out by cyclic voltammetry (CV) of the anolyte, respectively, water collected from the control-grown plants. The CV of the anolyte has shown oxidation and a reduction peak, which are missing in the CV of the control. Identical peaks have been observed after the filtration of the anolyte through a sterile filter, which supposes that the observed electrochemical activity is not due to existing bacteria. The linear dependence of both cathodic and anodic peak currents on the square root of the scan rate, typical for a diffusion-limited Nernstian reaction, supports the presence of soluble redox couple(s) in the medium (Hubenova and Mitov 2012). An indirect mechanism of EET implemented by electron shuttling compounds has been suggested.

9.2.3 The Generated Current by P-BFC Depends on the Light Source

The second duckweed species explored toward exoelectrogenicity was *L. valdiviana* (Hubenova and Mitov 2015). The duckweeds were grown in P-BFCs, situated in a lab under constant temperature and mixed natural and artificial lightening, and the cell voltage across an external resistor was recorded over time. During the first two days, the generated current sharply increased reaching steady-state values of c. 200 mA/m^2, which are eight times lower than that generated by *L. minuta*–DPPFCs under natural sunlight (Hubenova and Mitov 2012). Comparing current densities generated by biofuel cells with different duckweed species (Figure 9.4a) but under controlled laboratory lightening, the values vary between 100 and 220 mA/m^2 and show the same trend over time. It has been confirmed that as the sun rises and the lighting in the room increases, the current increases until around noon (Figure 9.4b), as it was also found for *L. minuta* (Hubenova and Mitov 2012).

To establish the possible origin of the electrochemical activity in *L. valdiviana*-based P-BFC, CV of the used anolyte with different scan rates has been performed (Hubenova and Mitov 2015). The linear dependence of both anodic and cathodic peak currents on the square root of the scan rate indicates a diffusion-controlled electrochemical reaction with the participation of an electrochemically active soluble compound in the medium (Hubenova and Mitov 2015). The cyclic voltammograms clearly show that a redox-active substance is secreted as a result of the polarization, suggesting that it could serve as an electron shuttle. Figure 9.4c demonstrates the CV pattern of an exploited anolyte. The midpoint (formal) potential $E^{0\prime}$ is calculated to be around −70 mV (versus SHE) for *L. valdiviana*, while $E^{0\prime}$ of *L. minuta* is around −100 mV (versus SHE), which suggests that the same metabolic pathway could respond to the polarization and synthesizes closely related secondary metabolites. This hypothesis is supported also by the lack of redox activity in the control-grown plants.

As far as the two representatives of the Lemna genus – *L. minuta* and *L. valdiviana* have roots contacting the anode, a mixed EET mechanism has been proposed – a direct EET, implemented by the duckweeds' roots and a mediated EET by the secreted redox-active compound.

9.2.4 The Rootles *Wolffia globosa* in Plant-based Biofuel Cell

P-BFCs using Lemna species as biocatalysts are characterized by the fact that duckweed leaves float on the water surface (anolyte) and are in constant contact with the anode through their roots. The duckweeds belonging to the family Lemnacea possess roots, which differentiate them from Wolffia. Wolffia water lentils do not touch the anode because they are rootless and thus the possibility of transferring electrons to the anode remains only by an indirect mechanism.

The characterization of the *Wolffia globose*-based P-BFC was made by polarization and power curves after the stabilization and adaptation of the plants (Hubenova et al. 2023). The results reveal that the developed BES is capable of generating a current reaching density of 1 A/m^2 and a power density of c. 90 mW/m^2 (Hubenova et al. 2023). When the voltage of *W. globosa*-based BFC is recorded through a permanently connected resistor, the current density is around 170 mA/m^2 (Hubenova et al. 2023), which is comparable to the current generated by *L. minuta* - biofuel cell operated under artificial lighting (Figure 9.4a).

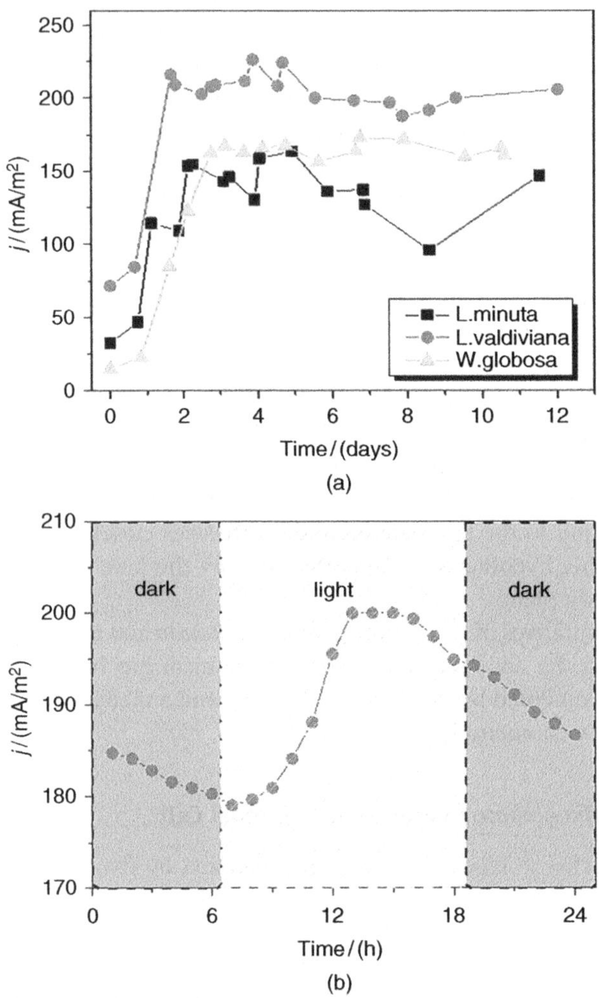

Figure 9.4 Electrical characteristics: (a) current density recorded by P-BFC across 1 kΩ resistor with three different duckweed species over time; (b) with *L. valdiviana* for 24 hours window; (c) CV of the exploited anolyte from *L. valdiviana*- P-BFC. The CV is carried out in a three-electrode electrochemical cell, where both working and counter electrodes are Pt-wires, and the reference electrode is Ag/AgCl, scan rate of 50 mV/s^1. Source: Adopted from Hubenova and Mitov (2015).

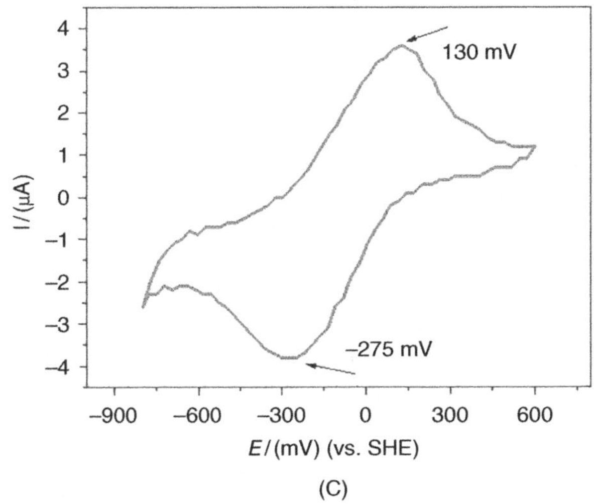

Figure 9.4 (Continued)

The possible mechanism of electron transfer from *W. globosa* to the anode has been verified by CV of the exploited anolyte. The observed oxidation peak at +90 mV (versus SHE) and reduction peak at −200 mV (versus SHE) reveal that *W. globosa* is capable of producing and secreting electrochemically active compounds when grown under polarization conditions. It has been suggested that this redox substance participates in the current generating processes as an EnM, shuttling electrons from the plants to the anode. This mediated electron transfer mechanism is in good agreement with previously proven EnMs for both aquatic plants (Hubenova and Mitov 2012; Hubenova and Mitov 2015) and yeast cells (Hubenova and Mitov 2015; Hubenova et al. 2014). It has to be mentioned that only a few EnMs have been identified till now (Hubenova and Mitov 2015). Among the examples are flavin molecules secreted by *Shewanella oneidensis* and nitrogen-containing heterocyclic phenazine metabolites such as pyocyanin produced by *Pseudomonas aeruginosa* (Sevda et al. 2018). Phenazines are derived from the shikimic acid pathway that is highly conservative in most organisms including plants (Pierson and Pierson 2010), and it is supposed that they participate in the EET and redox state maintenance of the duckweeds under polarization. The estimated formal potential E'^0 of −55 mV (versus SHE) from the CV of *W. globosa* (Hubenova et al. 2023) is close to that of *L. valdiviana* (Hubenova and Mitov 2015) and shifted in a positive direction from that of *L. minuta* (Hubenova and Mitov 2012). Although the nature of the EnM in duckweeds remains unclear and needs further clarification, the established redox potentials overlap for example that of the 2-cyano phenazine (Nakagawa and Nishina 2021), suggesting that it could be a phenazine derivate.

9.3 Influence of the Electrode Polarization on the Plants' Metabolism

The metabolic pathways are formally divided into primary and secondary pathways. It is supposed that the primary pathways support the cells with food substrate, energy, and reproduction, while the secondary pathways condition the adaptation of plants to environmental changes. Since survival in various conditions caused by both climatic and anthropogenic factors is directly dependent on secondary pathways, recently the term active metabolites is preferred to be used instead of secondary metabolites. So to understand the relationship between photosynthesis, substrate oxidation, respiration, and the caused polarization conditions in the biofuel cell, basic biochemical parameters have been analyzed for the duckweeds grown as biocatalysts and as controls (Hubenova and Mitov 2012; Hubenova and Mitov 2015; Hubenova et al. 2023) and compared.

Analyzes begin with the estimation of the growth and development of the water lens. Macroscopically, it has been observed that both the control and those grown in a biofuel cell divide vegetatively. It has been established that under polarization conditions the increase of the duckweed biomass exceeds up to 45% of that of the control plants. This means that the duckweeds not only upregulate the secondary pathways to communicate adequately with the anode but also intensify their anabolic and catabolic processes.

9.3.1 The Change in the Protein Content

A more precise estimation of the influence of the polarization conditions on the duckweed metabolism was achieved by the determination of the total protein content. After the mechanical disintegration of the duckweeds, the homogenates have been fractionated by centrifugation (12 000rpm, 10 minutes) to obtain a crude organelle-enriched fraction. The chloroplasts are sedimented in this pellet (Hubenova and Mitov 2012; Hubenova et al. 2023). The organelles of *L. valdiviana* were further purified and separated by centrifugation in a sucrose density gradient (Hubenova and Mitov 2015). The summary of the content of the determined protein concentration is presented in Table 9.1. The calculated data for the wet biomass is also presented. The analyses show that all kinds of duckweed have higher protein content when grown in biofuel cells. The increase varies for the single species from 25% for *W. globosa* to over 100% for *L. minuta*, suggesting that the polarization affects the biochemical processes in the duckweeds' chloroplasts and mitochondria.

Table 9.1 Summary of the intracellular protein quantity determined by Bradford method in duckweeds grown in biofuel cell under polarization and as control grown plant. The fractionation approach to obtain an organelle-enriched fraction is marked. The procedure for differential centrifugation followed by centrifugation in sucrose density gradient is described in detail in Hubenova and Mitov (2015).

An organelle-enriched fraction from duckweed species	Obtained by centrifugation	Protein (mg/ml)	Protein (mg/g wet biomass)	Reference
L. minuta in biofuel cell	In one step	2.90	621	(Hubenova and Mitov 2012)
L. minuta, control	In one step	1.38	321	(Hubenova and Mitov 2012)
L. valdiviana in biofuel cell	35% sucrose density gradient fraction	2.01	503	(Hubenova and Mitov 2015)
L. valdiviana, control	35% sucrose density gradient fraction	1.37	343	(Hubenova and Mitov 2015)
W. globosa in biofuel cell	In one step	1.98	495	(Hubenova et al. 2023)
W. globosa, control	In one step	1.57	393	(Hubenova et al. 2023)

9.3.2 Synthesis, Accumulation, and Degradation of Carbohydrates

Plants are the primary organic matter producers on Earth. The glucose is synthesized from the intermediate sugar phosphates (glyceraldehyde-3-phosphate) produced in the Calvin cycle in the chloroplasts after fixation and reduction of carbon dioxide. Glyceraldehyde-3-phosphate can be used for the production of polysaccharides in the form of starch, which is accumulated in granules as an energy reserve inside the chloroplasts. The enzyme amylase catalyzes the hydrolysis of glycoside bonds of the starch, producing dextrins, maltose, and glucose and thus participating in the regulation of carbohydrate synthesis and degradation. To understand how the carbohydrate synthesis and degradation is influenced by the polarization, the quantity of the glucose and starch quantity of duckweeds was deteremined and compared.

A twice higher quantity of reducing sugars (Hubenova et al. 2023) and 1.7 times more starch have been established in the organelle-enriched fraction of *W. globosa* (Figure 9.5) collected from P-BFC. The processes of accumulating polysaccharides as reserves for maintaining the positive energy balance required for the

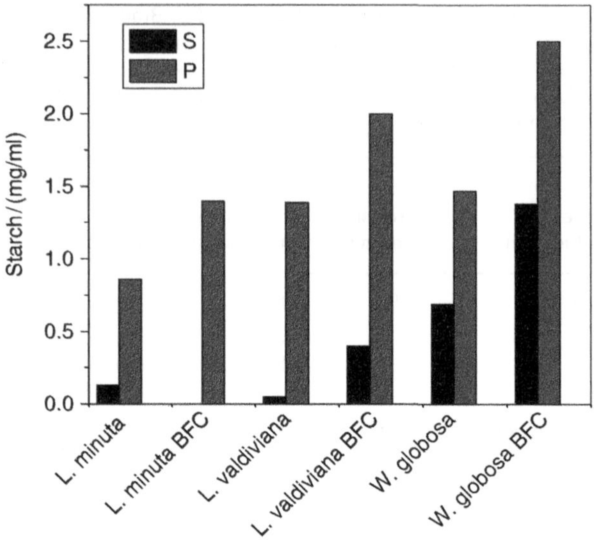

Figure 9.5 Summary of the starch content determined in the supernatant (S) and pellet (P) fractions, the latter containing the cellular organelles of three duckweed species grown as control and in P-BFC.

plants' growth under polarization are also enhanced. At the same time, the determined specific amylase activity of 70 U/g wet biomass was 1.82 times higher than that of the control, indicating not only that more starch is produced under polarization but also that more reserve carbohydrates have been hydrolyzed to cover the plant's energy demands (Hubenova et al. 2023). The highest amylase activity was expressed by *W. globosa* compared to the amylase activity of Lemna species (Hubenova and Mitov 2012; Hubenova and Mitov 2015; Hubenova et al. 2023). These results demonstrate that the whole metabolism of *W. globosa* grown under polarization is boosted. The comparison of the glucose content in *W. globosa* (Hubenova et al. 2023) and *L. valdiviana* (Hubenova and Mitov 2015) grown under polarization leads to the summary for a common tendency in the increase of the monosaccharides' content. For all investigated species, the accumulation of reserve carbohydrate starch also increases almost twice. The lower glucose content in the fraction of *L. minuta* grown in a biofuel cell is probably the reason for the increased amylase activity compared to the control. It has been suggested that this is a consequence of the adaptive response of the duckweed to the enhanced energy requirements due to the partial conversion of the produced chemical energy into electricity (Hubenova and Mitov 2012).

In summary, duckweed produces twice as much reducing sugars and accumulates twice as much starch when grown in a biofuel cell. The increased amylase activity indicates that the entire plant metabolism (synthesis and degradation) is intensified under these conditions.

9.3.3 Formation and Degradation of the Phytates

Metabolic processes of the duckweeds are enhanced under polarization. This statement is further supported by the reduced amounts of inorganic phosphate found in the chloroplast- and mitochondria-enriched fractions of Lemna spp. grown in P-BFC (Table 9.2). Lower amounts suggest that inorganic phosphate has been converted to organic, and used in oxidative phosphorylation to produce ATP as well as in phosphorylation to tag molecules, for example, to be incorporated into glycolysis as glucose-6-phosphate. In parallel, glucose-6-phosphate is the precursor of myo inositol 3-phosphate, which phosphorylated gives inositol (3,4,6) 3-phosphate (Hubenova et al. 2014). Myo-inositol hexakisphosphates (phytates) are the phosphorous storage form in plants. Like other plant species, duckweeds contain phytates (Hubenova et al. 2014; Idriss et al. 2002), and the determination of the phytase activity would direct the degree of the degradation of myo-inositol hexakisphosphates. The results presented in Table 9.2 show that the decreased amount of inorganic phosphate was inversely proportional to the measured phytase activity, which suggests that compensation of the faster inorganic phosphate decrease under polarization occurs.

The presence of phosphate buffer in the anolyte of *W. globose* (Hubenova et al. 2023) causes an increase in the inorganic phosphates, which differs from the above two cases. From the initial inorganic phosphate buffer (67 mM, pH 7),

Table 9.2 Summary of quantity inorganic phosphate and the phytase activity determined in the organelle fractions of three duckweed species.

An organelle-enriched fraction from duckweed species	Anolyte	PO_4^{3-}, mM	Phytase activity (U/g wet biomass)	Reference
L .minuta, biofuel cell	Water	80	-	(Hubenova and Mitov 2012)
L. minuta, control	Water	152	2.5	(Hubenova and Mitov 2012)
L. valdiviana, biofuel cell	Water	61	4.1	(Hubenova and Mitov 2015)
L. valdiviana, control	Water	74	2.3	(Hubenova and Mitov 2015)
W. globosa, biofuel cell	Phi	161	4.3	(Hubenova et al. 2023)
W. globosa, control	Phi	22	2.8	(Hubenova et al. 2023)

40 mM phosphates are sufficient to supply the 1g plants with the required amount of phosphorus (Hubenova et al. 2023). In total, 20 mmol/L phosphates have been measured in the organelle fraction of the control, while these values were more than seven times higher in the respective fraction of the plants grown under polarization. However, comparably high phytase activity was found in the organelle fraction of *W. globosa* with that of *L. valdiviana* (Table 9.2). This indirectly specifies that the presence of inorganic phosphates in the buffer does not cause hydrolysis of phytate to a greater or lower degree. The phytase activity of all control plants is almost identical and two times lower than those determined for the plants in biofuel cells.

One of the ecological problems is related to the eutrophication of water. Phytates contaminate composts and eutrophicate waters in regions with intensive stock farming because they are indigestible by nonruminant animals (Hubenova et al. 2014). If phytates get into foods, they reduce their value and appear as inhibitors for the assimilation of valuable trace elements such as zinc and iron, that is why phytate in food products must be degraded. Intensive investigations concerning the development of effective methods for phytate bioremediation have been carried out (Hubenova et al. 2014; Hubenova et al. 2014; Georgiev et al. 2013). The increase of the specific phytase activity by polarization offers a new alternative to the fermentation approach for decreasing the phytate content.

9.4 Components of Photosynthetic Systems Involved in the Direct EET to the Anode

After the establishment of the electrical current generation (summarized electrochemical parameters are presented in Table 1 of (Hubenova et al. 2023)) and the increased carbohydrate and protein content of the duckweeds under polarization, the next step was to verify the contribution of the photosynthetic processes to the electricity generation by the duckweeds–based biofuel cells by trying to figure out which plant photosystems, absorbing light and transferring energy and electrons in the chloroplasts, could be involved. For this purpose, a special experimental setup (Figure 9.6) for growing the duckweed in fuel cells under controlled light access and a constant temperature has been developed (Hubenova et al. 2018). Four identical fuel cells have been wrapped with a black light-proof folio and light-tight lid. A coherent light source with an illuminance of 60 kLx has been transmitted through optical fibers, 3 fibers in a package of which has been illuminating a single P-BFC. A filter for polychromatic or monochromatic light was put at the end of the fibers after a Polaroid sheet and illuminated the entire surface of the anode covered with 1 g wet biomass *Lemna minor*, equal to c. 100 plants (Hubenova et al. 2018). Monochromatic red light with a wavelength of 650 nm and measured illuminance

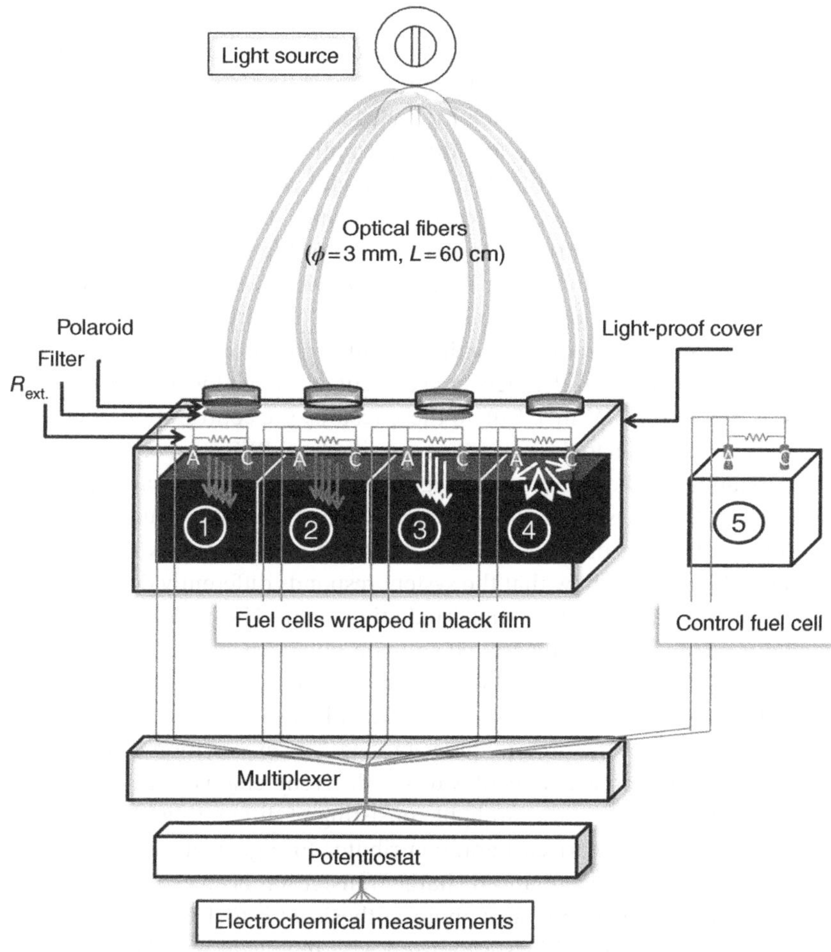

Figure 9.6 Schematic presentation of the developed experimental set-up for controlled irradiation of photosynthesizing plant fuel cells and real-time electrochemical measurements. Source: Reproduced from Hubenova et al. (2018).

of 40 ± 5 Lx, as well as blue (wavelength of 450 nm, measured illuminance 500 ± 100 Lx) polarized light, have been used for periodic irradiation of the P-BFC and the recorded electrical outputs have been compared to those obtained when visible polarized (280 ± 60 Lx) or non-polarized visible light (1050 ± 300 Lx) was used. The red and blue lights have been chosen because their wavelengths are near the characteristic absorption peaks of chlorophyll *a* and *b* playing role in the light-harvesting system of green plants. An additional sample grown under mixed laboratory lightening (1870 ± 200 Lx) has been performed as well (Figure 9.6).

During the irradiation, the current has been recorded and the change of the charge transfer has been evaluated. All irradiated samples after incubation in the dark give higher current values than the control kept under natural light during the day/night cycle. As expected, when irradiated with visible non-polarized light, the current difference (Δj) before and during irradiation has been smaller than that obtained with duckweeds-P-BFC irradiated with polarized (waves of the visible spectrum directed in one direction, irradiance $0.41\,W/m^2$) visible light (Hubenova et al. 2018). Right after the irradiation of the water lentils, grown in P-BFC, with monochromatic red or blue light, however, the generated current increases and with small oscillations is stable up to three hours (Hubenova et al. 2018). The periodic irradiation with red light (irradiance equal to $0.06\,W/m^2$) leads to the highest values of the registered current, although the irradiance of the blue light is higher ($0.73\,W/m^2$). Due to these results, the contribution of chlorophyll *a* and chlorophyll *b* to the current generation by P-BFC has been proposed.

To understand the photo-induced charge transfer, electrochemical impedance spectroscopy (EIS) analyses have been implemented in real time – during the irradiation (*operando*). The obtained impedance spectra, presented as Nyquist and Bode plots in Figure 9.7, show that the system responds differently depending on the type and the wavelength of the irradiating light. The impedance spectra obtained during the irradiation with red and non-polarized visible light represent one semi-arc, well fitted to a simplified Randel's model but the diameter of the latter was 6 times lower. The peaks on the Bode plots show a one-time constant reaction that appears in the high-frequency range. The explanation is that plants are adapted to irradiation with a set of waves of different lengths and energies (the whole spectrum). In nature, the sunlight consisting of different light colors, corresponding to the non-polarized light, contributes to higher photosynthesis. In the case of P-BFC growing, enhanced charge transfer has been estimated at non-polarized visible light, which relates photosynthesis with EET to the anode. The impedance patterns of the P-BFCs, irradiated with blue light or polarized visible light, differ from those of the red light and fit an equivalent circuit model consisting of two successive time constants. Both express negative phase angles in the low-frequency range.

The similarity of the impedance spectra obtained with the monochromatic red light and the non-polarized light allows the authors to hypothesize that complexes of the plants' photosystems, capable of absorbing red light, and could contribute to the direct electron transfer from duckweeds' fronds/roots to the fuel cell anode (Hubenova et al. 2018). Although the absorption maxima of chlorophyll *a* are at about 450 nm for blue light and c. 650 nm for a red light, the impedance data obtained *operando* reveal that red light is more effective in the charge transfer to the anode. This could be related to the photosystems (PS I, II) absorbing light of wavelengths in the red region. When photosystem II absorbs and transfers light, the electrons in the chlorophyll reaction centers are excited to a higher energy level (Appenroth et al. 2010). Photo-excited electrons travel through the cytochrome b6f complex to photosystem I via an electron transport chain set in the thylakoid membrane (Appenroth et al. 2010). This energy fall is harnessed to

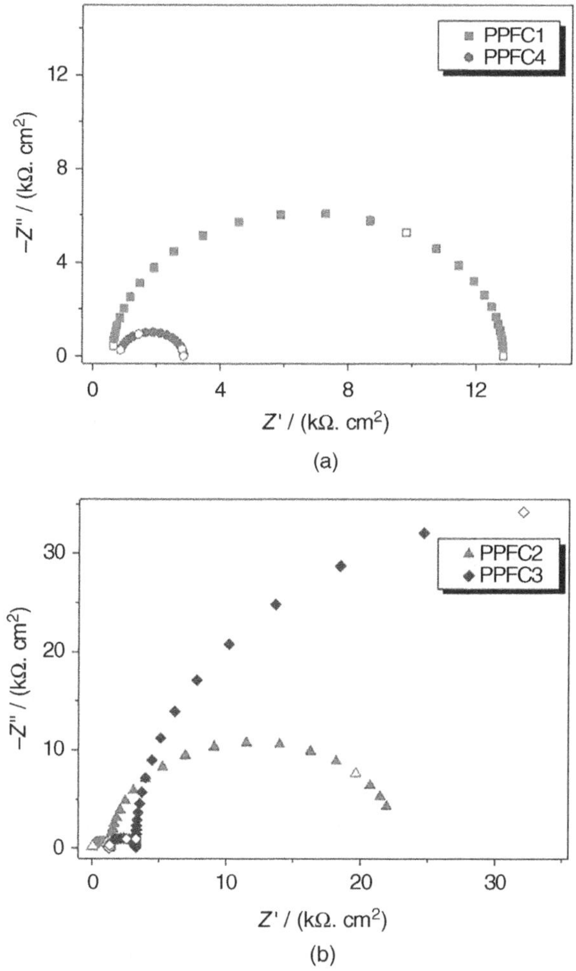

Figure 9.7 The charge transfer to the anode was evaluated by EIS, *operando* (during irradiation) and presented as (a) Nyquist plot of PPFC1 (red light – 650 nm, squares) and PPFC4 (non-polarized light, random mixture of waves circles); b) Nyquist plot of PPFC2 (blue light – 450 nm, triangles) and PPFC3 (polarized light – 390 to 700 nm, diamonds); (c) Bode plots of all PPFCs. Source: Reproduced from Hubenova et al. (2018). Abbreviation PPFC used in original (Hubenova et al. 2018) is preserved in the graphs, but meaningfully coincides with the unified abbreviation P-BFC.

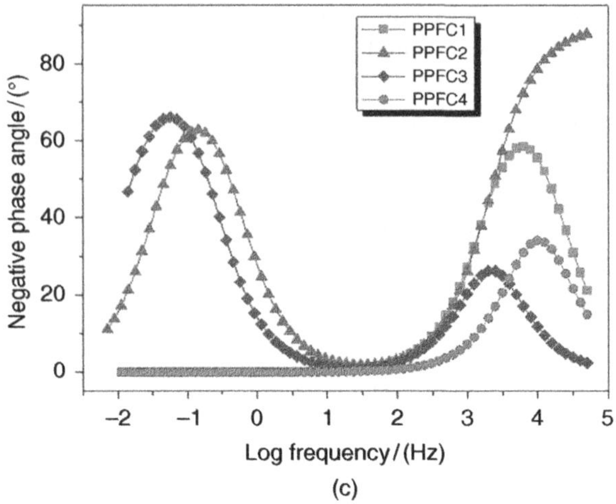

Figure 9.7 (Continued)

transport protons through the membrane. Another explanation could be related to facilitated electron transport driven by light-induced changes in the plant cell membrane potential.

9.5 Future Challenges and Concluding Remarks

So far, four types of duckweeds have been studied as biocatalysts in P-BFC, and all of them show the ability to transfer electrons extracellularly and generate electric currents. Although high values of 1.6 A/m have been reported (Hubenova and Mitov 2012), for the practical application of duckweeds-based P-BFC and for powering small consumers the generated current and achieved power have to be further increased. This is possible not only by improving the construction of the fuel cells utilized but also by a deeper understanding of the plant processes contributing to the current generation. So far it is proposed that the EET in P-BFC can be realized by: (i) direct mechanism EET – direct contact of plant cells (fond, root) with the anode, in which components related to photosynthesis are involved in the transfer during days and/or respiration processes during nights; (ii) indirect or mediated EET – with the participation of self-produced redox-active molecules shuttling electrons from the plant to the fuel cell anode; and (iii) mixed EET – direct and mediated. The EnMs are believed to be secondary metabolites, the nature of which, however, remains unknown and it is a challenge to identify.

The duckweeds have a cuticle that acts as a protective skin that covers the aerial surfaces of plants and serves as the interface between plant tissues and the environment (Borisjuk et al. 2018). Although known that cuticles consist of various types of free fatty acids and phytosterols (Borisjuk et al. 2018), it is not established how the EnM goes through them and if the polarization causes changes in the chemical composition of the protective layer of fronds and roots.

When the duckweeds have been grown in P-BFC under natural sunlight, its aseptic properties lead to a very low bacterial number. It has to be considered that among the Lemnaceae family grown in natural habitats bacteria belonging to the different families such as Comamonadaceae, Caulobacteraceae, Methylophilaceae, Oxalobacteraceae, Sphingomonadaceae, Rhizobiaceae, and Rhodospirillaceae et cetera are often spread (Inoue et al. 2022). Nevertheless, nothing is known about the electrochemical properties of these bacteria. Duckweed-associated bacteria have been investigated using V3-V4 16S rRNA amplicon sequencing (Bunyoo et al. 2022). The most abundant bacteria determined by Bunyoo et al. (2022) belong to the genera *Rhodobacter*, *Acinetobacter*, *Allorhizobium-Neorhizobium-Pararhizobium-Rhizobium*, and *Pseudomonas*. Under stress conditions, however, the authors proved that the duckweed microbiome changes (Bunyoo et al. 2022). A challenge remains the identification of new bacterial species living in symbiosis with the duckweeds in the laboratory and the field and if the polarization conditions also cause changes in the microbiome.

It has been found that growing the water lenses in a fuel cell leads to an intensification of plant metabolism. This, in turn, leads to faster growth, increased biomass, and increased protein and carbohydrate amounts. The latter could be useful in duckweed farms, which are being developed to produce not only bioethanol but also valuable nutrients. By boosting the duckweed's growth by P-BFC polarization one can yield duckweed with double the protein and carbohydrate content in half the time. A challenge for future research remains the scaling up of these systems and their application in natural settings with not only mixed duckweed species but also a rich variety of other macro and microscopic species.

Acknowledgments

The presented results have been implemented in the Biochemical laboratory of Plovdiv University, which has become a part of the scientific infrastructure "Energy Storage and Hydrogen Energetics," funded by a contract DO1-160/2018. The authors thank the Bulgarian Chemical Communication editors for the kind permission to reproduce figures from Hubenova et al. (2018) in this chapter.

References

Appenroth, K.-J., Krech, K., Keresztes, A. et al. (2010). Effects of nickel on the chloroplasts of the duckweeds *Spirodela polyrhiza* and *Lemna minor* and their possible use in biomonitoring and phytoremediation. *Chemosphere* 78 (3): 216–223. https://doi.org/10.1016/j.chemosphere.2009.11.007.

Bombelli, P., Iyer, D.M.R., Covshoff, S. et al. (2013). Comparison of power output by rice (*Oryza sativa*) and an associated weed (*Echinochloa glabrescens*) in vascular plant bio-photovoltaic (VP-BPV) systems. *Applied Microbiology and Biotechnology* 97 (1): 429–438. https://doi.org/10.1007/s00253-012-4473-6.

Bombelli, P., Dennis, R.J., Felder, F. et al. (2016). Electrical output of bryophyte microbial fuel cell systems is sufficient to power a radio or an environmental sensor. *Royal Society Open Science* 3 (10): 160249. https://doi.org/10.1098/rsos.160249.

Borisjuk, N., Peterson, A.A., Lv, J. et al. (2018). Structural and biochemical properties of duckweed surface cuticle. *Frontiers in Chemistry* 6 (317): https://doi.org/10.3389/fchem.2018.00317.

Bunyoo, C., Roongsattham, P., Khumwan, S. et al. (2022). Dynamic alteration of microbial communities of duckweeds from nature to nutrient-deficient condition. *Plants* 11 (21): 2915. https://doi.org/10.3390/plants11212915.

De Schamphelaire, L., Van Den Bossche, L., Hai, S.D. et al. (2008). Microbial fuel cells generating electricity from rhizodeposits of rice plants. *Environmental Science and Technology* 42: 3053–3058.

Georgiev, D., Gotcheva, V., Angelov, A. et al. (2013). Phytase production by *Candida melibiosica* 2491 alkalophilic strain. *Emirates Journal of Food and Agriculture* 25: 342–348.

Helder, M., Strik, D.P.B.T.B., Hamelers, H.V.M. et al. (2010). Concurrent bio-electricity and biomass production in three Plant-Microbial Fuel Cells using *Spartina anglica, Arundinella anomala* and *Arundo donax. Bioresource Technology* 101: 3541–3547.

Hubenova, Y. and Mitov, M. (2011). Bacterial mutualism in the mosses roots applicable in Bryophyta-microbial fuel cell. *Communications in Agricultural and Applied Biological Sciences* 76 (2): 63–65.

Hubenova, Y. and Mitov, M. (2012). Conversion of solar energy into electricity by using duckweed in Direct Photosynthetic Plant fuel Cell. *Bioelectrochemistry* 87: 185–191. https://doi.org/10.1016/j.bioelechem.2012.02.00.

Hubenova, Y. and Mitov, M. (2015). Extracellular electron transfer in yeast-based biofuel cells: a review. *Bioelectrochemistry* 106A: 177–185.

Hubenova, Y. and Mitov, M. (2015). Enhanced metabolic and redox activity of vascular aquatic plant *Lemna valdiviana* under polarization in Direct

Photosynthetic Plant Fuel Cell. *Bioelectrochemistry* 106A: 226–231. https://doi.org/10.1016/j.bioelechem.2014.07.007.

Hubenova, Y., Georgiev, D., and Mitov, M. (2014). Enhanced phytate dephosphorylation by using Candida melibiosica yeast-based biofuel cell. *Biotechnology Letters* 36: 1993–1997. https://doi.org/10.1007/s10529-014-1571-9.

Hubenova, Y., Georgiev, D., and Mitov, M. (2014). Stable current outputs and phytate degradation by yeast-based biofuel cell. *Yeast* 31: 343–348. https://doi.org/10.1002/yea.3027.

Hubenova, Y., Ivanov, G., Hubenova, E., and Mitov, M. (2018). Photo-induced charge transfer between *Lemna minor* and anode of photosynthesizing plant fuel cell. *Bulgarian Chemical Communications* 50 (B): 141–146.

Hubenova, Y., Hubenova, E., Manasiev, Y. et al. (2022). Draft genome sequence of *Paenibacillus profundus* YoMME, the new exoelectrogenic Gram-positive bacterium. *Microbiology Resource Announcements* https://doi.org/10.1128/mra.00235-22.

Hubenova, Y., Borisov, G., Slavcheva, E., and Mitov, M. (2022). Gram-positive bacteria covered bioanode in a membrane-electrode assembly for use in bioelectrochemical systems. *Bioelectrochemistry* 144: 108011. https://doi.org/10.1016/j.bioelechem.2021.108011.

Hubenova, E., Mitov, M., and Hubenova, Y. (2023). Electrochemical performance of *Paenibacillus profundus* YoMME encapsulated in alginate polymer. *Bioelectrochemistry* 150: 108354. https://doi.org/10.1016/j.bioelechem.2022.108354.

Hubenova, Y., Hubenova, E., and Mitov, M. (2023). *Wolffia globosa* as a biocatalyst in plant-based biofuel cells. *Journal of Electrochemical Science and Engineering* http://dx.doi.org/10.5599/jese.1547.

Idriss, E., Makarewicz, O., Farouk, A. et al. (2002). Extracellular phytase activity of *Bacillus amyloliquefaciens* FZB45 contributes to its plant-growth-promoting effect. *Microbiology* 148: 2097–2109.

Inoue, D., Hiroshima, N., Ishizawa, H., and Ike, M. (2022). Whole structures, core taxa, and functional properties of duckweed microbiomes. *Bioresource Technology Reports* 18: 101060. https://doi.org/10.1016/j.biteb.2022.101060.

Jassby, A.D. and Platt, T. (1976). Mathematical formulation of the relationship between photosynthesis and light for phytoplankton. *Limnology and Oceanography* 21: 540–547.

McCormick, J., Bombelli, P., Bradley, R.W. et al. (2015). Biophotovoltaics: oxygenic photosynthetic organisms in the world of bioelectrochemical systems. *Energy & Environmental Science* 8: 1092. https://doi.org/10.1039/c4ee03875d.

Nakagawa, R. and Nishina, Y. (2021). Simulating the redox potentials of unexplored phenazine derivatives as electron mediators for biofuel cells. *Journal of Physics: Energy* 3: 034008. https://doi.org/10.1088/2515-7655/abebc8.

Pierson, L.S. III and Pierson, E.A. (2010). Metabolism and function of phenazines in bacteria: impacts on the behavior of bacteria in the environment and biotechnological processes. *Applied Microbiology and Biotechnology* 86: 1659–1670. https://doi.org/10.1007/s00253-010-2509-3.

Santoro, C., Arbizzani, C., Erable, B., and Ieropoulos, I. (2017). Microbial fuel cells: from fundamentals to applications. A review. *Journal of Power Sources* 356: 225–244. https://doi.org/10.1016/j.jpowsour.2017.03.109.

Sevda, S., Sarma, P.J., Mohanty, K. et al. (2018). Microbial fuel cell technology for bioelectricity generation from wastewaters, Chapter 11. In: *Waste to Wealth, Energy, Environment, and Sustainability* (ed. R.R. Singhania et al.). Singapore: Springer https://doi.org/10.1007/978-981-10-7431-8_11.

Strik, D.P.B.T.B., Hamelers, H.V.M., Snel, J.F.H., and Buisman, C.J.N. (2008). Green electricity production with living plants and bacteria in a fuel cell. *International Journal of Energy Research* 32: 870–876.

Strik, D.P.B.T.B., Hamelers, H.V.M., Helder, M. et al. (2011). Plant-microbial fuel cells: matching results and model predictions to show the technological and economical perspectives of PlantPower. In: *Proceedings of the 3rd International Microbial Fuel Cell Conference*, 21. Leeuwarden, The Netherlands.

Timmers, R.A., Strik, D.P.B.T.B., Hamelers, H.V.M., and Buisman, C.J.N. (2010). Long-term performance of a plant microbial fuel cell with *Spartina anglica*. *Applied Microbiology and Biotechnology* 86: 973–981.

10

Low Power Voltage Acquisition System for Photosynthesis-Based Microbial Fuel Cells

Victor A. Maldonado-Ruelas[1], Raúl A. Ortiz-Medina[1], Sathish-Kumar Kamaraj[2], Wilgince Apollon[3], and Marco A. Vázquez-Gutierrez[4]

[1] *Dirección de Posgrado e Investigación, Universidad Politécnica de Aguascalientes, Aguascalientes, México*
[2] *Instituto Politécnico Nacional (IPN)-Centro de Investigación en Ciencia Aplicada y Tecnología Avanzada, Unidad Altamira (CICATA-Altamira), Altamira, Tamps., México*
[3] *Department of Agricultural and Food Engineering, Faculty of Agronomy, Autonomous University of Nuevo León, General Escobedo, Nuevo León, México*
[4] *Tecnológico Nacional de México, Campus El Llano Aguascalientes, Aguascalientes, México*

10.1 Low Power Sources

In recent decades, different solutions have been sought to the environmental problem that afflicts the whole world. The solutions are varied: solar irradiation, wind energy, and biofuels, among others, all of them with a certain degree of success in their application. However, the increase in electricity generated from this type of source is still limited, contributing only 12.6% of all energy used in 2020 worldwide (Renewables 2022). For this reason, work continues on alternatives for the generation of electricity through sources other than fossil fuels. One of the emerging alternative energies that have been extensively studied in recent years is microbial fuel cells (MFCs). These are the subject of study due to their growing potential in terms of electricity generation, their low cost, and their applications as a source of low-power energy. This chapter will provide a profile of the PMFCs and the basic electrical parameters measured in them.

Photosynthesis-Assisted Energy Generation: From Fundamentals to Lab Scale and In-Field Applications, First Edition. Edited by Sathish-Kumar Kamaraj and Iryna Rusyn.

10.1.1 Photosynthesis-Based Microbial Fuel Cells

MFCs based on photosynthesis are energy sources considered low power, whose operation is focused on plants that perform photosynthesis. Its operation is based on the biological interactions between plant roots and electrochemically active bacteria (EAB). The PMFC photosynthesis-based MFC uses exudates released by plant roots to nourish EABs found at the anode and generate bioelectricity (Rusyn 2021; Apollon et al. 2021). There is a great variety of PMFC arrangements, according to (Kabutey et al. 2019) there are four basic types of configuration: the PMFC with the cell integrated into vascular plants, in it the living plant grows with a surrounding anode that generates bioelectricity thanks to the organic matter released by the plant in its natural process of photosynthesis. This type of cell is very advantageous because the plant generates its own organic matter that is oxidized by the large number of microbes found in the rhizosphere (Pamintuan and Sanchez 2019; Apollon et al. 2020). The PMFC is based on macrophyte, these cells are built from these plants whose ecosystem is formed in bodies of water, have great influence on their ecosystems with respect to fauna, hydrology, and dynamics of the sediments where they are found (Guadarrama-Pérez et al. 2022; Liu et al. 2020). PMFC takes advantage of the photosynthesis carried out by the macrophyte using rhizodeposits and exudates as organic matter for bioelectricity generation. PMFCs based on constructed wetlands have basically the same principle as those based on macrophyte, but unlike these, in addition to the organic matter produced by plant photosynthesis, they have the load of organic matter from wastewater, which makes PMFC based on constructed wetlands a hybrid system of biogeneration of electrical energy and a water pollution control system (Liu et al. 2020; Huang et al. 2021). Finally, there are the PMFCs based on bryophyte, these cells are built with the anode surrounding the rhizoids of the bryophyte, since it is a plant that lacks roots and uses these rhizoids to adhere to surfaces. The organic matter generated by photosynthesis and the carbon and nitrogen cycles is what contributes to the biogeneration of electrical energy in this type of PMFC (Bombelli et al. 2016; Rusyn and Hamkalo 2020).

10.1.2 Potential Difference

The potential or voltage difference, as is widely known, is the work per unit charge needed to move a charged particle through an electric field. The PMFCs generate that potential difference between the anode and the cathode, thus, when connecting a charge between the electrodes there is a movement of electrons from the cathode to the anode, passing through the charge in question, this movement of electrons is known as electric current. The voltage of the cells is underreported, but it turns out to be a great indication of the cycles of the PMFC. In the comparison of (Sarma and Mohanty 2020), the maximum reported voltages of the different PMFC range from 300 to 700 mV. Although in other works maximums of up to 750 mV are reported, showing the changes in the voltage magnitudes that are in daily cycles (Apollon et al. 2020; Apollon et al. 2022).

The PMFCs deliver their energy in direct current, that is, the shape of the signals coming from the cell electrodes are straight lines, which remain with the same magnitude without variation in time (Hayt et al. 2019). From what is known with respect to PMFC, this definition of direct current is not entirely adequate, since the parameters of the PMFC do undergo changes in their magnitude over time; but the definition can be assumed valid because the changes that occur in the electrical parameters of the PMFCs respond to conditions of humidity and temperature, whose cycles usually occur throughout a day. As an example, the electrical energy used conventionally is considered alternative energy, since it has changes in its magnitude that respond to a sinusoidal nature, whose signals, depending on each country, have frequencies of 50 or 60 Hz, which implies having cycles of the sinusoidal signals completed every 20 or 16 ms, respectively; through the same analysis, the frequency of the electrical parameters of the PMFC would be 0.00001157 Hz, which implies having a cycle with a duration of one day, then, due to its proximity to 0 Hz, the signals coming from the PMFC are considered as direct current (Apollon et al. 2020; Hayt et al. 2019).

Electrical energy in conventional direct current is commonly used in electronic systems of low or high voltage, its value depends on the application that is going to be given, since this is generally obtained from the controlled transformation of alternating current to direct current. In the case of PMFCs, they deliver a voltage in direct current to which a load that makes use of this type of energy can be subjected, such as a diode, which in a simple way and without any major arrangement or transformation can be powered by PMFCs. There is a simple way by which the voltage involved in the combination of cells connected in series can be raised. In a configuration with two cells, these are connected to each other so that the anode of the first cell is connected to the cathode of the second cell, in this way the cathode of the first cell is free and the anode of the second is free and these free terminals become the output terminals of the cell, in this way the voltage of both PMFC is added and the resulting array has higher voltage (Hayt et al. 2019; Rusyn et al. 2021; Veerubhotla et al. 2019). Another way to increase the voltage of a cell is the use of DC–DC converters or direct current to direct current converters which are electronic circuits capable of increasing the voltage of a voltage source in direct current. The advantage of voltage transformation by the DC–DC converter is that you can have a control stage that keeps the output voltage at a desired and constant level, such as for 3.3 or 5 V applications that are nominal voltage levels of many electronic equipments, and the disadvantages are that the increase in output voltage implies a reduction in the output current and that the DC–DC converter it also consumes energy, reducing efficiency from the PMFC (Osorio de la Rosa et al. 2019).

10.1.3 Power and Energy

The power in direct current under stable conditions is defined by the product between the magnitudes of voltage and current, and energy is defined by the integral of the power by its time differential, so in a power curve against time, the area under the power curve is the energy, with power units per time. In the bio-generation of electrical energy, as well as in conventional electrical energy, voltage, current, power, and energy are electrical parameters that provide basic information on energy sources. In PMFC, the values of power and current are taken according to the surface area of the electrodes as densities, both surface density of current and surface density of power, although in some cases volumetric densities are used.

In the literature you can find the magnitudes of the maximum parameters that are reached with the different types of PMFC (Apollon et al. 2021; Kabutey et al. 2019; Sarma and Mohanty 2020; Maddalwar et al. 2021), with variations in plants, substrate, membranes, and electrodes, where their parameters in reported power density range from $0.001\,\mu W/m^2$ to $679\,mW/m^2$ maximum reported in a PMFC with plant *Spartina anglica*, and graphite felt used as an electrode. There is also information on the reported current density (Apollon et al. 2021; Kabutey et al. 2019; Sarma and Mohanty 2020; Maddalwar et al. 2021), ranging from $3.96\,mA/m^2$ to the maximum of $2080\,mA\,m^2$ reported in a PMFC with *S. anglica* plant, and graphite felt used as electrode

10.2 Voltage Acquisition System

This section will introduce the structure and electronic design of the voltage acquisition or voltage monitoring system for PMFC. It will delve into the type of analog configuration to be designed and the differences that exist in using a system in parallel with respect to a multiplexed configuration. The basic electronic instrumentation required and the reason for each of the stages are analyzed in detail. Likewise, its design and implementation, its functionality, and the method of information validation are presented. The programming for each of the configurations is analyzed, and the selection of the digital signal platform is justified. Finally, the general scheme of the voltage monitoring system is presented, as well as, the current result of the system implemented in the voltage acquisition.

10.2.1 Analog Instrumentation

A fundamental part of the acquisition system is the treatment of the voltage signal generated by the PMFC, of which its value must be guaranteed within a digital signal processor (DSP), either digitally or analog. According to the literature reviewed, it is considered for design that PMFCs generate a potential difference in the order of maximum 2V with a current in the order of μA; so its impedance value Z_c will be of the order of giga-ohms. For the value of the very low power signal, the acquisition can be done in two ways:

10.2.1.1 In Parallel (Independent PMFC)

This configuration allows each PMFC to be measured by means of an analog-to-digital converter (ADC) of the DSP using one channel for each PMFC. The high impedance of the cells can cause the input voltage to temporarily drop and an incorrect value to be recorded (ATMega328p 2016); therefore, the instrumentation of each channel is necessary for the treatment of the signal between the PMFC and the DSP, to couple the impedances of the cell with respect to the analog input.

10.2.1.2 Multiplexed (PMFC Dependent)

This configuration allows you to measure the voltage of an N number of PMFCs with a single ADC channel of the DSP, taking one voltage signal at a time. The chosen signal is dependent on the digital selection of the desired channel. As in the parallel configuration, there is a signal impedance treatment between the PMFC and the analog input of the DSP. For this configuration, the DSP must have sufficient digital outputs for the selection of the desired channel, and the number of channels to be read is obtained with Eq. (10.1).

$$2^n = \#\text{channels} \tag{10.1}$$

where

$N = 0, 1, 2, \ldots$ (Digital Outputs).

Table 10.1 presents the advantages and disadvantages of using one configuration over the other. The variables or characteristics that were considered to be compared were chosen based on the experimental systems that have been implemented over the last four years, both electrical, electronic, and physical.

10.2.2 Acquisition System Design

Figure 10.1 presents the diagram of the signal processing circuit used for impedance coupling, which is independent of the configuration used, the only difference

Table 10.1 Feature of the different acquisition configurations.

Feature	Parallel	Multiplexed
Energy consumption	Minimum (Only one circuit per N channels)	High with more than 4 channels
Impedance coupling circuit	One per N channels	One for each channel
DSP using	One analog channel and N digital outputs of the DSP	N analog channels
Electronic design	Compact on a single card	Modular, on various cards
Software design	Complex, due to the selection of the channel to choose	Simple, channel-by-channel reading
Voltage monitoring	Only one channel selected at a time	All channels "at the same time"
Maximum number of PMFC per DSP	Until 88 for an Arduino® Uno	Until 6 for an Arduino® Uno
Physical Desing	Suitable for field work	Complex for field work as it is modular
Cost-benefit	More PMFC reads for less cost	Fewer PMFC reads for higher cost

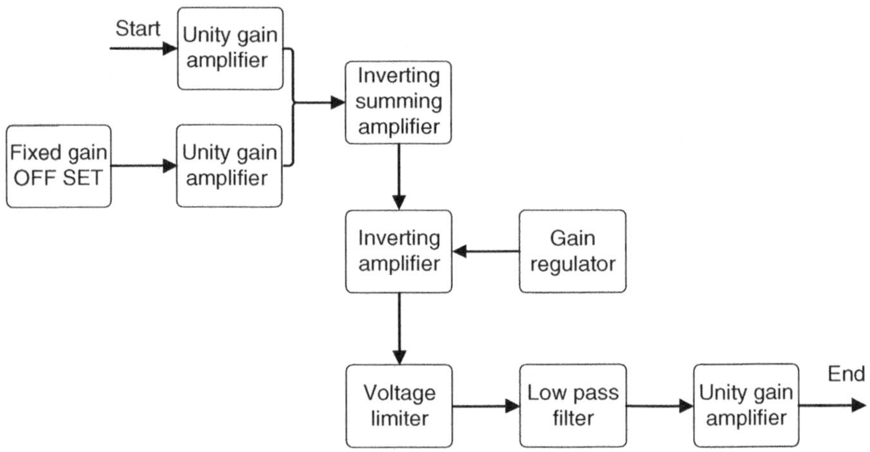

Figure 10.1 Schematic diagram of signal processing for PMFC.

being the number of circuits that one carries with respect to the other. The instrumentation used is analog, since the signal generated by the PMFC is maintained continuously and without loss of information. The active elements used are instrumentation operational amplifiers; which are more accurate and with greater impedance coupling margins with external electronic circuits (Huijsing 2017; Franco 2015). The general scheme of the procurement system is divided into 6 stages:

10.2.2.1 Input Impedance Coupler

The voltage generated by the MCC is connected to an amplifier in its non-inverter tracker configuration. This allows the input impedance to be coupled to the electronic circuit, ensuring the voltage magnitude of the cell signal.

10.2.2.2 OFF-SET Added to Suppress Input Signal Polarity

In this stage, the signal of the cell of the previous stage is located, at a level of 2.5 V as a zero reference; since the input voltage range of a DSP is commonly 0–5 V. The reference is made by means of a voltage divider by means of two resistors in series connected to a voltage source as shown in Eq. (10.2).

$$V_{\text{out}} = \frac{R_2}{R_1 + R_2} V_{\text{in}} \tag{10.2}$$

where

V_{in} = Input voltage source, greater than output voltage;
R_2 = Output resistance, connected to ground
R_1 = Limiting resistor, connected to V_{in}
V_{out} = Output voltage = 2.5 V

Therefore, when connecting the polarized PMFC directly, the voltage levels will be 2.5–5 V and when connecting it inversely, the voltage levels will be 2.5 to 0 V. This ensures that regardless of how the PMFC electrodes are connected, the voltage values will be within the range allowed by the DSP, thus ensuring that the analog input is not damaged when the reverse-biased PMFC is connected. The amplifier in the inverter adder configuration is used to join the PMFC signal of the non-inverter tracker with the reference of the 2.5 V OFF-SET. The gain of the adder is unitary, since it is sought not to affect the input signal at this stage.

10.2.2.3 Adjustable Gain Amplifier

An amplifier is used in inverter configuration, which can increase or decrease the value of the input depending on the resistance value that is adjusted in its reference (Precision Potentiometer). The inverter configuration counteracts the adder inversion of the previous stage so that the resulting signal continues to have the initially chosen polarity. The signal amplification is desired not to exceed 4.5 V or decrease from 0.5 V in reverse polarity; since at the limits of the ADC input of the DSP the signal can be attenuated and information lost.

10.2.2.4 Output Voltage Limiter

To ensure that the output voltage of the instrumentation system does not exceed the allowable voltage range of an ADC input of the DSP, protection is used by means of a 5.1 V zener diode connected in reverse bias in series with a current limiting resistance of 1 kΩ. This setting is known as a signal trimmer, which applies to the voltage of the zener diode being used. It is for DSP protection, and information above this level will be lost by limiting the signal.

10.2.2.5 Bass-Pass Filter to Eliminate High-Frequency Noise

For this stage, a first-order filter is built by means of an R resistor and a C capacitor connected in series. The cutoff frequency Fc resulting from 6 kHz is calculated with Eq. (10.3). Due to the nature of the filter, it will begin to cut the magnitude of the signal, 2 decades before the cutoff frequency, i.e. at 60 Hz. Therefore, any noise generated by the movement of the wires or some type of magnetic signal over the PMFC voltage greater than 60 Hz will be reduced in magnitude (Jung 2002).

$$F_{\mathrm{C}} = \frac{1}{RC} \tag{10.3}$$

10.2.2.6 Output Impedance Coupler

To ensure impedance coupling and avoid voltage losses of the signal processing system with the DSP; It is occupied as in the input an amplifier in its non-inverter follower configuration.

10.2.3 Physical Implementation and Operation Tests

To validate that the stages selected from the previous section meet the desired, it is simulated with the PROTEUS software. This program allows simulation tests with real values because the same components are used that are physically and not with ideal components; as well as, it allows to carry out the electronic design for its implementation in a phenolic table. In order for the electronic design to allow the acquisition of voltages reliably, the following considerations are taken into account:

- Design of a ground plane that eliminates scattered noise and prevents the generation of voltage curls.
- Decoupling capacitors in each integrated circuit between supply and ground voltages.
- Passive elements with tolerance or precision of 1%, to avoid variations of gains that occur when working with very low power.

Table 10.2 Data acquired from the first eight channels.

	Voltages per cells (V)							
Sample	**C1**	**C2**	**C3**	**C4**	**C5**	**C6**	**C7**	**C8**
0	0.7	0.6	0.5	0.6	0.5	0.6	0.7	0.7
1	0.7	0.6	0.6	0.6	0.5	0.6	0.7	0.7
2	0.8	0.7	0.7	0.7	0.6	0.7	0.7	0.7
3	0.8	0.7	0.7	0.7	0.6	0.7	0.8	0.7
4	0.8	0.7	0.7	0.7	0.6	0.7	0.8	0.7
5	0.8	0.7	0.7	0.7	0.6	0.7	0.8	0.7
6	0.8	0.7	0.7	0.7	0.6	0.7	0.8	0.7
7	0.8	0.7	0.7	0.7	0.6	0.7	0.8	0.7
8	0.8	0.8	0.7	0.7	0.6	0.7	0.8	0.7
9	0.8	0.8	0.7	0.7	0.6	0.7	0.8	0.7
10	0.8	0.8	0.7	0.7	0.6	0.7	0.8	0.8

- Circuit symmetry to prevent scattered currents and hot spots on some electronic components.
- Implementation of easily accessible connectors; for the connections of each PMFC; as well as power supplies and DSP.

For the operation test of the analog instrumentation board, 8 channels with 10 samples at different measurement times were considered. Each channel is connected to an equal PMFC under construction. Table 10.2 presents the values acquired by the voltage acquisition system.

The measurement of the values was carried out by means of a Fluke multimeter, calibrated, and verified by the Anova method for the study of repeatability and reproducibility of a measurement system (Mikulová and Plura 2018).

10.2.4 Programing of Digital Platform

The choice of digital platform must contain at least the following hardware architecture:

- **Analog inputs:** depending on the acquisition configuration, the parallel configuration must have the same number of inputs as the channels to be read. For the multiplexed system, one input for any number of channels is sufficient.

- **Digital outputs:** for the multiplexed system, the outputs are based on the channels to be selected. In the parallel system, they are not necessary for channel control.
- **16 kB or larger processor:** The monitoring and acquisition process has no sampling times less than 1 s.
- **Serial communication port Tx and Rx:** necessary for the processing of information and/or saving of data.
- **Any extra accessory for data display:** helps to see the acquired data in situ, but is not necessary for the monitoring and acquisition mode to operate.

Today, there are a large number of commercial DSPs on the market, therefore, the choice was based on the minimum functionality of operation, size, software, and price. For the different acquisition schemes, they have been tested on 2 Arduino platforms: UNO and MEGA with ATMEL microcontroller. One difference between these platforms is the number of channels that can be monitored, but both meet the necessary characteristics for proper monitoring and voltage acquisition.

The program for the monitoring system with parallel configuration is based on the independent reading of the analog signals of each PMFC. Equation (10.4) realizes linearity through least squares with at least 15 samples of each signal. This provides a better estimate, in the face of random and unbiased variations.

$$m = \frac{n\sum(xy) - \sum x \sum y}{n\sum x^2 - \left|\sum x\right|^2} \quad y \quad b = \frac{\sum y \sum x^2 - \sum x \sum(xy)}{n\sum x^2 - \left|\sum x\right|^2} \tag{10.4}$$

where

m = slope of the line;
b = intercept;
x = independent variable;
y = dependent variable;
n = number of samples;

The sampling time is established considering times between samples of five minutes, since the frequency of the voltage of the PMFC is very low. Data values are collected in a spreadsheet with a date and time for each channel. Likewise, the data are written to a .txt file for storage with little memory and constantly, in case there is a power outage or system failure in general, the data can be recovered.

The program for the monitoring system with multiplexed configuration is described through a flowchart in Figure 10.2. The process of linearity, sampling time, and printing or saving data is set in the same way as in the parallel configuration.

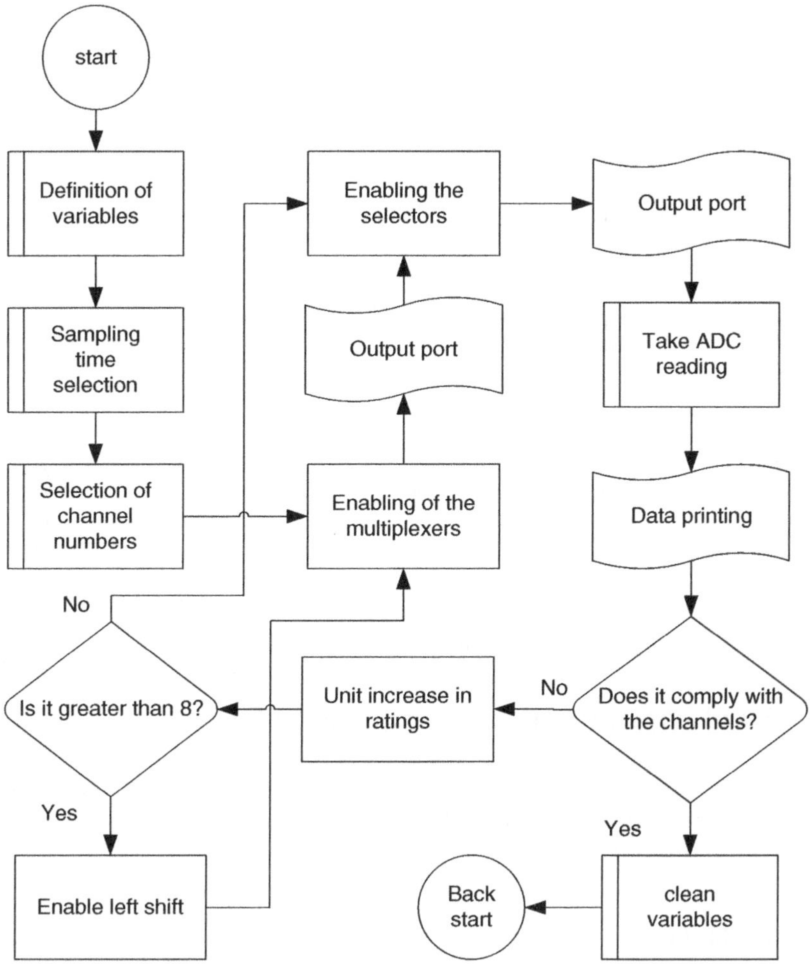

Figure 10.2 Flowchart of the multiplexed system.

10.2.5 Operation of the Integrated Voltage Acquisition System

Figure 10.3 presents the block diagram of the current system implemented for 16 PMFC voltage monitoring with multiplexed configuration. Two 8-1 multiplexers are used, with two ratings and three selectors controlled by the DSP, which in this case is an Arduino UNO. The outputs of the multiplexers are joined and connected to a single instrumentation and impedance coupling circuit. The output of the system is connected to an analog input of the Arduino UNO, where the voltage reading programming of each of the 16 channels is made. Table 10.3 presents the reading of 10 samples from the 16 channels in a total time of 50 minutes.

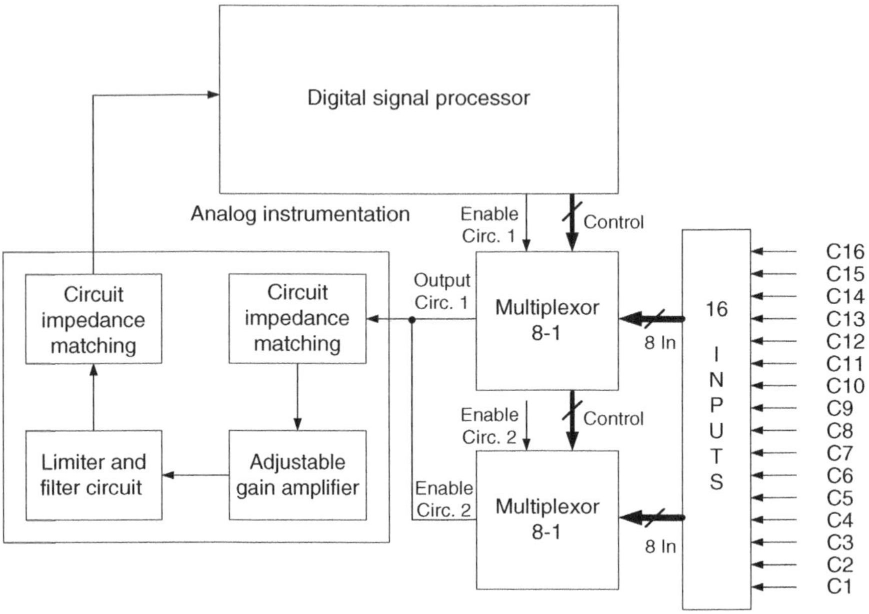

Figure 10.3 Block diagram of the voltage monitoring system.

10.3 Field Application of the Acquisition System

This section shows the voltage measurements made with the implemented acquisition system. The plants used in the PMFCs were Prickly pear, *Plectranthus hadiensis* (Vaporub), and *Stevia*.

10.3.1 Prickly Pear

Prickly pear is a very common plant in the semiarid regions of México. The low presence of water and energy in this climate prompted interest in the use of PMFC with the local flora. Figure 10.4 shows 1 day of the results of the voltage acquisition of 8 PMFC, using 8 channels, 4 of them belong to the species *Opuntia robusta* and 4 to *Opuntia albicarpa*. These measurements are part of the research work found in Apollon et al. (2020), which were performed with the measurement system exposed in this work.

10.3.2 *Plectranthus hadiensis* (Vaporub)

The cells based on *P. hadiensis*, also known as vaporub, are part of a research project at the Polytechnic University of Aguascalientes that combines power

Table 10.3 Monitoring system data for 16 PMFC.

Sample	C1	C2	C3	C4	C5	C6	C7	C8	C9	C10	C11	C12	C13	C14	C15	C16
1	1.07	1.14	1.09	1.11	1.11	1.12	1.09	1.05	1.14	2.45	1.1	1.1	0.92	1.12	1.16	1.09
2	1.07	1.14	1.09	1.11	1.11	1.12	1.09	1.05	1.14	2.25	1.1	1.1	0.92	1.12	1.16	1.09
3	1.08	1.14	1.09	1.11	1.11	1.12	1.09	1.05	1.14	2.1	1.1	1.1	0.93	1.12	1.16	1.09
4	1.08	1.14	1.09	1.11	1.11	1.12	1.09	1.05	1.14	2.46	1.1	1.1	0.93	1.12	1.16	1.09
5	1.08	1.14	1.09	1.1	1.11	1.12	1.09	1.05	1.14	1.92	1.1	1.1	0.93	1.12	1.16	1.09
6	1.08	1.14	1.09	1.1	1.11	1.12	1.09	1.05	1.14	2.3	1.1	1.1	0.93	1.12	1.16	1.09
7	1.09	1.14	1.09	1.1	1.11	1.12	1.09	1.05	1.14	1.98	1.1	1.1	0.93	1.12	1.16	1.09
8	1.09	1.14	1.09	1.1	1.11	1.12	1.09	1.05	1.14	1.37	1.1	1.1	0.93	1.12	1.16	1.09
9	1.09	1.14	1.09	1.1	1.11	1.12	1.09	1.06	1.14	1.98	1.1	1.09	0.93	1.12	1.16	1.09
10	1.09	1.14	1.09	1.1	1.11	1.12	1.1	1.06	1.14	1.99	1.1	1.1	0.93	1.12	1.16	1.09

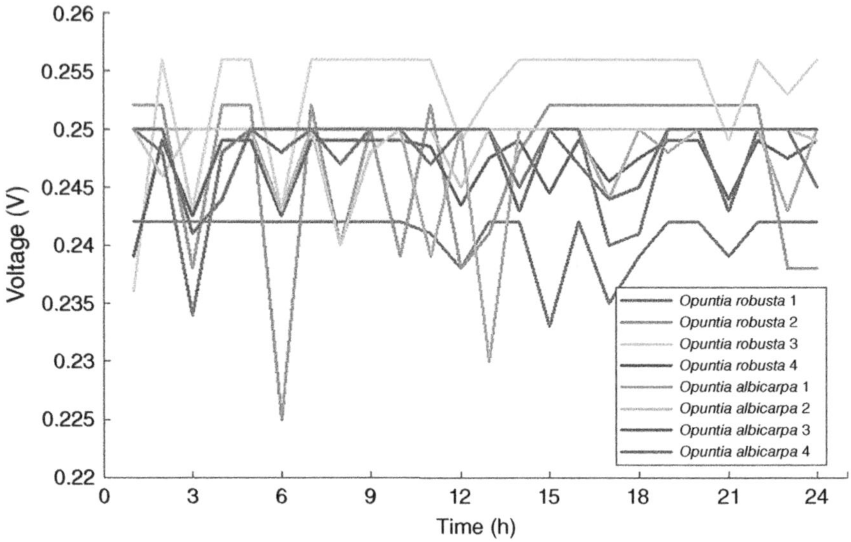

Figure 10.4 Voltage of 8 channels, corresponding to 8 PMFC, 4 of them belong to the species *Opuntia robusta* and 4 to *Opuntia albicarpa*.

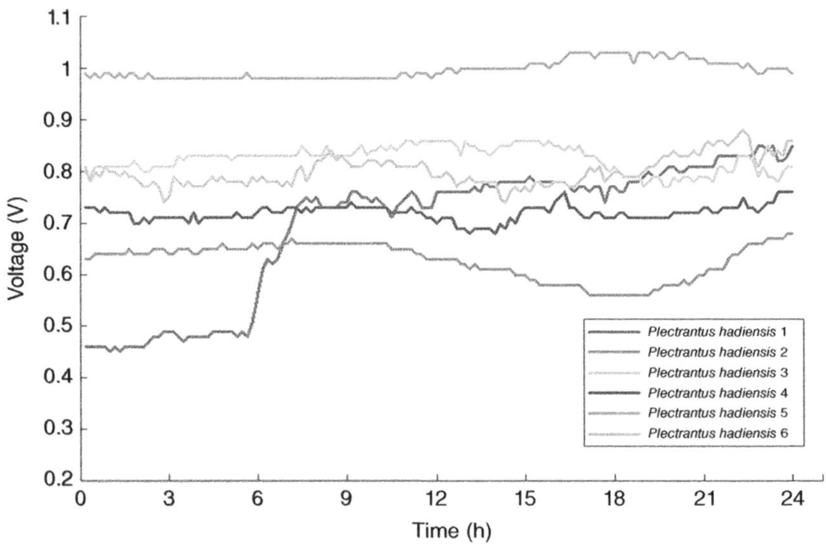

Figure 10.5 Voltage of 6 channels, corresponding to 6 PMFC, all of them belong to the species *Plectranthus hadiensis* (Vaporub).

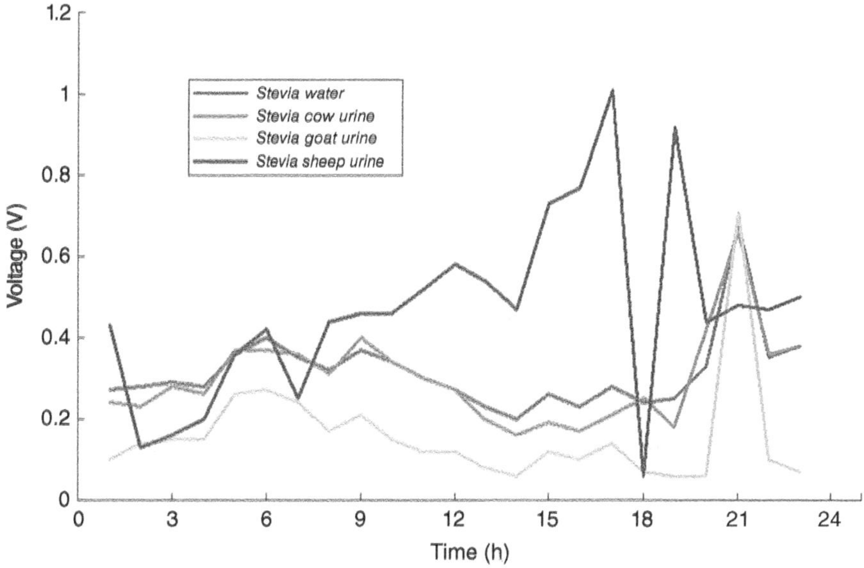

Figure 10.6 Voltage of 4 channels, corresponding to 4 PMFC based on *Stevia*, added with water, cow urine, goat urine, and sheep urine.

generation with plants used for the extraction of essential oils. Figure 10.5 shows a day of measurement of the voltages of 6 PMFC based on *P. hadiensis*, the six signals correspond to 6 voltage channels, one for each PMFC.

10.3.3 *Stevia*

Stevia-based PMFCs were tested with different types of soil enrichment. Figure 10.6 shows a day of measurement of four *Stevia*-based PMFC voltage signals, each corresponding to a different type of irrigation: water, cow urine, goat urine, and sheep urine. These measurements are part of the research work of (Apollon et al. 2022) carried out with the measurement system exposed in this work.

10.4 Conclusions

The measurement and acquisition of the electrical variables of the PMFC are essential to obtain their characteristics, such as polarization curves and it is also important to know their behavior over time. The system designed and implemented in its multiplexing configuration is capable of measuring voltage up to

88 MFCs simultaneously, with automatic polarity correction and adjustable sample rate. Due to its characteristics, sensors of current, voltage, environmental and soil humidity, temperature, and other parameters that can be electrical and nonelectrical, which contribute to the enrichment of research in PMFC can be added to the same system.

References

Apollon, W., Kamaraj, S.K., Silos-Espino, H. et al. (2020). Impact of Opuntia species plant bio-battery in a semi-arid environment: demonstration of their applications. *Applied Energy* 279: 115788, ISSN 0306-2619.

Apollon, W., Luna-Maldonado, A.I., Kamaraj, S.K. et al. (2021). Progress and recent trends in photosynthetic assisted microbial fuel cells: a review. *Biomass and Bioenergy* 148: 106028. ISSN 0961-9534.

Apollon, W., Vidales-Contreras, J.A., Rodríguez-Fuentes, H. et al. (2022). Livestock's urine-based plant microbial fuel cells improve plant growth and power generation. *Energies* 15: 6985.

ATMega328p (2016). Atmel Microchip, datasheet, p. 244.

Bombelli, P., Dennis, R.J., Felder, F. et al. (2016). Electrical output of bryophyte microbial fuel cell systems is sufficient to power a radio or an environmental sensor. *Royal Society Open Science*.

Franco, S. (2015). *Design with Operational Amplifiers and Analog Integrated Circuits*. NY, USA: McGraw-Hill.

Guadarrama-Pérez, O., Bahena-Rabadan, K.Y., Dehesa-Carrasco, U. et al. (2022). Bioelectricity production using shade macrophytes in constructed wetlands-microbial fuel cells. *Environmental Technology* 43 (10): 1532–1543.

Hayt, W.H., Kemerly, J.E., Philips, J., and Durbin, S.M. (2019). *Engineering Circuit Analysis*. NY, USA: Mc Graw Hill.

Huang, X., Duan, C., Duan, W. et al. (2021). Role of electrode materials on performance and microbial characteristics in the constructed wetland coupled microbial fuel cell (CW-MFC): a review. *Journal of Cleaner Production* 301: 126951, ISSN 0959-6526.

Huijsing, J. (2017). *Operational Amplifiers*. Switzerland: Springer Cham.

Jung, W. (2002). *OP AMP Application Handbook*. USA: Newnes.

Kabutey, F.T., Zhao, Q., Wei, L. et al. (2019). An overview of plant microbial fuel cells (PMFCs): configurations and applications. *Renewable and Sustainable Energy Reviews* 110: 402–414, ISSN 1364-0321.

Liu, F., Sun, L., Wan, J. et al. (2020). Performance of different macrophytes in the decontamination of and electricity generation from swine wastewater via an integrated constructed wetland-microbial fuel cell process. *Journal of Environmental Sciences* 89: 252–263, ISSN 1001-0742.

Maddalwar, S., Nayak, K.K., Kumar, M., and Singh, L. (2021). Plant microbial fuel cell: opportunities, challenges, and prospects. *Bioresource Technology* 341: 125772, ISSN 0960-8524.

Mikulová, P. and Plura, J. (2018). Comparison of approaches to gauge repeatability and reproducibility analysis. In: *MATEC Web Conference*, Poland, vol. 183, 03015.

Osorio de la Rosa, E., Vázquez-Castillo, J., Carmona-Campos, M. et al. (2019). Plant microbial fuel cells–based energy harvester system for self-powered IoT applications. *Sensors* 19: 1378.

Pamintuan, K.R.S. and Sanchez, K.M. (2019). *IOP Conference Series: Materials Science and Engineering*, Bali, Indonesia, vol. 703, 012037.

Renewables (2022). Gobal Status Report. In: *REN21 Renewables Now*. Paris: France.

Rusyn, I. (2021). Role of microbial community and plant species in performance of plant microbial fuel cells. *Renewable and Sustainable Energy Reviews* 152: 111697, ISSN 1364-0321.

Rusyn, I.B. and Hamkalo, K. (2020). Electro-biosystems with mosses on green roofs. *Environmental Research, Engineering and Management* 76: 20–31.

Rusyn, I.B., Medvediev, O.V., and Valko, B.T. (2021). Enhancement of bioelectric parameters of multi-electrode plant–microbial fuel cells by combining of serial and parallel connection. *International journal of Environmental Science and Technology* 2021 (18): 1323–1334.

Sarma, P.J. and Mohanty, K. (2020). An insight into plant microbial fuel cells, Chap. 8. In: *Bioelectrochemical Interface Engineering* (ed. R.N. Krishnaraj and R.K. Sani). Wiley.

Veerubhotla, R., Nag, S., and Das, D. (2019). Internet of Things temperature sensor powered by bacterial fuel cells on paper. *Journal of Power Sources* 438: 226947, ISSN 0378-7753.

Part III

Lab-Scale and Infield Application of Photosynthesis-Based Microbial Fuel Cells

The chapter highlights progress and recent trends in photosynthetic assisted microbial fuel cells to power different kinds of low-energy-consuming devices and IoT: as a power supply for real-time LED and digital clocks, IoT-based WSN sensor systems, powering batteries, and partially loading a mobile phone, etc., based on various types of PMFCs. There are considered problems in improving the removal of organic and nutrient pollutants, and in particular, ammonium and phosphorus treatment with hybrid photoautotrophic algal-assisted MFCs due to their synergistic interrelations, which presents an effective sustainable technology for the elimination of organic matters and the removal/recovery of nutrients from different waste with simultaneously zero waste bioenergy recovery. Research on photo-bioelectrochemical fuel cells for antibiotic and dye removal providing simultaneous power generation is just beginning to develop in recent years but shows significant progress. PBFC systems demonstrate high power generation capacity and almost complete up to 99–100% antibiotic and dye elimination, which is the subject of the section. The part also presents challenges related to the possibility of obtaining bioelectricity on green roofs in different climatic conditions such as in northern countries with frosty winters and in southern countries with arid climates, which simultaneously provide biosensing and climatic advantages, reducing urban heat island temperatures using different types of photosynthesis-based microbial fuel cells.

11

Plant-Based-Microbial Fuel Cells for Bioremediation, Biosensing, and Plant Health Monitoring

Roshan Regmi[1,2], *Vinh Nguyen*[1,3], *and Ranjita Sapkota*[4,5]

[1] *Thammasat University, School of Bio-Chemical Engineering and Technology, Sirindhorn International Institute of Technology, Thailand*
[2] *CSIRO, Microbiome for One Systems Health (MOSH), Australia*
[3] *Korea University, School of Civil, Environmental and Architectural Engineering, College of Engineering, Seoul, South Korea*
[4] *TERI School of Advance Studies, Department of Policy Studies, New Delhi, India*
[5] *University of South Australia, Department of Business, Centre for Markets, Value and Inclusion, Adelaide, South Australia, Australia*

11.1 Introduction

Achievements of sustainable energy sources and a clean environment are the top prioritized goals in society nowadays (Nepal et al., 2021). Industrialization and population growth have caused overexploitation of fossil fuels, leading to continuous environmental threats and energy crises (Fadil et al., 2023). This demands the search for alternative sources and technologies to address the shortage of fossil energy (Kralova & Sjöblom, 2010). On the one hand, the global electricity demand is increasing by 3.6% annually and the price of electricity has surged by 8% (Kan et al., 2021), and on the other hand, arable land is decreasing (Zhang & Cai, 2011). Therefore, a sustainable source of energy that concurrently produces electricity and plant biomass is a suitable candidate for upcoming generations. Renewable energies such as ocean, geothermal, biomass, wind, and solar are sustainable with low environmental impacts. Plant-based microbial fuel cell (PMFC) has gained

Photosynthesis-Assisted Energy Generation: From Fundamentals to Lab Scale and In-Field Applications, First Edition. Edited by Sathish-Kumar Kamaraj and Iryna Rusyn.

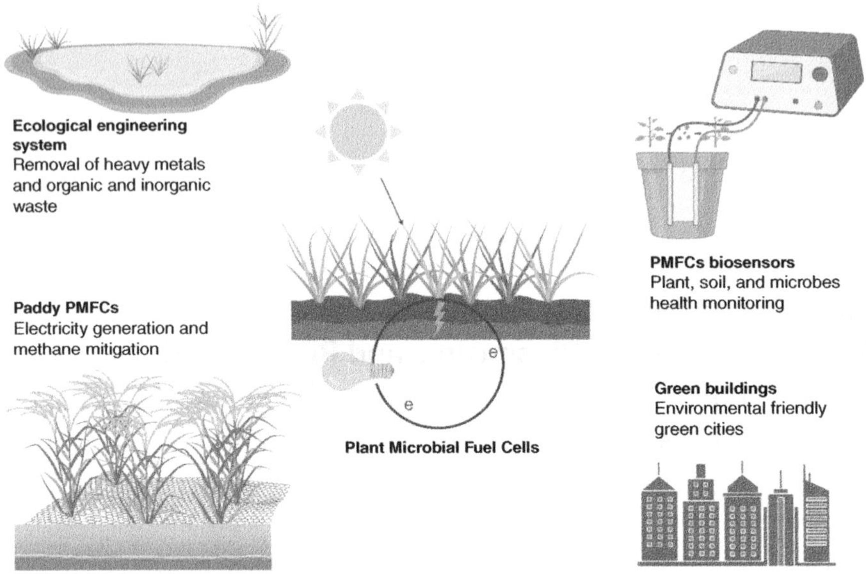

Figure 11.1 Diversity of plant-based microbial fuel cells and their real-world applications. The figure is designed in biorender.com.

increasing attention as a source of sustainable bioenergy sources that exploits the unique plant and microbe relationship at the rhizosphere region of a plant. Over the decades, different forms of PMFC have been developed and assessed in terms of their configuration and utilization. In this chapter, we present fundamental knowledge and applications that plant-based microbial fuel cells may apply in producing energies and bioremediations. (Figure 11.1)

PMFCs could play key roles in obtaining some of sustainable development goals (SDGs). The SDGs were devised to ensure economic, environmental, and social balance. PMFCs is aligned to SDG 1 (No Poverty), SDG 6 (Clean Water and Sanitation), SDG 7 (Affordable and clean energy), and SDG 13 (Climate actions). The progress of PMFCs and their role in achieving SDG goals has been reviewed recently in Ref. (Sayed et al., 2021).

11.2 Bioelectricity Generation Using a Plant-based Microbial Fuel Cell

In PMFCs, living plants use sunlight and release numerous exudates and organinc matter in the root zones (rhizosphere). Microbial communities that live beneath the rhizosphere oxidize the root exudates and organic matter into electrons, protons, biogas, and other by-products. This oxidation process occurs under anaerobic conditions called as anode. Electrons are transported through an external circuit, and protons simultaneously pass through a cation exchange membrane to the cathode. At the cathode, electrons, protons, and electron acceptors are combined and create an entire circuit thereby producing electricity (Nitisoravut & Regmi, 2017). PMFCs are therefore considered a promising green technology that can convert solar energy into electricity via bioelectrochemical interactions without any environmental footprints (Strik et al., 2008). Moreover, PMFCs are obtained without any food competition for land and non-destruction of plants thereby harnessing concurrent biomass and bioelectricity production (Helder et al., 2010; Strik et al., 2008). Various types of plants have been extensively investigated in PMFCs such as semi-arid green roof ecosystems (Tapia et al., 2017), wetland plants (Zhou et al., 2018), paddy fields (Kaku et al., 2008), fresh marshy species and salt marsh species (Helder et al., 2010), weeping alkali-grass (Md Khudzari et al., 2018), floating MFCs (Schievano et al., 2016), herbal plants (Li et al., 2020), hydroponic plants (Paucar, 2022), and wheat (Arulmani et al., 2021). PMFCs have been conducted from lab to field scale for bioenergy productions (biogas, biomass, and electricity), contaminant remediation, and PMFC-based sensors. Over the time, PMFCs have been diversified in terms of their usage and design. This chapter will discuss different uses and configurations of PMFCs.

11.3 PMFCs for Bioremediation

Initially, PMFC systems were tested for electricity generation; however, the output performances were relatively low (Takanewa et al., 2010; Kouzuma et al., 2014; De Schamphelaire et al., 2008). PMFCs are then shifted toward environmental remediation (Chen et al., 2015; Gustave et al., 2018). PMFC can remove a variety of contaminants, including organics and inorganics. The bioremediation of several PMFC systems is illustrated in Table 11.1.

Heavy metals are common contaminants in the soil, mainly arsenic (As), copper (Cu), zinc (Zn), chromium (Cr), cadmium (Cd), lead (Pb), mercury (Hg), and nickel (Ni). These pollutants are highly toxic, nonbiodegradable, and persistent in soil environments (He et al., 2013). Plant such as rice has high As accumulation due to the high bioavailability of As released from soil solid to the porewater of flooded soils. PMFC systems have been applied for the removal of heavy metals existing in both plants and soils. The application of PMFCs decreased the As accumulation in rice tissues from 44.7% to 67.9% with a maximum power output of 22.2 mW/m^2 compared to the control, which was operated without any resisters (open circuit conditions) (Gustave et al., 2019). This may have been attributed to the fact that the PMFCs limited the migration of As to rice plants and the

Table 11.1 Use of plant-based microbial fuel cells for waste removal.

Plant types	Soil types	Electrode materials	Targeted treatment	Removal efficiency	Outputs	References
Rice (*Oryza sativa L.*)	Paddy soil	Carbon felts	As in stems, leaves, husks, and rice grains	53.4–67.9%	22.2 mW/m^2	Gustave et al. (2019)
Ryegrass (*Lolium perenne*)	—	Graphite granules	Cr(VI)	99%	30–70 mA/m^2	Habibul et al. (2016)
Canna indica	Wetland soil	Granular activated carbon	COD NH4+-N	83.19–96.82 %	92.05 mW/m^3	Wen et al. (2021)
– *O. sativa* – *A.calamus* – *S. petite* – *C. elegans*	Sediment soil	Graphite rods	Sulfide	29.4–61.6%	0.0278–0.0256 V	Liu et al. (2019)
– *Pennisetum (Pennisetum alopecuroides)* – Reeds (*Phragmites communis)*	Wetland soil	Carbon felts	Total Cr, Cr(VI), Cu, Ni, Pb, Zn	14.6–73.9%	162.47 mV 128.01 mV	Guan et al. (2019)
V. spiralis	Sediment soil	Graphite felts	NH4+	57.2–58.7%	3.16 mW/m^2	Xu et al. (2017)
Aglaonema commutatum	—	Graphite felts	Phenanthrene Pyrene Crude oil	48.4–82.3%	184.9 mV	Zhao et al. (2019)
Wild rice (*Oryza rufipogon*) Cattail (*Typha Orientalis*)	Wetland	Carbon felts	Cd	22.8–30.2 %	350.50 mV 137.12 mV	Tongphanpharn et al. (2021)
Canna indica	Wetland	Granularactivated carbon	Sulfamethoxazole (SMX) Tetracycline (TC)	99.66–100%	256.7–301.5 mV	Wen et al. (2020)

mobilization of As into the soil porewater (Gustave et al., 2019). The anode of PMFCs might act as an alternative electron acceptor, thereby limiting As and Fe reduction (Gustave et al., 2018; Gustave et al., 2019). This led to less As being released into the porewater of paddy soil, hence impeding As uptake and translocation by rice plants (Gustave et al., 2018). Cr(VI) is greatly removed by 99% using PMFCs. The maximum current density of their system reached 55 mA/m^2 at Cr(VI) concentration of 19 mg/l (Habibul et al., 2016). The PMFC systems improved the removal of Cr(VI) because Cr(VI) was reduced directly by electrons generated from rhizospheres according to the cathode and anode potentials.

Constructed wetlands (CWs) combined with MFCs could simultaneously achieve pollutant removal and energy generation (Doherty et al., 2015). The planted CW-MFCs had higher achievements in terms of nitrate, ammonium, and total nitrogen removals. The removal efficiency of $NH4^+$-N could reach 95.91–96.82% in planted CW-MFCs compared with 56.54–59.95% removal of unplanted CW-MFCs. The power output of planted CW-PMFCs was 92.05 mW/m^3 which was higher than 3.29 mW/m^3 from unplanted systems (Wen et al., 2021).

PMFCs also have been investigated for the removal of polyaromatic hydrocarbons (PAHs), which are contaminated in soils due to leakage from refining of crude oil. PAHs are persistent and harmful substances. PAHs can threaten human health and damage the ecological systems. Degradation of PAHs (phenanthrene and pyrene) was significantly obtained using enhanced PMFCs with the addition of β-cyclodextrin (Zhao et al., 2019). The addition of β-cyclodextrin to PMFCs increased the degradation of PAHs since it solubilizes the organic petroleum molecules and increases the bioavailability such as nutrients and thereby accelerates the degradation of PAHs. The voltage production with plants increased rapidly and reached a maximum voltage of 184.9 mV while unplanted systems showed a lower maximum voltage output (117.3 mV) (Zhao et al., 2019). Those studies demonstrated that PMFC systems can remove a variety of pollutants from both plant tissues and soil environments. The main mechanism of pollutant removal by PMFCs is attributed to (i) activity of anodic zones (rhizospheres) – contaminants such as organic matters are oxidized by anaerobic microbial communities; (ii) anode reductions – pollutants retrieve electrons from the anode; (iii) metabolism involvements – increase nutrient uptake of plants; (iv) chemical interactions – increase the interaction between species donating and receiving electrons within soil matrix.

11.4 PMFCs for Control of Biogas Emission

Paddy fields and wetlands are the major sources of methane emissions. Rice cultivation accounted for 7% of global methane emissions in 2005 resulting in the second largest source of CH_4 emission in the agriculture (EPA, 2013). PMFCs can be utilized to control CH_4 emissions due to the out competition of methanogens by exogenic bacteria for organic substrates (Arends & Verstraete, 2012). Biochar anode was used to mitigate methane from PMFCs with a reduction of 39% accommodated with the highest power value of 11.11 mW/m^2 compared to carbon felt anode (Md Khudzari et al., 2019). However, the actual mechanisms for the reduction of methane emissions of MFCs with biochar were not clearly determined (Md Khudzari et al., 2019). CWs integrated with MFCs using *Canna indica* reduced the emissions of greenhouse gases (GHGs) by 7.2–38.7% for N_2O, 5.9–32.4% equivalents for CO_2, and 17.9–36.9% for CH_4. Organic matter in rhizospheres is oxidized to produce electrons that flow to the cathode (Puig et al., 2012). The microbes can receive the electrons to reduce nitrite and nitrate to nitrogen gas (Naga Samrat et al., 2018; Puig et al., 2012), promoting the removal of N_2O during nitrification and denitrification processes. PMFC systems also promote the degradation of organic substrates in the anodes, thus improving the production of CO_2 (Doherty et al., 2015). These were proved by higher average total nitrogen and chemical oxygen demand (COD) removal in closed-circuit conditions for 84.3–91.4% and 81.4–91.6%, respectively (Wang et al., 2019). Another study also showed that regulations of carbon cathode materials could control GHG emissions (Liu et al., 2022). Among stainless-steel mesh, carbon fiber felt (CFF) and carbon cloth, CFF was superior compared to carbon cloth with an average emission flux of 0.77 ± 0.04 and 130.78 ± 13.08 mg/m^2/h for CH_4 and N_2O, respectively. CFF in CW-MFC obtained the maximum power density of 2.99 W/m^3. The reason behind this prohibition might be associated with the proliferation of aerobic microbes, which can catalyze O_2 reduction into H_2O. Porous carbon materials possess a larger specific surface area and high cathode activities, enhancing the voltage output and proliferation of aerobic bacteria. Methanotrophs as aerobic bacteria could be enriched on CFF cathode than others, accounting for less CH_4 emission. Produced N_2O at lower surface CFF could be used as electron acceptors and reduced to N_2, thereby lowering the emissions flux of N_2O (Liu et al., 2022).

PMFCs can be simultaneously applied for multiple purposes including pollutant removal, bioelectricity, and biogas production. Previous studies in PMFCs mainly focused on either electricity generation coupled with controlling reducing gas emissions or/and pollutant removals. However, there is limited research investigating the capability for biogas production as well as the integration of multiple functions of PMFC systems. In the study of T.E. Kuleshova et al., lettuce (*Lactuca sativa* L.) planted in PMFC systems delivered about 50 mW/m^2 with a hydrogen yield of 0.2 mmol/day per m^3 in the root environment. This finding reveals that PMFC systems can be used for harvesting energy such as biogas production. Further studies are therefore suggested to interpret the maximum energies that can be harvested in PMFCs.

11.5 PMFCs-based Sensors

Sensors are widely used for detecting the properties of surrounding events and transferring the sensed information into signals (Do Kim et al., 2007). Sensors can be deployed to sense the soil water content, Internet-of-Things (IoT) application, flora health, and rain event indicators in PMFCs. PMFCs provide a power source for attaining an energy-neutral monitoring sensor. The performance of PMFCs depends on many factors including flora health, unhealthy plants can reduce the performance of PMFC systems. The activity of flora can be monitored through the measurement of power output. This relation is clarified by Davide Brunelli et al (Brunelli et al., 2016), sub-milliwatt harvested power from PMFCs is enough for obtaining an smart sensor. This smart sensor can monitor both the flora's health and environmental conditions. Their results showed that the lower output voltage was proportional to the decrease of cell performance due to the depletion of nutrients. Bench and pilot-scale experiments in PMFCs using green roof species were carried out to observe the relationship between substrate/soil water content and current generation. The outcomes indicated that current generation and soil water content had a tight relationship, which is reflected as higher soil moisture (around 27% vs. 17.5% v/v) accelerated larger output power (114.6 and 82.3 $\mu W/m^2$ vs. 32.5 $\mu W/m^2$). Their results could illustrate the ability to develop a water-content biosensor using bioelectrochemical technology and stimulate further studies as well as irrigation indicators for plants.

IoT-based wireless sensor networks (WSNs) are used to monitor real-time data such as humidity, water stress, salinity, and temperature (Zanella et al., 2014). Traditional IoT-based WSN systems operate on batteries and need to be replaced when energies are exhausted, those batteries are then disposed causing serious environmental risks. The combination of energy harvesting – WSNs (EH-WSN) and PMFCs could overcome this issue by providing a self-powered supply while eliminating environmentally harmful batteries. PMFC-Energy Harvesting Sensor Nodes were designed with an autonomous energy operation of sensor node with the energy gained from PMFC. The energy autonomy of the sensor node is obtained with a mean consumption value of only 1.27 mW while PMFC could generate up to 3.5 mW/cm^2 (De La Rosa et al., 2019).

11.6 PMFCs for Plant Health Monitoring

Proper monitoring of plant health is crucial for food and agricultural industries. Given PMFCs performance is linked to plant, soil, and microbes, the power generation from PMFCs could be linked to monitor the health of these interconnected components in the system. Root and shoot biomass along with high photosynthetic activities can enhance the voltage generation in PMFCs (Regmi et al., 2018). For example, rough-haired sedge *C. Hirta*, that was planted in PMFCs produced the highest power output of $970 \pm 22\,mW/m^2$ with the largest accumulated total dry leaf/steam and root mass. This was due to the higher development of the root system and the stronger photosynthesis activities compared to other species and non-vascular plants (Rusyn et al., 2022). One of the pioneer papers in PMFCs reported a decline in the voltage generation due to the decreased vitality of plants with aphid infestations (Strik et al., 2008). Regmi, 2017 reported that diseased/wilted tomato plants produced almost five times less voltage than healthy tomato plants. Therefore, PMFCs could be a useful wireless monitoring system to know the plant health (Brunelli et al., 2016). A WSNs based on PMFCs were successfully investigated for monitoring rice plant health by tracking photosynthetic activity (Puengsungwan, 2022). Interconnection of plant growth and power generation of PMFCs was studied with the amendment of different livestock's urine (Apollon et al., 2022). Therefore, amendment that concurrently enhances plant biomass and bioelectricity is a viable option in future. Apart from the plant, the electric signal from a PMFC could be useful to detect different stresses imparted on soil organism in rhizosphere region as reported in MFCs (Deng et al., 2015; Shen et al., 2012). Plant microbial diversity is directly related to its health (Berg et al., 2017). As rhizosphere microbes are the drivers of electricity generation, investigating their diversity and quantity is one of the potential applications of PMFCs (Rusyn, 2021). Recently, it is reported that open circuit potential generation from the PMFC is due to plant health and affected by the cultivation environment (Yoon et al., 2018). Varying level of electric potential is reported from healthy, dead, and no plant PMFC systems. Diversity of bacterial strain was also differed from live plant rhizosphere and heat-shocked plant rhizosphere (Yoon et al., 2018). These evidence suggest the impact of plant health in power generation of PMFCs and corroborate the idea presented previously as PMFCs for trios assessment of plant, microbes, and soil health (Regmi et al., 2018).

11.7 Design Criteria for Plant-based Microbial Fuel Cells

11.7.1 Anode

Anode electrodes functioned as electron acceptors in the PMFCs. A good electrode should possess the following qualities: high surface area, electrical conductivity, non-toxicity, chemical stability, biocompatibility, and low cost (Goswami &

Mishra, 2018; Munoz-Cupa et al., 2021; Slate et al., 2019). Electrodes are commonly made of carbon materials such as carbon felt, graphite felt/rod, carbon cloth, and carbon brush (Nitisoravut & Regmi, 2017). As shown in Table 11.1, graphite is one of the most common materials used in PMFCs due to its durability, low cost, and easy handling (ter Heijne et al., 2008). Anode carbon material can also be designed from naturally available substances like biochar which is a source of biomass that is abundantly available in forests, agricultural residues, manure, and marginal lands. Biochar anode PMFCs could produce a power density of 11.11 mW/m^2 of cathode area, which was lower than that of 41.41 mW/m^2 carbon felt anode MFCs (Md Khudzari et al., 2019). Although biochar anode generated lower power as compared to carbon felt anode PMFCs, methane emissions in biochar anode were reduced by 39% compared to carbon felt anode. Meanwhile, the price between 51 and 381 dollars per ton of biochar was much cheaper than 500–800 dollars per ton of graphite granules. In addition, anode materials could influence the microbial communities in the rhizospheres. The mixture of activated carbon (AC) and marine sediment PMFC outperformed with a single PMFC since a higher diversity of microbial communities was recorded with 42.5–69.7% *Proteobacteria*. Their study also indicated that the mixture of AC and marine sediment is worthy of consideration as bioanodes and useful for PMFCs in a real wetland. These studies suggest that the utilization of natural products for electrode materials such as biochar and/or sediment could be an appealing approach, which could reduce the cost, save energy, control GHGs and other applications might be further exploited.

11.7.2 Cathode

PMFC performance is also affected by electrode at the cathode and reduction reaction. Oxygen permanganate or ferricyanide is widely applied for terminal electron acceptors which occurred in cathode area (Kouam Ida & Mandal, 2022; Md Khudzari et al., 2019). Unlike MFCs, which are mainly exploited for wastewater treatment and electricity production, PMFCs commonly utilize the upper layers as cathode zones in single chamber configuration. Electrodes are placed on the topsoil surface, where contacted directly with the air while accepting electrons and protons from the anode.

11.7.3 Single Chamber PMFCs

In a single chamber, supply of oxygen for cathodic chamber is not required, as cathode is directly exposed to the air (Obileke et al., 2021). As shown in Table 11.1, single chamber design is more popular than dual chamber. Single chamber configuration is considered as cost-effectiveness and energy reduction since soil itself can function as a cation exchange membrane, and protons are spontaneously moved from the anode to the cathode within the soil matrix. However, there are several drawbacks that are inevitable in this system such as oxygen diffusion into the anode, low electricity production, and occurrence of short-circuit.

11.7.4 Dual Chamber PMFCs

Dual chamber is a conventional design of MFCs for wastewater treatment and electricity production. This design is less used for PMFCs since it requires the use of catholytes and membranes for the operations which pose difficulty while operating in rhizosphere, although it can produce higher electricity compared to a single PMFC. With an engineering modification, it is possible to use the dual chamber PMFCs *in situ* to reduce the energy needs. Integration of dual chamber PMFC with an algal-assisted cathode was reported to generate a high voltage output of up to 399 mV (Srivastava et al., 2019). This integration utilizes algal photosynthesis at the cathode to produce oxygen that acts as terminal electron acceptors. This design has some advantages such as low-cost cathode, operating in self-sustainable mode coupled with natural processes while producing high voltage output. Another cost-effective dual chamber PMFC has been designed using a novel earthen pot-PMFC using Vetiver and rice for wastewater treatment and electricity production (Nitisoravut & Regmi, 2017; Regmi et al., 2018). Vetiver PMFC conducted with low-cost double chamber earthen membrane achieved a maximum average current density of $242 \pm 10.5\,mA/m^2$ while organic removal rate could reach up to 99% while in rice PMFC double chamber PMFCs increased current generation by 87%. However, this design is difficult to integrate into the natural bodies but has exciting potential to be used as roof top and garden PMFCs.

11.7.5 Reactor Design

PMFCs can be designed either as a flat plate or tubular configuration. In the design of tubular PMFC, the anode is made in the form of a tube with a membrane at the bottom and the cathode underneath it. However, this design requires protons to pass a large distance across the membrane (Kuleshova et al., 2021). This drawback is overcome through the development of a flat-plate PMFC reactor, in which the membrane is vertically placed between anode and cathode, significantly reducing the resistance of proton transportation. In this reactor, anodes were placed at different positions, thus building a relationship between plant growth and power generation (Kuleshova et al., 2022). Wetland plants PMFC operated with AC mixed with marine sediment under a flat-plate configuration was reported to deliver a power density of $1.04\,mW/m^2$ (Sudirjo et al., 2019), while a tubular under the same configuration was also recorded to produce up to $22\,mW/m^3$ (Sudirjo et al., 2020).

Connection of multiple PMFCs and multielectrode in parallel and in series can also improve voltage and current of larger scale systems. While connecting three cells in series and parallel, increased voltage and current generation in *Ipomoea aquatica* and *Pistia stratiotes* operated PMFCs. However, power production was not significantly different either in series or parallel (Pamintuan et al., 2018). In other studies, *Lobelia Queen Cardinalis* PMFC operated in series delivered an equal open circuit potential compared to all single open circuit voltage from individual pots (Bataillou et al., 2022).

Furthermore, designs of mono and multielectrode systems were carried out using *Caltha palustris* plant (Rusyn et al., 2021). Mono-electrode MFC system was conducted with different electrode pairs including graphite/aluminum, graphite/zinc steel, graphite/copper, graphite/steel, aluminum/zinc steel, and aluminum/copper. Meanwhile, multielectrode MFCs were conducted with anodes and cathodes including 12 zinc-galvanized steel plates/11 graphited, 10 cathodes/9 anodes, 8 cathodes/9 anodes, and 8 cathodes/8 anodes. Their results showed that average voltage potentials were 1009.8 and 1014.6 mV for graphite/aluminum and graphite/zinc-galvanized steel electrode pairs, respectively. Electrode pairs of aluminum/copper, graphite/stainless steel, and aluminum/zinc-galvanized steel had a lower performance with electric potential ranging from 519.3 to 783.4 mV. The higher performance of graphite/aluminum and graphite/zinc-galvanized steel electrodes was attributed to the increase in surface properties of electrodes. Zinc and aluminum coated surfaces protected the electrodes from corrosion and improved the electrical conductivities of electrodes. Besides, it is notable that their work utilized graphite as waste of electric transport to reduce the cost of configuration installations. Their study also showed that multielectrode PMFCs and serial connection of multielectrode PMFCs increased the bioelectricity potentials and suggested the feasibility of enhancing PMFC performance by combining different configuration setups in bioelectrical systems. The configuration and desgins of PMFCs in different studies is shown in Table 11.2.

11.8 Conclusion and Recommendation

Although PMFCs have great potential to be a sustainable source of energy, there are different technological and feasibility challenges such as unstable electricity, low voltage, and short life span of plants that restrict their real-world application (Maddalwar et al., 2021). Given the multiple uses of diverse PMFC types, each system needs to be optimized for the desired objectives inspired by geographical locations and client needs. In rural area where electricity is a major constrain, research on maximizing power output with longer duration need to be focused. For methane mitigation, configuration design should be directed to control methane emission. Currently, there is a lack of standardization in materials used,

Table 11.2 Configurations and designs in PMFCs.

Configurations	Plant types	Anode material	Cathode material	Membranes	Power outputs	References
Flat-plate reactors	*Spartina anglica*	Graphite rods	Graphite felt	Fumasep FKD-PK-75 PEEK-reinforces	1.04 mW/m^2	Sudirjo et al. (2019)
Single chamber	Rice plant	Carbon felt/biochar	Manganese-based catalyzed carbon	—	41.41 mW/m^2	Md Khudzari et al. (2019)
Dual chamber	*C. phormidium*	Graphite rod	Carbon cloth	Nafion117	29.78 mW/m^3	Srivastava et al. (2019)
Single chamber	*Oryza sativa*	Graphite felt	Graphite felt	—	49 mW/m^2	Goto et al. (2015)
Single chamber	Rice	Graphite felt	Graphite felt	—	130 mW/m^2	Matsumoto et al. (2020)
Flat-plate MFCs	plants	Polyurethane foams/activated carbon	Graphite felt	Fumasep FKD-PK-75	22 mW/m^3	Sudirjo et al. (2020)
Single chamber	*Lactuca sativa L.*	Graphite felt	Graphite felt	—	170 mV	Kuleshova et al. (2021)
Single chamber	Wetland plants	Graphite felt	Graphite felt	—	12.82 mW/m^2	Zhou et al. (2018)
Dual chamber	*Vetiver*	Graphite fiber	Graphite fiber	Earthen	68 mW/m^2	Regmi et al. (2018)
Dual chamber	Herbal plant	Carbon brushes	Carbon brushes	Cation exchange membrane	18.67 mW/m^3	Li et al. (2020)
Single chamber	*Agapanthus africanus*	Carbon fiber	Carbon fiber	—	690 mV	Gómora-Hernández et al. (2020)
Single chamber	Food plants	Graphite rod	Carbon cloth	—	194.45 mW/m^2	Arulmani et al. (2021)
Single chamber	Wet plants	Granule activated carbon	Platinum coated carbon	—	229 ± 52 mW/m^3	Oodally et al. (2019)
Single chamber	Weeping alkali grass	Carbon felt	Manganese-based catalyzed carbon	—	8.59–83.7 mW/m^2	Md Khudzari et al. (2018)

configuration, design, and microbes in PMFCs which poses further difficulties in commercialization of PMFCs (Jadhav et al., 2021). Scaling-up PMFCs with stacking of modular units need to be done for usable electricity generation. Value-added by-products recovery from PMFC operations like biogas would be another exciting opportunity to investigate.

References

Apollon, W. et al. (2022). Livestock's Urine-Based Plant Microbial Fuel Cells Improve Plant Growth and Power Generation. *Energies* 15 (19). https://doi.org/10.3390/en15196985.

Arends, J.B.A. and Verstraete, W. (2012). 100 years of microbial electricity production: three concepts for the future. *Microbial Biotechnology* 5 (3): 333–346. https://doi.org/10.1111/j.1751-7915.2011.00302.x.

Arulmani, S.R.B., Gnanamuthu, H.L., Kandasamy, S. et al. (2021). Sustainable bioelectricity production from *Amaranthus viridis* and *Triticum aestivum* mediated plant microbial fuel cells with efficient electrogenic bacteria selections. *Process Biochemistry* 107 (April): 27–37. https://doi.org/10.1016/j.procbio.2021.04.015.

Bataillou, G., Haddour, N., and Vollaire, C. (2022). Bioelectricity production of PMFC using Lobelia Queen Cardinalis in individual and shared soil configurations. *E3S Web of Conferences* 334: https://doi.org/10.1051/e3sconf/202233408001.

Berg, H., Ekman Söderholm, A., Söderström, A.S., and Tam, N. (2017). Recognizing wetland ecosystem services for sustainable rice farming in the Mekong Delta, Vietnam. *Sustainability Science* 12 (1): 137–154. https://doi.org/10.1007/s11625-016-0409-x.

Brunelli, D., Tosato, P., and Rossi, M. (2016). Flora health wireless monitoring with plant-microbial fuel cell. *Procedia Engineering* 168: 1646–1650. https://doi.org/10.1016/j.proeng.2016.11.481.

Chen, Z., Zhu, B.K., Jia, W.F. et al. (2015). Can electrokinetic removal of metals from contaminated paddy soils be powered by microbial fuel cells? *Environmental Technology and Innovation* 3: 63–67. https://doi.org/10.1016/j.eti.2015.02.003.

De La Rosa, E.O., Edith, J.V.C., Campos, M.C. et al. (2019). Plant microbial fuel cells-based energy harvester system for self-powered IoT applications. *Sensors (Switzerland)* 19 (6): 1–16. https://doi.org/10.3390/s19061378.

De Schamphelaire, L., Rabaey, K., Boeckx, P. et al. (2008). Outlook for benefits of sediment microbial fuel cells with two bio-electrodes. *Microbial Biotechnology* 1 (6): 446–462. https://doi.org/10.1111/j.1751-7915.2008.00042.x.

Deng, H., Jiang, Y.B., Zhou, Y.W. et al. (2015). Using electrical signals of microbial fuel cells to detect copper stress on soil microorganisms. *European Journal of Soil Science* 66 (2): 369–377. https://doi.org/10.1111/ejss.12215.

Do Kim, J., Kim, D.J., Byun, H.G. et al. (2007). The definition of basic TEDS of IEEE 1451.4 for sensors for an electronic tongue and the proposal of new template TEDS for electrochemical devices. *Talanta* 71 (4): 1642–1651. https://doi.org/10.1016/j.talanta.2006.07.049.

Doherty, L., Zhao, Y., Zhao, X. et al. (2015). A review of a recently emerged technology: constructed wetland – microbial fuel cells. *Water Research* 85: 38–45. https://doi.org/10.1016/j.watres.2015.08.016.

Doherty, L., Zhao, Y., Zhao, X., and Wang, W. (2015). Nutrient and organics removal from swine slurry with simultaneous electricity generation in an alum sludge-based constructed wetland incorporating microbial fuel cell technology. *Chemical Engineering Journal* 266: 74–81. https://doi.org/10.1016/j.cej.2014.12.063.

EPA, (2013). Global Mitigation of Non-CO2 GHGs Report.

Fadil, D.A., Al-Bahadili, R.J., and Abdullah, M.N. (2023). Energy harvesting schemes for internet of things: a review. *Indonesian Journal of Electrical Engineering and Computer Science* 29 (2): 1088–1094. https://doi.org/10.11591/ijeecs.v29.i2.pp1088-1094.

Gómora-Hernández, J.C., Serment-Guerrero, J.H., Carreño-de-León, M.C., and N. F.-A. (2020). Voltage production in a plant-microbial fuel cell using *Agapanthus africanus*. *Revista Mexicana de Ingeniería Química* 12 (3): 505–511. http://www.redalyc.org/articulo.oa?id=62029966013.

Goswami, R. and Mishra, V.K. (2018). A review of design, operational conditions and applications of microbial fuel cells. *Biofuels* 9 (2): 203–220. https://doi.org/10.1080/17597269.2017.1302682.

Goto, Y., Yoshida, N., Umeyama, Y. et al. (2015). Enhancement of electricity production by graphene oxide in soil microbial fuel cells and plant microbial fuel cells. *Frontiers in Bioengineering and Biotechnology* 3: 1–8. https://doi.org/10.3389/fbioe.2015.00042.

Guan, C.Y., Hu, A., and Yu, C.P. (2019). Stratified chemical and microbial characteristics between anode and cathode after long-term operation of plant microbial fuel cells for remediation of metal contaminated soils. *Science of the Total Environment* 670: 585–594. https://doi.org/10.1016/j.scitotenv.2019.03.096.

Gustave, W., Yuan, Z.-F., Sekar, R. et al. (2018). Arsenic mitigation in paddy soils by using microbial fuel cells. *Environmental Pollution* 238: 647–655. https://doi.org/10.1016/j.envpol.2018.03.085.

Gustave, W., Yuan, Z.F., Ren, Y.X. et al. (2019). Arsenic alleviation in rice by using paddy soil microbial fuel cells. *Plant and Soil* 441 (1–2): 111–127. https://doi.org/10.1007/s11104-019-04098-7.

Gustave, W., Yuan, Z.F., Sekar, R. et al. (2019). The change in biotic and abiotic soil components influenced by paddy soil microbial fuel cells loaded with various resistances. *Journal of Soils and Sediments* 19 (1): 106–115. https://doi.org/10.1007/s11368-018-2024-1.

Habibul, N. et al. (2016). Bioelectrochemical chromium(VI) removal in plant-microbial fuel cells. *Environmental Science and Technology* 50 (7): 3882–3889. https://doi.org/10.1021/acs.est.5b06376.

He, B., ZhaoJun, Y.U.N., JianBo, S.H.I. et al. (2013). Research progress of heavy metal pollution in China: sources, analytical methods, status, and toxicity. *Chinese Science Bulletin* 58 (2): 134–140. https://doi.org/10.1007/s11434-012-5541-0.

ter Heijne, A., Hamelers, H.V., Saakes, M. et al. (2008). Performance of non-porous graphite and titanium-based anodes in microbial fuel cells. *Electrochimica Acta* 53 (18): 5697–5703. https://doi.org/10.1016/j.electacta.2008.03.032.

Helder, M., Strik, D.P.B.T.B., Hamelers, H.V.M. et al. (2010). Concurrent bio-electricity and biomass production in three plant-microbial fuel cells using *Spartina anglica*, *Arundinella anomala* and *Arundo donax*. *Bioresource Technology* 101 (10): 3541–3547. https://doi.org/10.1016/j.biortech.2009.12.124.

Jadhav, D.A., Mungray, A.K., Arkatkar, A. et al. (2021, 2020). Recent advancement in scaling-up applications of microbial fuel cells: from reality to practicability'. *Sustainable Energy Technologies and Assessments* 45: 101226. https://doi.org/10.1016/j.seta.2021.101226.

Kaku, N., Yonezawa, N., and Kodama, Y. (2008). Plant/microbe cooperation for electricity generation in a rice paddy field. *Applied Microbiology and Biotechnology* 79: 43–49. https://doi.org/10.1007/s00253-008-1410-9.

Kan, X., Reichenberg, L., and Hedenus, F. (2021). The impacts of the electricity demand pattern on electricity system cost and the electricity supply mix: a comprehensive modeling analysis for Europe. *Energy* 235: 121329. https://doi.org/10.1016/j.energy.2021.121329.

Kouam Ida, T. and Mandal, B. (2022). Microbial fuel cell design, application and performance: a review. *Materials Today: Proceedings* https://doi.org/10.1016/j.matpr.2022.10.131.

Kouzuma, A., Kaku, N., and Watanabe, K. (2014). Microbial electricity generation in rice paddy fields: recent advances and perspectives in rhizosphere microbial fuel cells. *Applied Microbiology and Biotechnology* 98 (23): 9521–9526. https://doi.org/10.1007/s00253-014-6138-0.

Kralova, I. and Sjöblom, J. (2010). Biofuels-renewable energy sources: a review. *Journal of Dispersion Science and Technology* 31 (3): 409–425. https://doi.org/10.1080/01932690903119674.

Kuleshova, T.E., Gall, N.R., Galushko, A.S. et al. (2021). Electrogenesis in plant–microbial fuel cells in parallel and series connections. *Technical Physics* 66 (3): 496–504. https://doi.org/10.1134/S1063784221030142.

Kuleshova, T., Rao, A., Bhadra, S. et al. (2022). Plant microbial fuel cells as an innovative, versatile agro-technology for green energy generation combined with wastewater treatment and food production. *Biomass and Bioenergy* 167 (September): 106629. https://doi.org/10.1016/j.biombioe.2022.106629.

Li, T., Yang, X.L., Chen, Q.L. et al. (2020). Enhanced performance of microbial fuel cells with electron mediators from anthraquinone/polyphenol-abundant herbal plants. *ACS Sustainable Chemistry and Engineering* 8 (30): 11263–11275. https://doi.org/10.1021/acssuschemeng.0c03058.

Liu, Y., Zhang, H., Lu, Z. et al. (2019). Decreasing sulfide in sediment and promoting plant growth by plant–sediment microbial fuel cells with emerged plants. *Paddy and Water Environment* 17 (1): 13–21. https://doi.org/10.1007/s10333-018-0679-2.

Liu, S., Xue, H., Wang, M. et al. (2022). The role of microbial electrogenesis in regulating methane and nitrous oxide emissions from constructed wetland-microbial fuel cell. *International Journal of Hydrogen Energy* 47 (63): 27279–27292. https://doi.org/10.1016/j.ijhydene.2022.06.063.

Maddalwar, S., Nayak, K.K., Kumar, M., and Singh, L. (2021). Plant microbial fuel cell: opportunities, challenges, and prospects. *Bioresource Technology* 341: 125772. https://doi.org/10.1016/j.biortech.2021.125772.

Matsumoto, A., Nagoya, M., Tsuchiya, M. et al. (2020). Enhanced electricity generation in rice paddy-field microbial fuel cells supplemented with iron powders. *Bioelectrochemistry* 136: 107625. https://doi.org/10.1016/j.bioelechem.2020.107625.

Md Khudzari, J., Kurian, J., Gariépy, Y. et al. (2018). Effects of salinity, growing media, and photoperiod on bioelectricity production in plant microbial fuel cells with weeping alkaligrass. *Biomass and Bioenergy* 109: 1–9. https://doi.org/10.1016/j.biombioe.2017.12.013.

Md Khudzari, J., Gariépy, Y., Kurian, J. et al. (2019). Effects of biochar anodes in rice plant microbial fuel cells on the production of bioelectricity, biomass, and methane. *Biochemical Engineering Journal* 141 (June 2018): 190–199. https://doi.org/10.1016/j.bej.2018.10.012.

Munoz-Cupa, C., Hu, Y., Xu, C., and Bassi, A. (2021). An overview of microbial fuel cell usage in wastewater treatment, resource recovery and energy production. *Science of the Total Environment* 754: 142429. https://doi.org/10.1016/j.scitotenv.2020.142429.

Naga Samrat, M.V.V., Rao, K.K., Ruggeri, B. et al. (2018). Denitrification of water in a microbial fuel cell (MFC) using seawater bacteria. *Journal of Cleaner Production* 178: 449–456. https://doi.org/10.1016/j.jclepro.2017.12.221.

Nepal, R., Phoumin, H., and Khatri, A. (2021). Green technological development and deployment in the association of southeast Asian economies (ASEAN)—At crossroads or roundabout? *Sustainability (Switzerland)* 13 (2): 1–19. https://doi.org/10.3390/su13020758.

Nitisoravut, R. and Regmi, R. (2017). Plant microbial fuel cells: a promising biosystems engineering. *Renewable and Sustainable Energy Reviews* 76: 81–89. https://doi.org/10.1016/j.rser.2017.03.064.

Obileke, K.C., Onyeaka, H., Meyer, E.L. et al. (2021). Microbial fuel cells, a renewable energy technology for bio-electricity generation: a mini-review. *Electrochemistry Communications* 125: 107003. https://doi.org/10.1016/j.elecom.2021.107003.

Oodally, A., Gulamhussein, M., and Randall, D.G. (2019). Investigating the performance of constructed wetland microbial fuel cells using three indigenous South African wetland plants. *Journal of Water Process Engineering* 32 (February): https://doi.org/10.1016/j.jwpe.2019.100930.

Pamintuan, K.R.S., Clomera, J.A.A., Garcia, K.V. et al. (2018). Stacking of aquatic plant-microbial fuel cells growing water spinach (*Ipomoea aquatica*) and water lettuce (*Pistia stratiotes*). *IOP Conference Series: Earth and Environmental Science* 191 (1): https://doi.org/10.1088/1755-1315/191/1/012054.

Paucar, N. E. (2022). Electricity generation, organic removal, and nutrient.

Puengsungwan, S. (2022). Wireless sensor networks (WSNs) based on the existing perception for monitoring rice-paddy health. In: *ITC-CSCC 2022 – 37th International Technical Conference on Circuits/Systems, Computers and Communications*, 298–301. https://doi.org/10.1109/ITC-CSCC55581.2022.9895072.

Puig, S., Coma, M., Desloover, J. et al. (2012). Autotrophic denitrification in microbial fuel cells treating low ionic strength waters. *Environmental Science and Technology* 46 (4): 2309–2315. https://doi.org/10.1021/es2030609.

Regmi, R., Nitisoravut, R., Charoenroongtavee, S. et al. (2018). Earthen pot–plant microbial fuel cell powered by vetiver for bioelectricity production and wastewater treatment. *Clean – Soil, Air, Water* 46 (3): https://doi.org/10.1002/clen.201700193.

Regmi, R. (2017). Examining different classes of plants under various operating conditions for bioelectricity production in plant microbial fuel cell. *Thammasat University.*

Regmi, R., Nitisoravut, R., and Ketchaimongkol, J. (2018). A decade of plant-assisted microbial fuel cells: looking back and moving forward. *Biofuels* 9: 1–8. https://doi.org/10.1080/17597269.2018.1432272.

Rusyn, I. (2021). Role of microbial community and plant species in performance of plant microbial fuel cells. *Renewable and Sustainable Energy Reviews* 152: 111697. https://doi.org/10.1016/j.rser.2021.111697.

Rusyn, I.B., Medvediev, O.V., and Valko, B.T. (2021). Enhancement of bioelectric parameters of multi-electrode plant–microbial fuel cells by combining of serial and parallel connection. *International Journal of Environmental Science and Technology* 18 (6): 1323–1334. https://doi.org/10.1007/s13762-020-02934-3.

Rusyn, I.B., Fihurka, O.M., and Dyachok, V.V. (2022). Effect of plants morphological parameters on plant-microbial feel cell efficiency. *Innovative Biosystems and Bioengineering* 6 (3-4): 161–168. https://doi.org/10.20535/ibb.2022.6.3-4.273108.

Sayed, E.T., Abdelkareem, M.A., Obaideen, K. et al. (2021). Progress in plant-based bioelectrochemical systems and their connection with sustainable development goals. *Carbon Resources Conversion* 4: 169–183.

Schievano, A., Pizza, F., Pino, C. et al. (2016). Experiences of floating microbial fuel cells, supplying on-line sensors for water quality. In: *Proceedings of the 13th IWA Leading Edge Conference on Water and Wastewater Technologies Spain*, vol. 1, 1–4.

Shen, Y.J., Lefebvre, O., Tan, Z., and Ng, H.Y. (2012). Microbial fuel-cell-based toxicity sensor for fast monitoring of acidic toxicity. *Water science and technology: a journal of the International Association on Water Pollution Research* 65 (7): 1223–1228. https://doi.org/10.2166/wst.2012.957.

Slate, A.J., Whitehead, K.A., Brownson, D.A., and Banks, C.E. (2019). Microbial fuel cells: an overview of current technology. *Renewable and Sustainable Energy Reviews* 101 (September 2018): 60–81. https://doi.org/10.1016/j.rser.2018.09.044.

Srivastava, P., Gupta, S., Garaniya, V. et al. (2019). Up to 399 mV bioelectricity generated by a rice paddy-planted microbial fuel cell assisted with a blue-green algal cathode. *Environmental Chemistry Letters* 17 (2): 1045–1051. https://doi.org/10.1007/s10311-018-00824-2.

Strik, D.P.B.T.B., Hamelers, H.V.M., Snel, J.F., and Buisman, C.J. (2008). Green electricity production with living plants and bacteria in a fuel cell. *Short communication Int. J. Energy* 32: 870–876. https://doi.org/10.1002/er.

Sudirjo, E., Buisman, C.J.N., and Strik, D.P.B.T.B. (2019). Activated carbon mixed with marine sediment is suitable as bioanode material for *Spartina anglica* sediment/plant microbial fuel cell: plant growth, electricity generation, and spatial microbial community diversity. *Water (Switzerland)* 11 (9): https://doi.org/10.3390/w11091810.

Sudirjo, E., Diaz, P.Y.C., and Cociancich, M. (2020). A thin layer of activated carbon deposited on polyurethane cube leads to new conductive bioanode for (plant) microbial fuel cell. *Energies* 13 (3): https://doi.org/10.3390/en13030574.

Takanewa, K., Nishio, K., Kato, S. et al. (2010). Factors affecting electric output from rice-paddy microbial fuel cells. *Bioscience, Biotechnology, and Biochemistry* 74 (6): 1271–1273. https://doi.org/10.1271/bbb.90852.

Tapia, N.F., Rojas, C., Bonilla, C.A., and Vargas, I.T. (2017). Evaluation of Sedum as driver for plant microbial fuel cells in a semi-arid green roof ecosystem. *Ecological Engineering* 108: 203–210. https://doi.org/10.1016/j.ecoleng.2017.08.017.

Tongphanpharn, N., Chou, C.H., Guan, C.Y. et al. (2021). Plant microbial fuel cells with *Oryza rufipogon* and *Typha orientalis* for remediation of cadmium contaminated soil. *Environmental Technology and Innovation* 24: 102030. https://doi.org/10.1016/j.eti.2021.102030.

Wang, X., Tian, Y., Liu, H. et al. (2019). The influence of incorporating microbial fuel cells on greenhouse gas emissions from constructed wetlands. *Science of the Total Environment* 656: 270–279. https://doi.org/10.1016/j.scitotenv.2018.11.328.

Wen, H., Zhu, H., Yan, B. et al. (2020). Treatment of typical antibiotics in constructed wetlands integrated with microbial fuel cells: roles of plant and circuit operation mode. *Chemosphere* 250: 126252. https://doi.org/10.1016/j.chemosphere.2020.126252.

Wen, H., Zhu, H., Yan, B. et al. (2021). Constructed wetlands integrated with microbial fuel cells for COD and nitrogen removal affected by plant and circuit operation mode. *Environmental Science and Pollution Research* 28 (3): 3008–3018. https://doi.org/10.1007/s11356-020-10632-3.

Xu, P., Xiao, E.R., Xu, D. et al. (2017). Internal nitrogen removal from sediments by the hybrid system of microbial fuel cells and submerged aquatic plants. *PLoS ONE* 12 (2): 1–20. https://doi.org/10.1371/journal.pone.0172757.

Yoon, T.H., Song, H.J., Jung, W.Y. et al. (2018). Monitoring plant health using a plant microbial fuel cell. *Bulletin of the Korean Chemical Society* 39 (10): 1193–1197. https://doi.org/10.1002/bkcs.11575.

Zanella, A., Bui, N., Castellani, A. et al. (2014). Internet of things for smart cities. *IEEE Internet of Things Journal* 1 (1): 22–32. https://doi.org/10.1109/JIOT.2014.2306328.

Zhang, X. and Cai, X. (2011). Climate change impacts on global agricultural land availability. *Environmental Research Letters* 6 (1): https://doi.org/10.1088/1748-9326/6/1/014014.

Zhao, L., Deng, J., Hou, H. et al. (2019). Investigation of PAH and oil degradation along with electricity generation in soil using an enhanced plant-microbial fuel cell. *Journal of Cleaner Production.* https://doi.org/10.1016/j.jclepro.2019.02.212.

Zhou, Y., Xu, D., Xiao, E. et al. (2018). Relationship between electrogenic performance and physiological change of four wetland plants in constructed wetland-microbial fuel cells during non-growing seasons. *Journal of Environmental Sciences (China)* 70: 54–62. https://doi.org/10.1016/j.jes.2017.11.008.

12

Progress and Recent Trends of Application of Low-energy Consuming Devices and IoT Based on Photosynthesis-assisted Power Generation

Edith Osorio-de-la-Rosa[1,*], *Mirna Valdez-Hernández*[2,*], *Rosa M. Woo-García*[3,4], *and Javier Vázquez-Castillo*[5]

[1] *División de Ciencia, Ingeniería y Tecnología, CONAHCYT-Universidad Autónoma del Estado de Quintana Roo, Chetumal, Quintana Roo, México*
[2] *Herbario, Departamento Conservación de la Biodiversidad, El Colegio de la Frontera Sur, Chetumal, Quintana Roo, México*
[3] *Maestría en Ingeniería Aplicada, Facultad de Ingeniería de la Construcción y el Hábitat, Universidad Veracruzana, Boca del Río, Veracruz, México*
[4] *Facultad de Ingeniería Eléctrica y Electrónica, Universidad Veracruzana, Boca del Río, Veracruz, México*
[5] *División de Ciencia, Ingeniería y Tecnología, Universidad Autónoma del Estado de Quintana Roo, Chetumal, Quintana Roo, México*

12.1 Introduction

The rapidly growing concerns about climate change, the excessive dependence on fuel-based power generation, and the environmental issues of traditional energy resources have led to a search for the generation of sustainable energy. Considering that the concentration of atmospheric CO_2 has reached ppm levels since 2011 and continues to increase (IPCC, 2021), it becomes necessary to use and develop new technologies to obtain energy without relying on fossil fuels.

Energy harvesting includes a range of technologies to capture freely available energy from the environment and convert it to more useful forms including electrical, thermal, and mechanical energies. The efficiency and performance of energy harvesting techniques are directly related to the characteristics of the employed materials.

*Corresponding authors

Photosynthesis-Assisted Energy Generation: From Fundamentals to Lab Scale and In-Field Applications, First Edition. Edited by Sathish-Kumar Kamaraj and Iryna Rusyn.

In this sense, energy harvesting is one of the key emerging technologies of the twenty-first century (Ku et al., 2016). It refers to the collection of energy from the environment; energy that would otherwise be lost to heat. To distinguish from renewable energy sources more generally, energy harvesting can be defined as the collection of locally naturally available energy for local use. Most often it involves small electronics systems to collect tiny amounts of power, in the range from nanowatts to hundreds of milliwatts.

This amount of energy harvesting can used to energize low-power devices used in the new era of the Internet of Things (IoT). IoT transforms real-world devices/objects into smart devices. So, IoT aims to unite every object in the work, industry, or any place, in such a way that people can control them via the Internet. IoT depicts the transactional relationship between things, people, and the Internet. Then, IoT is a vision of anything in the physical world can be digitally represented and connected for communication.

There are many architectures of IoT proposed due to its complexity of focusing on finer detailed aspects. Five-layer architecture is one of the simplest that includes the processing and business layers. (i) The perception layer is the physical layer, which has sensors (low-power consumption preferred) for sensing and gathering information about the environment. It senses some physical parameters or identifies other smart objects in the environment. (ii) The transport layer transfers the sensor data from the perception layer to the processing layer and vice versa through networks such as wireless, 5G, Bluetooth, Radio Frequency Identification, and others. Communication in the IoT technology requires special networking protocols and mechanisms. Therefore, new mechanisms and protocols have been proposed and implemented for each layer of the networking stack, according to the requirements imposed by IoT devices. (iii) Processing (known as fog computing applications) layer is also known as the middleware layer. It stores, analyzes, and processes huge amounts of data that come from the transport layer. It can manage and provide a diverse set of services to the lower layers. The technologies more used are cloud computing, databases, and big data processing modules. It is possible to say that this layer summarizes data before sending it on the network with a temporary storage, a small processing unit, and some security features. (iv) The application layer is responsible for delivering application-specific services to the user. It defines various applications in which the IoTs can be deployed, for instance, smart cities, and smart health, precision agriculture (PA). (v) The business layer manages the whole IoT system, including applications, business and profit models, and users' privacy (Sethi & Sarangi, 2017).

The leading communication technologies used in the IoT world are IEEE 802.15, low-power WiFi, Bluetooth, 6LoWPAN, LoRa, and other proprietary protocols for wireless networks.

This chapter presents a reviewer on photosynthesis-assisted power generation emphasizing the relation between type of photosynthetic rate and energy harvesting. Section three presents the energy harvesting concept applied to photosynthesis-assisted power generation. Section four presents the evolution of low-power devices in focus on the semiconductor devices and technology. Section five presents Bluetooth Low Energy (BLE) as support to IoT applications. Finally, section six talks about the trends in agriculture precision.

12.2 Promising Plants for Use as Energy Sources

Plants have developed the most efficient energy system known so far, transforming light energy into chemical energy to synthesize carbohydrates from CO_2 and water (Bloom & Lancaster, 2018; Volkov, 2012). Therefore, they constitute the main energy input to the terrestrial system and are the energy base. However, despite the large amount of energy that is generated on earth through plants, so far it has not been possible to develop efficient techniques to harvest this energy directly and massively (Kabutey et al., 2019).

An emerging technology is the plant microbial fuel cell (PMFC), which can use the photosynthetic reaction of plants to generate renewable and sustainable bioelectricity with plants as the only source of organic matter (Osorio-de-la-Rosa et al., 2021). PMFCs are made up of living plants, a support matrix, an anode, a cathode, and a membrane that separates two chambers; where the soil acts as a membrane while operating as a single sediment configuration (Osorio-de-la-Rosa et al., 2019).

In the most studied configuration, one of the electrodes is in contact with bacteria, capable of oxidizing organic molecules, releasing mainly protons and electrons. The released protons can reach the contact electrode through a salt bridge, while the electrons travel through an external circuit, causing a current flow.

PMFC performance is influenced by many factors (Jiang et al., 2021; Long et al., 2021) such as plant species used in PMFC (Valdez-Hernández et al., 2022), operation parameters, characteristics, and properties of wastewater (Jung & Park, 2021), soil microbial (He et al., 2021), electrode materials (Osorio-de-la-Rosa et al., 2019), and conformations of design (Jadhav et al., 2021).

The plants are a critical parameter, as plants are responsible for providing the organic matter to the system, with up to 70% composed of dead roots, leaf litter, and exudates. This organic matter is oxidized by bacteria that live in the rhizosphere, releasing CO_2, protons, and electrons (Ayala-Ruiz et al., 2019). In this sense, the choose plants in a PMFC technology, it is necessary to consider their lifetime, and photosynthetic pathway, that is, if they are of type C3, C4, or CAM. Metabolic pathways play a very important role in energy conversion with PMFCs. CAM plants grow very slowly, resulting in less biomass production in each period.

Several studies mention that C4 species have greater potential to produce energy, since they have higher net photosynthesis rates (Apollon et al., 2021; Rusyn, 2021) (Shrirang et al. 2021). Considering the assumption that a higher rate of photosynthesis will be directly related to higher energy generation (Apollon et al., 2021). However, few studies have determined physiological traits of the plant species used in PMFCs (Apollon et al., 2021; Valdez-Hernández et al., 2022). In most of the studies, only the photosynthetic pathway of the species used in the PMFC is mentioned, even sometimes the metabolic pathway assigned to the species can be difficult to determine if physiological measurements are not made to validate them. Currently, a greater number of species are reported with a type of facultative photosynthesis, which implies that they can change from one photosynthetic route to another depending on the prevailing environmental conditions (Srivastava et al., 2014). In this case, there are several of the species commonly used in PMFC: *Sedum album* which has been reported as facultative C3-CAM (Habibi & Hajiboland, 2012); *Hydrilla verticillata*, *Phragmites australis*, and *Phragmites communis* which have been reported as facultative C3-C4 (Magnin et al., 1997; Srivastava et al., 2014).

In PMFC systems, more than 70% of the integrated species are of type C3 (see Figure 12.1), which coincides with the dominance of this photosynthetic route in the diversity of plant species (Yamori et al., 2014). To analyze the relationship of the net photosynthesis rate in species commonly employed in PMFCs and power generation in terms of maximum power density, a literature review was conducted for more than 50 plant species employed in PMFCs (Apollon et al., 2022; Shrirang et al., 2021; Rusyn, 2021), for 25 species reports of net photosynthetic rates were found (see Figure 12.1). In the analysis, it was found that some of the species with the highest net photosynthetic rate do not generate more energy (Figure 12.1A); for example, *Ciperus papyrus* has a high net photosynthetic rate of 35 μmol $CO_2/m^2\ s^1$ (Jones, 1987) and a low maximum power density of 1083 mW/m^2 (Azri et al., 2018); *while Codiaeum variegatum* presents a low net photosynthetic rate of 2.6 μmol $CO_2\ m^2/s^1$ (Valdez-Hernández et al., 2022) and a high maximum power density of 189.97 mW/m^2 (Valdez-Hernández et al., 2022). This is reinforced by the results of a Spearman correlation that indicates that there is no correlation between the variables (see Figure 12.1B). Nor is it possible to determine which C4 species have higher energy generation; in fact, the two species that produce the highest power density are C3 (*Canna indica* and *Codiaeum variegatum*; see Figure 12.1A).

Although the photosynthetic pathway is very important for energy conversion in plants, it may not be directly linked to energy generation in PMFCs. Therefore, for the choice of the plant species that will be used in the PMFC system, it must

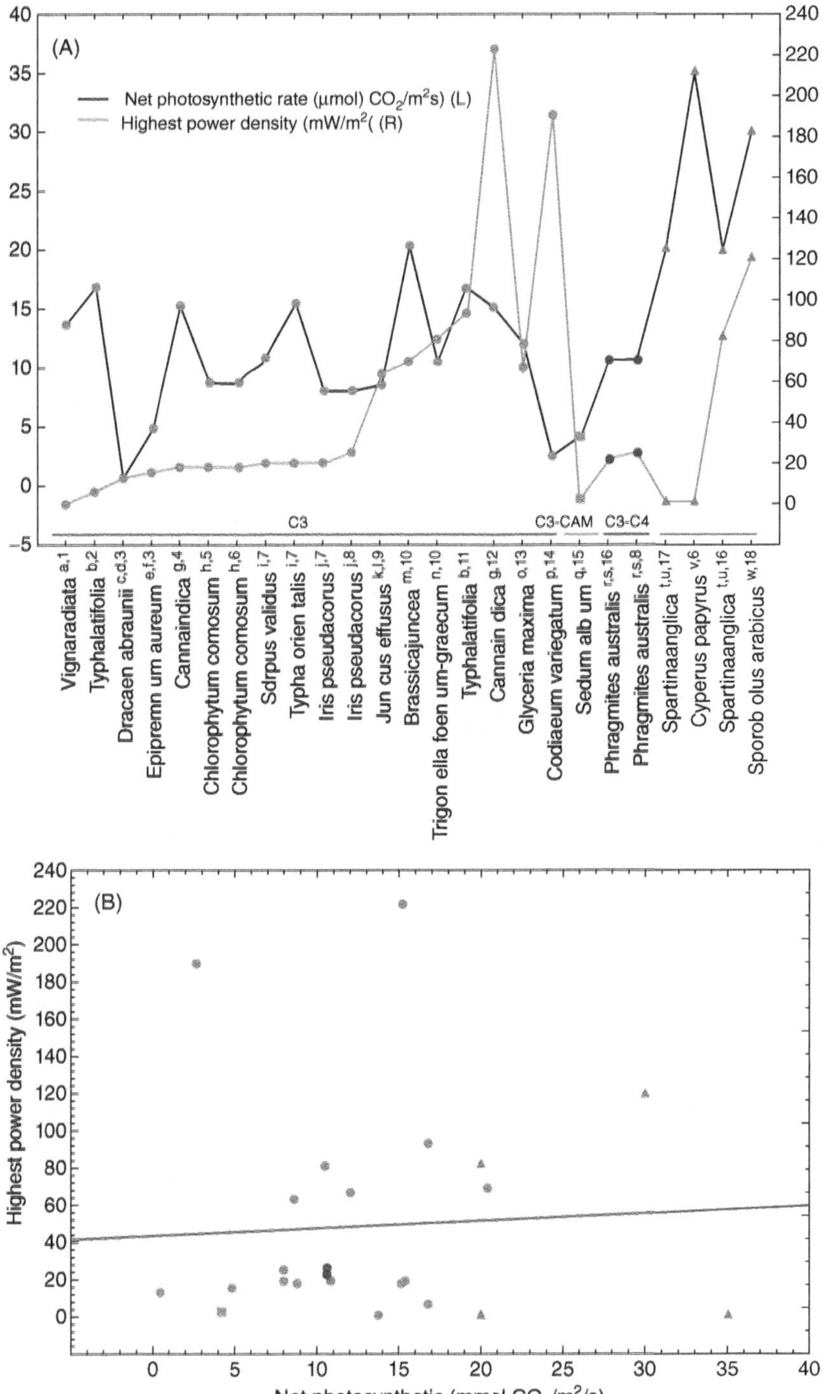
(A)
Net photosynthetic rate (μmol) CO2/m2s) (L)
Highest power density (mW/m2((R)
C3
C3-CAM
C3-C4
Vignaradiata a,1
Typhalatifolia b,2
Dracaen abraunii c,d,3
Epipremn um aureum e,f,3
Cannaindica g,4
Chlorophytum comosum h,5
Chlorophytum comosum h,6
Sdrpus validus i,7
Typha orien talis i,7
Iris pseudacorus j,7
Iris pseudacorus j,8
Jun cus effusus k,l,9
Brassicajuncea m,10
Trigon ella foen um-graecum n,10
Typhalatifolia b,11
Cannain dica g,12
Glyceria maxima o,13
Codiaeum variegatum p,14
Sedum alb um q,15
Phragmites australis r,s,16
Phragmites australis r,s,8
Spartinaanglica t,u,17
Cyperus papyrus v,6
Spartinaanglica t,u,16
Sporob olus arabicus w,18
(B)
Highest power density (mW/m2)
Net photosynthetic (mmol CO2/m2/s)

Figure 12.1 (A) Ratio of net photosynthesis rate and maximum recorded energy output for 25 plant species used in plant microbial fuel cells. On the *X* axis, the name of the species and the photosynthetic route reported in specialized literature on plant physiology are presented; on the left *Y* axis (L), the net rate of photosynthesis obtained by literature review is presented. The right *Y* axis (R) presents the maximum power density values reported in various studies. Letter superscripts refer to citations related to physiological traits (photosynthetic pathway and net photosynthesis rate): a) Fariduddin et al. (Fariduddin et al., 2003); b) McNaughton and Fullem, (McNaughton & Fullem, 1970); c) Gupta et al. (Gupta et al., 2016); d) Ju et al. (Ju et al., 2019); e) Gong et al. (Gong et al., 2019); f) Lee et al. (Lee et al., 2021); g) Imai, (Imai, 2008); h) Gupta et al. (Gupta et al., 2016); i) Xu et al. (Xu et al., 2018); j) Duan (Duan et al., 2022); k) Arslan et al. (Arslan et al., 2019); l) Carroll et al. (Mann & Wetzel, 1999); m) Ruhil et al. (Ruhil et al., 2015); n) Jain et al. (Jain et al., 2007); o) Mejdová et al. (Mejdová et al., 2021); p) Valdez-Hernández et al. (Valdez-Hernández et al., 2022); q) Habibi and Hajiboland, (Habibi & Hajiboland, 2012); r) Srivastava et al. (Srivastava et al., 2014); s) Gorai et al. (Gorai et al., 2011); t) Dunn et al. (Dunn et al., 1987); u) Matsuba et al. (Matsuba et al., 1997); v) Jones (Jones, 1987); w) Tisarum et al. (Tisarum et al., 2022). The superscripts in number refer to the citations related to the maximum energy power: 1) Pamintuan and Sanchez (Pamintuan & Sanchez, 2019); 2) Oon et al. (Oon et al., 2015); 3) Sarma and Mohanty (Sarma & Mohanty, 2018); 4) Lu et al. (Lu et al., 2015); 5) Md Khudzari et al. (Md Khudzari et al., 2018); 6) Azri et al. (Azri et al., 2018); 7) Di et al. (Di et al., 2020); 8) Yang et al. (Yang et al., 2020); 9) Guadarrama-Perez et al. (Guadarrama-Pérez et al., 2022); 10) Sophia and Sreeja, (Sophia & Sreeja, 2017); 11) Oon et al. (Oon et al., 2016); 12) Srivastava et al. (Srivastava et al., 2015); 13) Strik et al. (Strik et al., 2008); 14) Valdez-Hernandez et al. (Valdez-Hernández et al., 2022); 15) Tapia et al. (Tapia et al., 2017); 16) Wetser et al. (Wetser et al., 2017); 17) Sudirjo et al. (Sudirjo et al., 2019); 18) Gilani et al. (Gilani et al., 2016). (B) Spearman correlation between the highest power density versus the net rate of photosynthesis ($R = 0.089875, p = 0.669209$).

be considered that the environmental conditions in which the system will be established satisfy the basic environmental requirements of the species: active photosynthetic radiation (if the PMFCs will be established in outdoor or indoor environments), soil moisture, oxygen present (if the PMFCs will be established in flooded systems), nutrients (macro and micronutrients from the soil or sediments).

The type and level of development of the root system and of the above-ground photosynthetic surface of plant are an important prognostic factor of the PMFC performance. PMFC based on sedge *C. hirta* with the largest accumulated total dry leaf/stem and root mass and the rhizome-like and developed fibrous root system was shown the highest power output compared to other biosystems (Rusyn et al., 2022).

In addition, it is essential to integrate measurements of physiological traits in plants subjected to energy harvesting. The net photosynthetic rate indicates the amount of CO_2 fixed and is related to the yield and biomass gain of the plant; the concentration of chlorophyll will allow us to evaluate the nutritional status of the plant; Chlorophyll fluorescence can be used to reveal response mechanisms, quantification of stress responses, and identification of certain pollutants and their sources. The three mentioned physiological traits have the advantage of being nondestructive measurements that can be continuously measured in PMFC plants without causing damage to the system; knowing the physiological state of plants and its changes during the operation of PMFCs will help us answer questions about what, how, and how much plant physiology affects the PMFCs system and vice versa.

12.3 Understanding Energy Harvesting

Energy harvesting is the process by which small amounts of energy are gleaned from ambient (Dixit & Rangra, 2017), i.e. external sources such as solar power (Psomopoulos, 2013), thermal energy (Alhawari et al. 2013), wind energy, vibrations (Martínez-Cisneros et al., 2021), salinity gradients (Hsu et al., 2021), kinetic energy (Zhu & Beeby, 2011), and bioenergy. Multiple sources of natural can be harvested by a harvester device and converted to electrical energy, and each power provides a low power density.

This energy is captured by energy-harvesting devices and stored to power low-consumption devices small. Energy harvesting technologies have greatly improved in terms of their efficiencies, providing continuous power output from different energy sources, in addition, recent low-energy devices are designed to be extremely energy efficient, enabling them to operate using the harvested energy, for instance, wireless sensor networks (Castillo-Atoche et al., 2021; Osorio-de-la-Rosa et al., 2021), IoT devices (Osorio-de-la-Rosa et al., 2019), telecommunications (Ku et al. 2016), health care (Wang et al., 2021), Internet of Energy (IoE) (Wang et al., 2018), and smart city (Ayala-Ruiz et al., 2019).

Three components more common in energy harvesting devices are source ambiental, harvesting architecture, and load (devices to energizer) (Sanislav et al., 2021). So, it is possible to consider two great general architectures of the energy harvesting systems: harvest-load (HL) and harvest-store-load (HSL) (Sudevalayam & Kulkarni, 2011). In the HL model, the harvested energy is immediately consumed by the load. This model is appropriate when a steady supply of harvested energy can be guaranteed, and a battery-free circuit is desired. The major problem with this architecture is that, due to the possible random nature of the energy source, when the harvested energy is less than the minimum operational energy required by the load device, the device is disabled or not on.

In the HSL architecture, the harvested energy is first transferred to a storage element such as a supercapacitor or rechargeable battery. The consumer then accesses the harvested energy from the storage. By this method, it can ensure a steady supply of energy. Figure 12.2 shows a block diagram of the HSL architecture. Its power

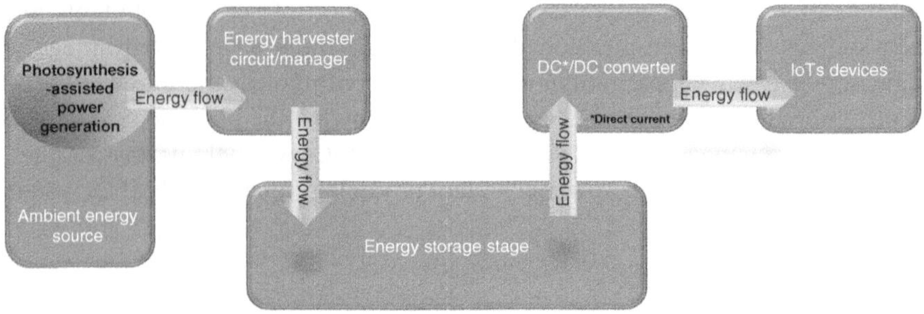

Figure 12.2 Block diagram of main components of HSL architecture, in this case, a photosynthesis assisted is considered. Source: Adapted from Osorio-de-la-Rosa et al. (Osorio-de-la-Rosa et al., 2019).

generation part includes a specific energy harvesting circuit (EHC) that converts ambient energy to direct current (DC) energy, power management units that increase the efficiency of power generation-use, and storage stage, which can store energy and power the electronics loads. All components, their associated characteristics, and actions are the subjects of research, which target the manufacturing of energy-autonomous wireless sensors when deployed in IoT devices, low-consumption devices.

STMicroelectronics®, Texas Instrument®, and Nordic Semiconductors® have developed several integrated circuits designed to work as energy harvesting boards.

12.4 Low-consumption Electronic Devices for IoT Applications

A low-consumption electronic device is made up of semiconductor materials. These materials have the ability that under certain conditions can act as either an electric conductor or an insulator. The main application of semiconductor materials is creating semiconductor devices, these devices can conduct electric currents in the solid state. One of the very commonly used semiconductor devices is the metal-oxide-semiconductor field-effect (MOSFET) transistors, which are used both for fast switching and for current amplification. Both MOSFET and semiconductor silicon serve as material for microchip or integrated circuits (ICs) development. The two main ICs are the logic chip (central processing unit, PC) and the memory chip (working memory or NAND Flash, DRAM). Logic chips process information to help electronic devices to complete their tasks.

Figure 12.3 shows the transition in semiconductor devices technology, together with a summary history of semiconductors. Since the invention of transistors in 1947, semiconductor devices have developed successfully for over 70.

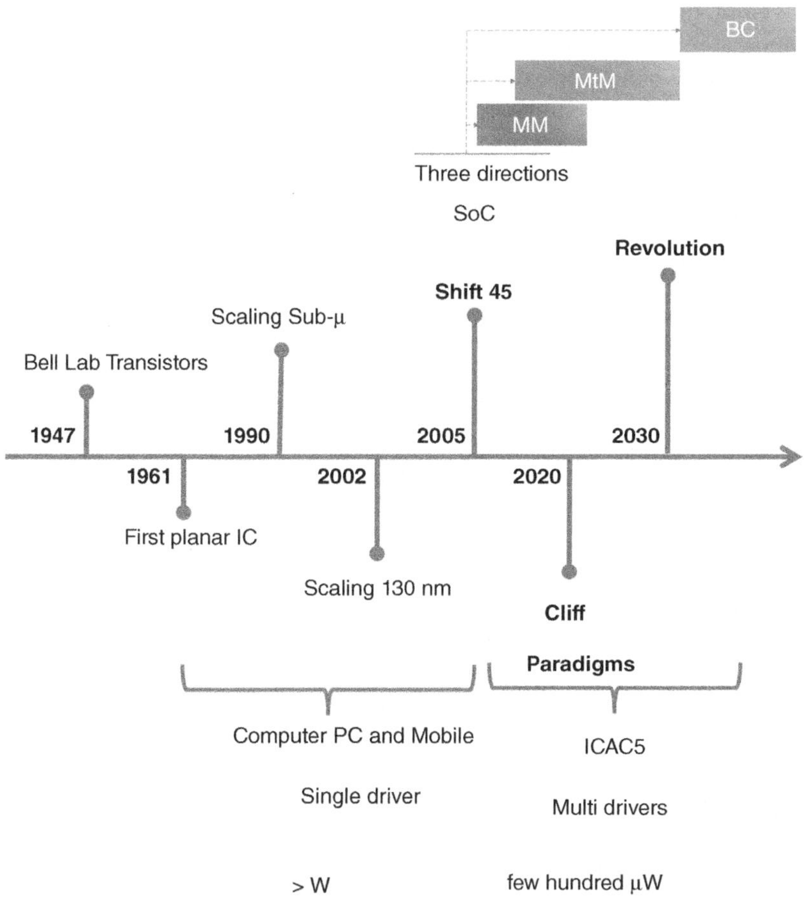

Figure 12.3 Summary history of semiconductors, Paradigms, and power consumption of technology. Source: Adapted from (Tsujimura, 2020).

This development followed Moore's and Dennard's scaling laws, established in 1965 and 1974, respectively. Miniaturization came next, with transistors shrinking to 130 nm.

By 2005, the semiconductor devices had entered an era of Paradigm Shift 45, so, the miniaturization allowed to unify the integration capacity of different stages within the same IC, i.e. the semiconductor devices, which formerly were driven by single-driver technology, including PCs and mobile phones, now, the devices are driven by the multi-driver devices represented by IoT, cloud, AI, cars, and 5G (ICAC5). Moreover, miniaturization has contributed to the development of ICs more its power consumption more efficient, consumption bellow the watts (see Figure 12.3), or has even managed to convert small amounts of energy into energy available for self.

In this point, the device has been developed in three directions (see Figure 12.3): More Moore (MM), More than Moore (MtM), and Beyond CMOS (BC) (Tsujimura, 2020). MM direction indicates successful scaling to 7 nm, allowing different stages to be coupled on the same chip. In MtM, the system on chip (SoC) has been developed, with theses which managed to develop CI for different areas such as embedded systems are small computers that form part of a larger device. Embedded systems that we commonly use include central heating systems, digital watches, fitness trackers, engine management systems in vehicles, and low-power consumption sensor to IoT applications.

By 2020, the technology development became Paradigm Cliff and by 2030 a Paradigm Revolution is expected, such as new transistors, spin transistors, and quantum computers. So, ultra-low-power consumption (up to tens of microamps) is a very important requirement to operate with small energy sources (to reduce size) and IoT device applications. Opening the vision of the Industry 4.0 to the introduction of novel communication and computation technologies.

12.4.1 Bluetooth® Low Energy

12.4.1.1 Bluetooth 4.0

BLE was developed by the Bluetooth Special Interest Group (SIG), this is innovative, and emerging technology device that operates under a standard for short-range, and low-cost wireless communication that uses radio technology (Gomez et al., 2012). This Wireless Personal Area Network (WPAN) technology is now an IEEE standard under the 802.15WPANs denomination.

The main kindness on top of the BLE protocol is the low-energy consumption of power from the device into which it is integrated. The BLE market has been introduced as an integral part of Bluetooth V4.0, transforming it into an excellent choice for smart devices increasing battery levels, and mainly fueling the new era of low-power IoTs (Palattella et al., 2016).

The BLE has become a pervasive wireless technology that is embedded in numerous types of personal devices and in diverse applications, such as in the medical field for e-health applications (Fan et al., 2021), blood flowmeter (Kuwabara et al., 2014), domotics for smart houses (Collotta & Pau, 2015), Hand Hygiene Tracking Sensor System for Prevention of Infections (Fan et al., 2021), and transmitting Internet Protocol Version 6 (IPv6) packets over Low power Wireless Personal Area Networks (6LowPAN) in health monitoring application (Tabish et al., 2013). Coupled with this, smart tags and asset tracking, and Proximity Marketing are exclusive applications of BLE that were made feasible by the low energy consumption of BLE modules and chipsets, since this protocol can locate misplaced personal equipment (Industryarc n.d.).

In ad-hoc mode, the connection via Bluetooth of various devices is called a piconet. A cluster of maximum of eight devices constitutes a piconet. In this area, a master device is designated and the rest as slaves. A point-to-point link is the arrangement of a single slave and master devices. When there is a single master controller linked to a maximum of seven active slaves, it has a point-to-multipoint. When two piconets are connected through a gateway or bridge Bluetooth device, it has the form of a scatter net.

BLE technology operates in the 2.4 GHz Industrial, Scientific, and Medical (ISM) frequency band, transmitting data over 37 channels with 2 MHz spacing, the bit rate is 1 Mbit/s, and the maximum transmit power is 10 mW, peak current consumption <15 mA, nominal maximum range <100 m (<330 ft), wake latency (from a non-connected state) typically 6 ms, and minimum total time to send data 3 ms. BLE is compatible with several provider ICs vendors, allowing single-mode chips to be more energy efficient. Therefore, BLE has an energy-efficient method of keeping connectivity while minimizing active radio usage.

Figure 12.4 shows the three primary components of the BLE protocol: (i) application, (ii) host, and (iii) controller and inside each of these blocks are distinct layers. The controller allows Bluetooth devices to locate each other and manage

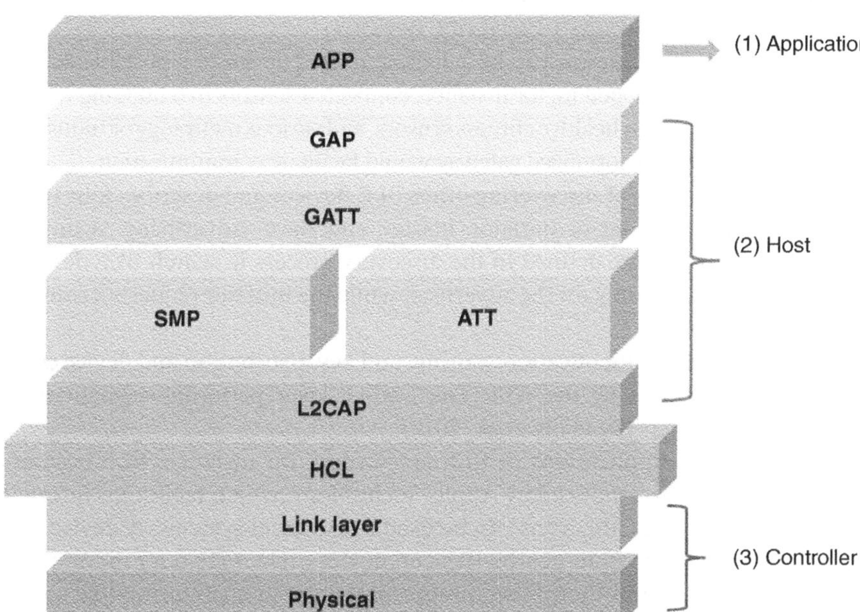

Figure 12.4 BLE protocol stack.

physical and logical links with higher-layer protocols and applications. The controller is implemented as a small System-on-Chip (SOC) with an integrated radio, it comprises the Physical Layer (PHY) and the Link Layer (LL).

The Host layer includes third-party and industry-standard protocols and Bluetooth SIG-developed protocols. The Host includes upper-layer functionalities, including the Logical Link Control and Adaptation Protocol (L2CAP), the Attribute Protocol (ATT), the Generic Attribute Profile (GATT), the Security Manager Protocol (SMP), and the Generic Access Profile (GAP). Communication between the Host and the Controller is standardized as the Host Controller Interface (HCI). Applications reside above the Host layer.

Data to be transmitted is kept in a GATT server database or client. The application dictates whether the GATT server/client is kept on the master or the client node. GATT specifies a framework for communication between GATT server and client databases.

Based on the hierarchy, the information is organized on the GATT server. Attributes may be defined as small, addressable GATT data entities. In GATT-based devices and services, devices are connected to a BLE device called a gateway. The devices implement one or more GATT-based services, offered to the GATT-based devices. The gateway interprets the service/profile and then exposes an "xyz" Internet Application Applications (APP) that sends and receives data from the Internet to the GATT-based BLE device.

In this use case, the gateway can be a dedicated fixed device or a portable device. The use case can be applied in, for instance, connected sensors in a building (home automation), body-worn health or fitness sensors, and various metrology or Industrial 4.0 devices, where the centralized gateway would facilitate communication.

GAP identifies ways of discovering other BLE devices and describes four functions: advertiser, scanner or initiator, master, and slave. Advertising, scanning, and initiating modes are defined in the discovery process to search BLE devices. The scanner searches only for the advertiser, while the initiator requests a connection with the advertiser.

SMP implements authentication, pairing, and key distribution and device communication is defined by the ATT. The L2CAP layer performs segmentation; reassembly of packets (Zeadally et al., 2019).

In terms of the Applications of BLE are several, for instance, BLE beacons are called iBeacon, introduced by Apple® technology, which facilitates internal positioning of devices. iBeacons can be used by Android systems. A device can get push notifications from nearby iBeacon devices. The IoTs refer to uniquely identifiable objects and their virtual representations in an Internet-like structure. iBeacon can register the nearness of an object and send an unambiguous universal identifier token from a compatible application to transform it into a physical location. The beacons come in different formats, including small coin cell-powered devices, flash drives, and software in embedded devices such as smart-bulbs, and so on (Zeadally et al., 2019).

12.4.1.2 Bluetooth 5.0

The adoption of the IoTs has contributed to an accelerated explosion in the number of connected devices, reaching 13.1 billion connected devices by 2022's end. Due to this, the number of IoT applications has significantly increased, including covering a wide range of applications. In this sense, Bluetooth 5.0 technology has been lately introduced for IoT purpose, this technology remains low-consumption and improvement over in its messaging capacity, achieving higher transferred speed (8 more times than BLE 4.0), increased transmission range (4 more times than BLE 4.0), improved Speed (two times the transmission speed of Bluetooth 4.0), and Bluetooth 5.0 can be deployed massively in many variety topologies. The architecture of Bluetooth 5.0 IoT can be divided into six layers as schematic in Figure 12.5.

Figure 12.5 shows the hardware Layer (orange stack), which is the bottom layer comprising the hardware components of IoT, such as sensors or sensors in smartphones, weather sensors, environment sensors, and other physical components. The hardware layer includes the abstraction to leverage information about its periphery to the above layer the microcontroller layer.

Microcontroller Layer (see Figure 12.5, yellow stack), this layer comprises the microcontrollers that are required for automatically controlled devices such as embedded systems. Few instances of what layer comprises are the Raspberry® Pi, AT Mega®, etc.

Bluetooth Connectivity Layer (Figure 12.5, navy blue), this layer contains the wireless standard for exchange, Bluetooth which uses a variety of protocols. The Bluetooth stack is divided into two segments. First, the controller – the controller is implemented in low-cost silicon devices that contain a time-critical radio interface that is the Bluetooth radio and a microprocessor. The second segment is the Host which deals with high-level data, the host is implemented as a package on top of an operating system or a part of an operating system itself.

However, the host stack also comprises the protocol for communication between the host and the controller. This stack is similar as Bluetooth 4.0 schematized in Figure 12.5 as Bluetooth Host Controller, where HCI is a pseudo protocol for standardized communication between the two stacks. LL is used to control the connections between the devices, which include creation, modification, and release of logical transports or logical links. As well as, to query any device capabilities and power control. The final protocol smart objects are creating a significant improvement in IoT for intelligent devices.

Connectivity layer are two pathways that can be taken in this layer, Internet Connectivity or Smart Phone OS. Internet connectivity is the central part of the

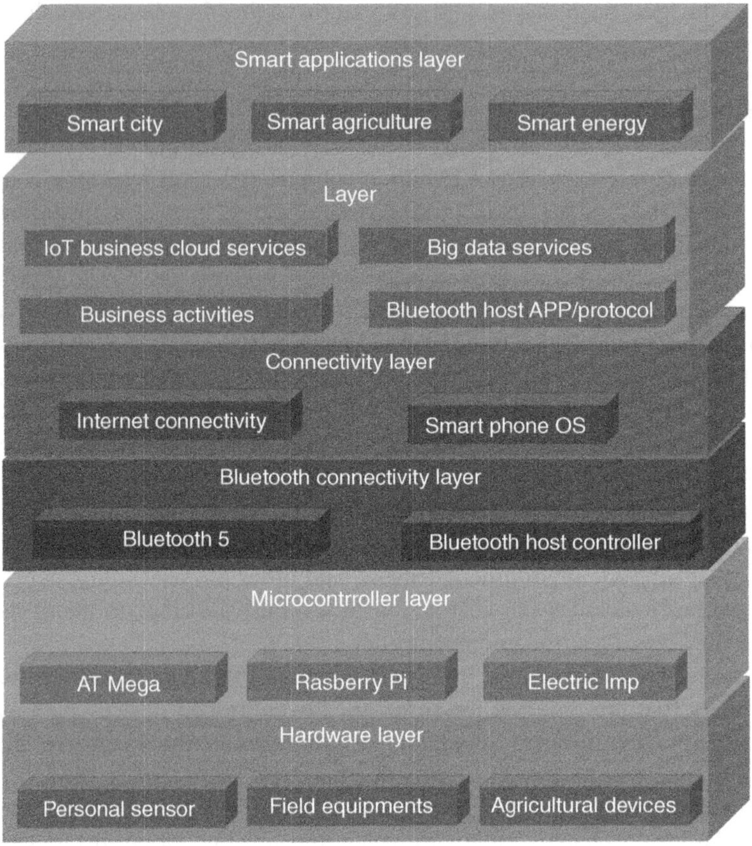

Figure 12.5 Bluetooth 5 protocol stack.

whole stack. IoT-specific communication protocols are added here for energy efficiency, resource constraints, and lightweight information processing. Message Queuing Telemetry Transport (MQTT), Constrained Application Protocol (CoAP), and Advanced Message Queuing Protocol (AMQP) protocols perform the following tasks: publish/subscribe messaging, multicast support, real-time messaging, packet-switched networking, message queue for middleware, lightweight local automation, and direct addressing publish/subscribe based communication for real-time embedded systems.

The other pathway is the Smart Phone OS: The WPAN technology which is designed for application in health care, fitness, security, and home automation. Current mobile devices are always released with hardware and software support for Bluetooth. There are already Bluetooth Profiles present in mobile devices, such as the Blood Pressure Profile (BLP), and many other Fitness profiles.

Infrastructure layer (Figure 12.5, gray stack) comprises four different compositions of IoT business cloud services, Big Data Services, Bluetooth Host APP/protocol, and Business Activities. The IoT Business Cloud – an abstraction for business-specific services. Various business transactions and activities are served here by Software as a Service (SaaS), Platform as a Service (PaaS), and Infrastructure as a Service (IaaS) cloud solutions. This modular architecture, as well as packet business APPs, eases the performance of any business-related operations. Parallelly also performs resource definition, abstraction, orchestration, and optimization of its external ecosystem. These business clouds reduce the organization's overhead of operational activities by providing a common layered approach with necessary supports provided. In Bluetooth Host, APP/Protocol has the host and controller stacks which are divided into various subcategories. The GAP defines the topology being used by the Bluetooth network stack. The one-to-one, one-to-many, or many-to-many mechanisms which are means to connect to the other devices fall under the mechanism of GAP. During a one-to-one connection, there should be an explicit connection and here handshake is required to transfer data. The GATT describes the detailed description of the transfer attributes once the devices are connected. The Object Exchange is a communication protocol that facilitates the exchange of binary objects between devices. For example, file transfers require simple data exchange. The OBEX is bound to Radio Frequency Communication, which is a set of transport protocol that is commonly used for its widespread support and publicly available APP on most operating systems. The low energy attribute protocol is adapted for BLE since this protocol allows the client to read and write certain attributes with a lot more ease. The Big Data Services make it possible for Big Data and IoT to work in conjunction. These services can be used for providing optimizations, statistical analysis, predictions, etc. By using these services, the network disks and computer power are impacted. The stack for big data processing must be done in consideration of the influx of data that IoT will deliver.

Finally, Application Layer (Figure 12.5, green stack) is designed to disseminate the user experience via the software in applications that are performed over Bluetooth Connectivity. Bluetooth 5.0 with IoT can help solve numerous problems in fields such as agriculture precision, sensors in smartphones, smart cities, and other more.

12.5 Precision Agriculture

At the end of 2022, the world population reached 8 billion inhabitants, and it is expected that by 2050 it will reach 9.7 billion. This population increase will demand a greater production of food in a smaller agricultural area, so it is expected that an improved production, management, and maintenance of agricultural food. In this sense, PA emerges as a science that improves crop yields and helps management decisions using analysis tools and low-energy consumption high-tech sensors.

PA is yet area emerging and gaining importance as an application of IoT to give the smart technological solution at agriculture processes, since it is required monitoring of soil and weather conditions for better decision making. IoT adopted for PA uses wireless communication technologies for the collection of agricultural parameters.

The major application areas of IoT in agriculture are identified as monitoring, control, and tracking. About 70% of the study of IoT in agriculture is focused on, monitoring, 25% on control, and only 5% of the total work on PA is focused on tracking. Major agricultural applications or activities being supported by IoT are monitoring of temperature, humidity, fertilization, disease development, animals, water, irrigation, and monitoring of air conditions.

The technical specifications of every technology have a great impact on the utilization of specific communication technology in IoT applications. For instance, communication range decides how far the radio devices (BLE, LoRa) can be placed for monitoring and control, power efficiency decides the operation period (energy harvesting stage).

There are also more special types of energy harvesting. For example, the use of PMFC for generating the energy required for powering wireless sensors has been investigated. These devices generate a power between 1.4 W during a day The plant-as-source power approach is meant to simplify the deployment of devices related in PA applications, where the measurement of soil moisture, ambient humidity, and transmitted using BLE technology to terminal devices such as smartphones (Castillo-Atoche et al., 2021).

In terms of the radio devices, many applications are developed using BLE or Bluetooth 5, it is preferred for portable devices with a communication distance of up to 100 m. The main advantages of using Bluetooth are low power, availability, and ease of use by countrymen. LoRa with specifications such as low power consumption, and long-distance communication range is also deployed in many agriculture applications such as smart irrigation, green area monitoring, fertilization improvement, and many more. The control and monitoring are done through a mobile application developed using MIT app inventor. LoRa offers a communication range of up to 5 km it works on an unlicensed frequency band with good power efficiency, BLE up to 10 m. LoRa is the best contender if the cost of deployment for agricultural applications is considered, thereafter comes the BLE.

The most common low-energy consumption wireless communication technologies used in agriculture are Bluetooth®, and LoRa® is the most suitable technology for agriculture applications, falls in short-distance communication technology, and offers the best power efficiency.

The data collection and its management through the underlying communications technologies and protocols, mainly application layer protocols of IoT architecture, are an important aspect of IoT. In terms of power consumption, CoAP and MQTT are the protocols with the least consumption present. When the latency is compared, irrespective of Quality of Service (QoS), CoAP showed less latency than MQTT. Latency of MQTT shows variance with change in QoS factor. As for the deployment of protocols in smart farming is considered, MQTT and Representational state transfer (REST) protocols are used for simple and small applications such as soil water content measurement, controlling of water pump, and measurement of atmospheric pressure, relative humidity, etc. For smart farming prototypes employed in medium and large applications MQTT, CoAP, REST, and sometimes MQTT and CoAP combinedly are used.

12.6 Conclusion and Future Perspectives

The development of low energy consumption devices for use in IoTs is related to the development of semiconductor devices, where the development of SoC, chips located in the MtM direction, allowing greater integration of more functionalities, such as sensors, radios: Bluetooth-LoRa, low energy harvesting board to give more support and autonomous energy to the IoT technology.

By improving the efficiency of the energy harvesting card, it is possible to harvest surrounding energy to provide continuous power output to power IoT devices. In some cases, it is possible to have a battery-free circuit that guarantees energy only through the environmental source.

The trends regarding the PMFC will be the monitoring of the photosynthetic path of the plant in situ with the energy harvesting process. Since it is essential to integrate measurements of physiological traits in plants subject to energy harvesting, in order to carry out the choice of plant species that will be used in the PMFC. Furthermore, knowing the physiological state of plants and their changes during PMFCs operation will help us to answer questions about what, how, and how much plant physiology affects the PMFCs system and vice versa.

The trend in the semiconductor technology, driven by multiple applications, by ICAC5, as well as PCs and mobile phones, will yet more grow explosively.

In this sense, the emerging trend is in the development of energy-harvesting ICs designed for each type of sources that generate energy through photosynthesis (ultra-low energy).

In terms of radio, the trend is to develop devices focused on the IoTs, there will be an increasing tendency to use short-range wireless connectivity for numerous types of devices powered by alternative sources. BLE aims to address the requirements for IoT applications. Such as range, more efficient transfer of large amounts of data, transmission speed, increased transmission power, and low power consumption.

These devices could contribute to future applications in PA where the plants are connected to the monitoring device as a power source through a collector card that supports low-consumption sensors and controllers, and thus can monitor the growth of plants in real time and control of conditions related to irrigation and lighting.

References

Alhawari, M., Baker, M., Hani, S., and Mohammed, I. (2013). A survey of thermal energy harvesting techniques and interface circuitry. In: *2013 IEEE 20th International Conference on Electronics, Circuits, and Systems (ICECS), Abu Dhabi, United Arab Emirates*, 381–384. https://doi.org/10.1109/ICECS.2013.6815434.

Apollon, W., Luna-Maldonado, A.I., Kamaraj, S.K. et al. (2021). Progress and recent trends in photosynthetic assisted microbial fuel cells: a review. *Biomass and Bioenergy* 148: 106028.

Apollon, W., Rusyn, I., González-Gamboa, N. et al. (2022). Improvement of zero waste sustainable recovery using microbial energy generation systems: a comprehensive review. *Science of the Total Environment* 817: 153055. https://doi.org/10.1016/j.scitotenv.2022.153055.

Arslan, M., Devisetty, U.K., Porsch, M. et al. (2019). RNA-Seq analysis of soft rush (*Juncus effusus*): transcriptome sequencing, de novo assembly, annotation, and polymorphism identification. *BMC Genomics* 20 (1): 489. https://doi.org/10.1186/s12864-019-5886-8.

Ayala-Ruiz, D., Castillo Atoche, A., Ruiz-Ibarra, E. et al. (2019). A self-powered PMFC-based wireless sensor node for smart city applications. *Wireless Communications and Mobile Computing* 2019, 8986302: 10. https://doi.org/10.1155/2019/8986302.

Azri, Y.M., Tou, I., Sadi, M., and Benhabyles, L. (2018). Bioelectricity generation from three ornamental plants: *Chlorophytum comosum*, *Chasmanthe floribunda* and *Papyrus diffusus*. *International Journal of Green Energy* 15.

Bloom, A.J. and Lancaster, K.M. (2018). Manganese binding to Rubisco could drive a photorespiratory pathway that increases the energy efficiency of photosynthesis. *Nature Plants* 4: 414–422. https://doi.org/10.1038/s41477-018-0191-0.

Castillo-Atoche, A., Vázquez-Castillo, J., Osorio-de-la-Rosa, E., Heredia-Lozano, J.C., Avilés-Viñas, J.F., Cetina, R.Q., and Estrada-López, J.J. (2021). An Energy-Saving Data Statistics-Driven Management Technique for Bio-Powered Indoor Wireless Sensor Nodes. *IEEE Transactions on Instrumentation and Measurement* 70: 10. http://doi.org/10.1109/TIM.2021.3063187.

Collotta, M. and Pau, G. (2015). A solution based on bluetooth low energy for smart home energy management. *Energies* 8: 11916–11938.

Di, L., Li, Y., Nie, L. et al. (2020). Influence of plant radial oxygen loss in constructed wetland combined with microbial fuel cell on nitrobenzene removal from aqueous solution. *Journal of Hazardous Materials* 394: 122542.

Dixit, N.K. and Rangra, K.J. (2017). A survey of energy harvesting technologies. In: *2017 International Conference on Innovations in Control, Communication and Information Systems (ICICCI)*, 1–7. https://doi.org/10.1109/ICICCIS.2017.8660812.

Duan, R., Xing, D., Chen, T., and Wu, Y. (2022). Effects of different inorganic nitrogen sources of *Iris pseudacorus* and *Iris japonica* on energy distribution, nitrogen, and phosphorus removal. *HortScience* 57 (6): 698–707.

Dunn, R., Thomas, S.M., Keyss, A.J., and Long, S.P. (1987). A comparison of the growth of the C4 grass *Spartina anglica* with the C3 grass *Lolium perenne* at different temperatures. *Journal of Experimental Botany* 38 (3): 433–441.

Fan, W., Qiu, C., Taiyang, W., and Yuce, M.R. (2021). Edge-based hybrid system implementation for long-range safety and healthcare IoT applications. *IEEE Internet of Things Journal* 8 (12): 9970–9980. https://doi.org/10.1109/JIOT.2021.3050445.

Fan, W., Taiyang, Z., Morfuni, D.C. et al. (2021). An autonomous hand hygiene tracking sensor system for prevention of hospital associated infections. *IEEE Sensors Journal* 13 (21): 4308–14319. https://doi.org/10.1109/JSEN.2020.3041331.

Fariduddin, Q., Ahmad, A., and Hayat, S. (2003). Photosynthetic response of *Vigna radiata* to pre-sowing seed treatment with 28-homobrassinolide. *Photosynthetica* 41 (2): 307–310.

Gilani, S.R., Yaseen, A., Zaidi, S.R.A. et al. (2016). Photocurrent generation through plant microbial fuel cell by varying electrode materials. *Journal of The Chemical Society of Pakistan* 38.

Gomez, C., Oller, J., and Paradells, J. (2012). Overview and evaluation of bluetooth low energy: an emerging low-power wireless technology. *Sensors (Basel)* 12 (9): 11734–11753. https://doi.org/10.3390/s120911734. PMCID: PMC3478807.

Gong, Y., Zhou, T., Wang, P. et al. (2019). Fundamentals of ornamental plants in removing benzene in indoor air. *Atmosphere* 10 (4): 221.

Gorai, M., Ennajeh, M., Khemira, H., and Neffati, M. (2011). Influence of NaCl-salinity on growth, photosynthesis, water relations and solute accumulation in *Phragmites australis*. *Acta Physiologiae Plantarum* 33 (3): 963–971.

Guadarrama-Pérez, O., Moeller-Chávez, G.E., Bustos-Terrones, V. et al. (2022). Identification of sugars as root exudates of the macrophyte species *Juncus effusus* and *Philodendron cordatum* in constructed wetland-microbial fuel cells during bioelectricity production. *Environmental Technology* https://doi.org/10.1080/09593330.2022.2121180.

Gupta, S.M., Agarwal, A., Dev, B. et al. (2016). Assessment of photosynthetic potential of indoor plants under cold stress. *Photosynthetica* 54 (1): 138–142.

Habibi, G. and Hajiboland, R. (2012). Comparison of photosynthesis and antioxidative protection in *Sedum album* and *Sedum stoloniferum* (Crassulaceae) under water stress. *Photosynthetica* 50 (4): 508–518.

He, J., Xin, X., Pei, Z. et al. (2021). Microbial profiles associated improving bioelectricity generation from sludge fermentation liquid via microbial fuel cells with adding fruit waste extracts. *Bioresource Technology* 337: 125452.

Hsu, W.-S., Preet, A., Lin, T.-Y., and Lin, T.-E. (2021). Miniaturized salinity gradient energy harvesting devices. *Molecules* 26: 5469. https://doi.org/10.3390/molecules26185469.

Imai, K. (2008). Edible canna: a prospective plant resource from South America. *Japanese Journal of Plant Science* 2 (2): 46–53.

Industryarc (n.d.) https://www.industryarc.com/ (accessed on 04 December 2022).

IPCC (2021). *Climate Change 2021: The Physical Science Basis*. Contribution of Working Group I to the Sixth Assessment Report of the Intergovernmental Panel on Climate Change (ed. V. Masson-Delmotte, P. Zhai, A. Pirani, et al.). Cambridge University Press In Press.

Jadhav, D.A., Ghosal, D., Chendake, A.D. et al. (2021). Plant microbial fuel cell as a biomass conversion technology for sustainable development. In: *Catalysis for Clean Energy and Environmental Sustainability* (ed. K.K. Pant, S.K. Gupta, and E. Ahmad). Cham: Springer https://doi.org/10.1007/978-3-030-65017-9_5.

Jain, V., Pal, M., Raj, A. et al. (2007). Photosynthesis and nutrient composition of spinach and fenugreek grown under elevated carbon dioxide concentration. *Biologia Plantarum* 51: 559–562. https://doi.org/10.1007/s10535-007-0122-9.

Jiang, J., Wang, H., Zhang, S. et al. (2021). The influence of external resistance on the performance of microbial fuel cell and the removal of sulfamethoxazole wastewater. *Bioresource Technology* 336: 125308.

Jones, M.B. (1987). The photosynthetic characteristics of papyrus in a tropical swamp. *Oecologia* 71: 355–359.

Ju, J.H., Son, H.M., Kim, W.T., and Yoon, Y.H. (2019). Effects of activated carbon on growth and physical responses of indoor plant *Dracaena braunii* to alleviate salt-induced stress in water culture. *Journal of Environmental Science International* 28 (3): 321–328.

Jung, K., K. and Park, B.-J. (2021). Efficiency of *Spathiphyllum* spp. as a plant-microbial fuel cell. *Ornamental Horticulture* 27: 173–182.

Kabutey, F.T., Zhao, Q., Wei, L. et al. (2019). An overview of plant microbial fuel cells (PMFCs): configurations and applications. *Renewable and Sustainable Energy Reviews* 110: 402–414. https://doi.org/10.1016/j.rser.2019.05.016.

Ku, M.L., Li, W., Chen, Y., and Liu, K.J.R. (2016). Advances in energy harvesting communications: past, present, and future challenges. *IEEE Communications Surveys & Tutorials* 18 (2): 1384–1412, 2nd Quart. https://doi.org/10.1109/COMST.2015.2497324.

Kuwabara, K., Higuchi, Y., Ogasawara, T. et al. (2014). Wearable blood flowmeter appcessory with low-power laser Doppler signal processing for daily-life healthcare monitoring. In: *Proceedings of the 2014 36th Annual International Conference of the IEEE Engineering in Medicine and Biology Society, Chicago, IL, USA, 26–30 August 2014*, 6274–6277.

Lee, J.K., Kim, D.Y., Park, S.H. et al. (2021). Particulate matter (PM) adsorption and leaf characteristics of ornamental sweet potato (*Ipomoea batatas* L.) cultivars and two common indoor plants (*Hedera helix* L. and *Epipremnum aureum* Lindl. & Andre). *Horticulturae* 8 (1): 26.

Long, S., Zhao, L., Chen, J. et al. (2021). Tetracycline inhibition and transformation in microbial fuel cell systems: performance, transformation intermediates, and microbial community structure. *Bioresource Technology* 322: 124534.

Lu, L.u., Xing, D., and Ren, Z.J. (2015). Microbial community structure accompanied with electricity production in a constructed wetland plant microbial fuel cell. *Bioresource Technology* 195: 115–121. https://doi.org/10.1016/j.biortech.2015.05.098.

Magnin, N.C., Cooley, B.A., Reiskind, J.B., and Bowes, G. (1997). Regulation and localization of key enzymes during the induction of Kranz-less, C4-type photosynthesis in *Hydrilla verticillata*. *Plant Physiology* 115 (4): 1681–1689. https://doi.org/10.1104/pp.115.4.1681.

Mann, C.J. and Wetzel, R.G. (1999). Photosynthesis and stomatal conductance of Juncus effusus in a temperate wetland ecosystem. *Aquatic Botany* 63 (2): 127–144. https://doi.org/10.1016/S0304-3770(98)00111-9.

Martínez-Cisneros, E. et al. (2021). Analytical modeling of the mechanical behavior of MEMS/NEMS-multilayered resonators with variable cross-sections for sensors and energy harvesters. *IEEE Access* 9: 81040–81056. https://doi.org/10.1109/ACCESS.2021.3084600.

Matsuba, K., Imaizumi, N., Kaneko, S. et al. (1997). Photosynthetic responses to temperature of phosphoenolpyruvate carboxykinase type C4 species differing in cold sensitivity. *Plant, Cell & Environment* 20 (2): 268–274.

McNaughton, S.J. and Fullem, L.W. (1970). Photosynthesis and photorespiration in *Typha latifolia*. *Plant Physiology* 45 (6): 703–707.

Md Khudzari, J., Kurian, J., Gariépy, Y. et al. (2018). Effects of salinity, growing media, and photoperiod on bioelectricity production in plant microbial fuel cells with weeping alkaligrass. *Biomass and Bioenergy* 109.

Mejdová, M., Dušek, J., Foltýnová, L. et al. (2021). Photosynthetic parameters of a sedge-grass marsh as a big-leaf: effect of plant species composition. *Scientific Reports* 11: 3723. https://doi.org/10.1038/s41598-021-82382-2.

Oon, Y.L., Ong, S.A., Ho, L.N. et al. (2015). Hybrid system up-flow constructed wetland integrated with microbial fuel cell for simultaneous wastewater treatment and electricity generation. *Bioresource Technology* 186: 270–275. https://doi.org/10.1016/j.biortech.2015.03.014.

Oon, Y.L., Ong, S.A., Ho, L.N. et al. (2016). Synergistic effect of up-flow constructed wetland and microbial fuel cell for simultaneous wastewater treatment and energy recovery. *Bioresource Technology* 203.

Osorio-de-la-Rosa, E., Vázquez-Castillo, J., Carmona-Campos, M., Barbosa-Pool, G.R., Becerra-Nuñez, G., Castillo-Atoche, A., and Ortegón-Aguilar, J. (2019). Plant Microbial Fuel Cells–Based Energy Harvester System for Self-powered IoT Applications. *Sensors* 19 (6): (1378): 1–16. https://doi.org/10.3390/s19061378.

Osorio-de-la-Rosa, E., Vázquez-Castillo, J., Castillo-Atoche, A. et al. (2021). Arrays of plant microbial fuel cells for implementing self-sustainable wireless sensor networks. *IEEE Sensors Journal* 21 (2): 1965–1974. https://doi.org/10.1109/JSEN.2020.3019986.

Palattella, M.R., Dohler, M., Grieco, A., Rizzo, G., Torsner, J., Engel, T., Ladid, L. (2016). Internet of Things in the 5G Era: Enablers, Architecture, and Business Models. *IEEE Journal on Selected Areas in Communications* 34 (3): 510–527. https://doi.org/10.1109/JSAC.2016.2525418.

Pamintuan, K.R.S. and Sanchez, K.M. (2019). Power generation in a plant-microbial fuel cell assembly with graphite and stainless-steel electrodes growing Vigna Radiata. In: *IOP Conference Series: Materials Science and Engineering*, 012037. IOP Publishing Ltd https://doi.org/10.1088/1757-899X/703/1/012037.

Psomopoulos, C.S. (2013). Solar energy: harvesting the sun's energy for sustainable future. In: *Handbook of Sustainable Engineering* (ed. J. Kauffman and K. Lee), 1065–1107. Dordrecht, The Netherlands: Springer.

Ruhil, K., Ahmad, A., Iqbal, M., and Tripathy, B.C. (2015). Photosynthesis and growth responses of mustard (*Brassica juncea* L. cv Pusa Bold) plants to free air carbon dioxide enrichment (FACE). *Protoplasma* 252 (4): 935–946.

Rusyn, I. (2021). Role of microbial community and plant species in performance of plant microbial fuel cells. *Renewable and Sustainable Energy Reviews* 152: 111697. https://doi.org/10.1016/j.rser.2021.111697.

Rusyn, I., Fihurka, O., and Dyachok, V. (2022). Effect of plants morphological parameters on plant-microbial feel cell efficiency. *Innovative Biosystems and Bioengineering* 6 (3-4): 161–168. https://doi.org/10.20535/ibb.2022.6.3-4.273108.

Sanislav, T., Mois, G.D., Zeadally, S., and Folea, S.C. (2021). Energy harvesting techniques for internet of things (IoT). *IEEE Access* 9: 39530–39549. https://doi.org/10.1109/ACCESS.2021.3064066.

Sarma, P.J. and Mohanty, K. (2018). *Epipremnum aureum* and *Dracaena braunii* as indoor plants for enhanced bio-electricity generation in a plant microbial fuel cell with electrochemically modified carbon fiber brush anode. *Journal of Bioscience and Bioengineering* 126.

Pallavi Sethi and Smruti R. Sarangi. (2017). "Internet of things: architectures, protocols, and applications", *Journal of Electrical and Computer Engineering*, vol. 2017, Article ID 9324035, 25 pages, https://doi.org/10.1155/2017/9324035

Shrirang, M., Nayak, K.K., Kumar, M., and Singh, L. (2021). Plant microbial fuel cell: opportunities, challenges, and prospects. *Bioresource Technology* 341: 125772. https://doi.org/10.1016/j.biortech.2021.125772.

Sophia, A.C. and Sreeja, S. (2017). Green energy generation from plant microbial fuel cells (PMFC) using compost and a novel clay separator. *Sustainable Energy Technologies and Assessments* 21: 59–66. https://doi.org/10.1016/j.seta.2017.05.001.

Srivastava, J., Kalra, S.J., and Naraian, R. (2014). Environmental perspectives of *Phragmites australis* (Cav.) Trin. Ex. Steudel. *Applied Water Science* 4 (3): 193–202.

Srivastava, P., Yadav, A.K., and Mishra, B.K. (2015). The effects of microbial fuel cell integration into constructed wetland on the performance of constructed wetland. *Bioresource Technology* 195.

Strik, D.P., Hamelers, H.V.M., Snel, J.F. et al. (2008). Green electricity production with living plants and bacteria in a fuel cell. *International Journal of Energy Research* 32 (9): 870–876. https://doi.org/10.1002/er.1397.

Sudevalayam, S. and Kulkarni, P. (2011). Energy harvesting sensor nodes: survey and implications. *IEEE Communications Surveys & Tutorials* 13 (3): 443461, 3rd Quart.

Sudirjo, E., Buisman, C.J.N., and Strik, D.P.B.T.B. (2019). Activated carbon mixed with marine sediment is suitable as bioanode material for *Spartina anglica* sediment/plant microbial fuel cell: plant growth, electricity generation, and spatial microbial community diversity. *Water (Switzerland)* 11 (9): 1810. https://doi.org/10.3390/w11091810.

Tabish, R., Mnaouer, A.B., Touati, F., and Ghaleb, A.M. (2013). A comparative analysis of BLE and 6LoWPAN for U-HealthCare applications. In: *Proceedings of the 2013 7th IEEE GCC Conference and Exhibition (GCC), Doha, Qatar*, 17–20 November 2013, 286–291.

Tapia, N.F., Rojas, C., Bonilla, C.A., and Vargas, I.T. (2017). Evaluation of Sedum as driver for plant microbial fuel cells in a semi-arid green roof ecosystem. *Ecological Engineering* 108: 203–210. https://doi.org/10.1016/j.ecoleng.2017.08.017.

Tisarum, R., Chaitachawong, N., Takabe, T. et al. (2022). Physio-morphological and biochemical responses of dixie grass (*Sporobolus virginicus*) to NaCl or Na_2SO_4 stress. *Biologia* 77 (11): 3059–3069.

Tsujimura, M. (2020). Lead the future: semiconductor evolution as seen by CMP manufacturers. In: *2020 China Semiconductor Technology International Conference (CSTIC)*. https://doi.org/10.1109/CSTIC49141.2020.9282449.

Valdez-Hernández, M., Acquaroli, L.N., Vázquez-Castillo, J., González-Pérez, O., Heredia-Lozano, J.C., Castillo-Atoche, A., Sosa-Vargas, L., and Osorio-de-la-Rosa, E. (2022). Plant/soil-microbial fuel cell operation effects in the biological activity of bioelectrochemical systems. *Energy Sources, Part A: Recovery, Utilization, and Environmental Effects* 44 (2): 2715–2729. https://doi.org/10.1080/15567036.2022.2059597.

Volkov, A.G. (2012). *Plant Electrophysiology: Methods and Cell Electrophysiology*. Berlin: Springer https://doi.org/10.1007/978-3-642-29119-7.

Wang, K., Jun, Y., Yan, Y. et al. (2018). A survey on energy internet: architecture, approach, and emerging technologies. *IEEE Systems Journal* 12 (3): 2403–2416. https://doi.org/10.1109/JSYST.2016.2639820.

Wang, B., Long, Z., Hong, Y. et al. (2021). Woodpecker-mimic two-layer band energy harvester with a piezoelectric array for powering wrist-worn wearables. *Nano Energy* 89: 106385.

Wetser, K., Dieleman, K., Buisman, C., and Strik, D. (2017). Electricity from wetlands: tubular plant microbial fuels with silicone gas-diffusion biocathodes. *Applied Energy* 185.

Xu, J.P., Sun, M., Wang, H. et al. (2018). Photosynthetic response of *Scirpus validus* and *Typha orientalis* to elevated temperatures in Dianchi Lake, Southwestern China. *Journal of Mountain Science* 15 (12): 2666–2675.

Yamori, W., Hikosaka, K., and Way, D.A. (2014). Temperature response of photosynthesis in C3, C4, and CAM plants: temperature acclimation and temperature adaptation. *Photosynthesis Research* 119: 101–117. https://doi.org/10.1007/s11120-013-9874-6.

Yang, Y., Zhao, Y., Tang, C. et al. (2020). Role of macrophyte species in constructed wetland-microbial fuel cell for simultaneous wastewater treatment and bioenergy generation. *Chemical Engineering Journal* 392: 123708.

Zeadally, S., Siddiqui, F., and Baig, Z. (2019). 25 years of bluetooth technology. *Future Internet* 11 (9): 194. https://doi.org/10.3390/fi11090194.

Zhu, D. and Beeby, S. (2011). Kinetic energy harvesting. In: *Energy Harvesting Systems* (ed. T. Kaźmierski and S. Beeby). New York, NY: Springer https://doi.org/10.1007/978-1-4419-7566-9_1.

13

Problems of Improving Organics, Ammonium and Phosphorus Treatment with Algal-assisted MFCs

Nguyen Trung Hiep

Research Institute for Sustainable Development, Ho Chi Minh University of Natural Resources and Environment, Ho Chi Minh City, Vietnam

13.1 Introduction

The world is now working to meet the universally agreed-upon sustainable development goals. However, the world population is growing rapidly leading to increased demand for energy and wastewater worldwide. Due to a large number of contaminants, wastewater treatment is an energy-intensive process. As a result, sustainable and energy-efficient wastewater treatment systems are a key item. Furthermore, valuable resources remain unexplored in conjunction with wastewater treatment. To address this issue, resource recovery and bioenergy generation via biological wastewater treatment have emerged as hot subjects.

The algal-assisted microbial fuel cell (A-MFCs) is a novel green device that may be used for both nutrient removal/recovery and energy generation from wastewater (Kannan & Donnellan, 2021; Yi et al., 2022). The A-MFCs combine algae phototrophs with bioelectrodes to enhance pollutant removal and bioenergy recovery from wastewater. In A-MFCs, oxygen produced by algae photosynthesis in the cathode compartment is used for cathodic reactions (He et al., 2014; Kakarla & Min, 2014; Kondaveeti et al., 2014). For conventional MFCs, applying mechanical aeration to the cathode accounts for about half of the total operating costs.

Photosynthesis-Assisted Energy Generation: From Fundamentals to Lab Scale and In-Field Applications, First Edition. Edited by Sathish-Kumar Kamaraj and Iryna Rusyn.

Therefore, oxygen supply from photosynthesis greatly reduces this cost (Del Campo et al., 2015). Algae produce oxygen during autotrophic cultivation by using light and inorganic carbon as energy and carbon sources, respectively. Besides, algae also absorb nutrients like nitrogen and phosphate during this process. Thus, in addition to treating wastewater in the anode chamber, A-MFCs may treat nutrient-rich wastewater in the cathode chamber.

Some research on the usage of various configurations of A-MFCs to remove nutrients from wastewater has been published in the literature in recent years (Aiyer, 2021; Das et al., 2022; Fu et al., 2010). The combined electrochemical/photochemical dynamics with microbial catalysis of A-MFCs has emerged as a sustainable platform for increased pollutant degradation and bioenergy recovery from wastewater. However, the development of A-MFCs is still at the laboratory scale. This chapter describes the present A-MFCs wastewater treatment technologies as a whole, as well as their overall performance constraints. Herein, the focus issues will be discussed including the configuration of A-MFCs, algae species, anode bacteria, environmental factors, hydraulic retention time (HRT), and organic loading rate (OLR). Furthermore, future prospects will also be discussed to make them more widely adopted.

13.2 Components and Designs of Algal-assisted MFCs

13.2.1 Components of A-MFCs System

13.2.1.1 Electrodes

The basic A-MFCs, similar to a convention MFC, comprise two major electrodes: a cathode and an anode. Normally, the electrodes will be connected through an external circuit. Depending on the purpose of research or application, there are different types of materials used as electrodes in A-MFC systems. The potential materials chosen as electrodes must have characteristics such as low cost, high conductivity, low resistance, biocompatibility, and high stability. In addition, the distribution, thickness, and stability of the electrode biofilm would be influenced by porosity and surface roughness. Therefore, nonmetallic materials such as graphite rods, carbon paper, carbon cloth, activated charcoal, and biochar are commonly employed as electrodes (Cao et al., 2022b; Zhou et al., 2011). However, the electrode materials are chosen based on MFC investigations. So far, no research has been done to find suitable electrode materials for the A-MFC system. Thus, the effect of electrode material and configuration in A-MFCs is a prospective research topic.

13.2.1.2 Proton Exchange Membrane (PEM)

In the original MFCs, there were separations between the anode and cathode using membranes or separators. As a fundamental component of any MFCs, a physical barrier between anode and cathode chambers plays a vital role in ion transfer from anode to cathode compartments. As a result, ion exchange membranes (IEM) are also utilized in the structure of A-MFC systems. Previous studies have shown that the quality IEM must have a high ion transfer rate, low gas permeability, strong thermal stability, and resistance to biofouling. The IEM has been divided into different categories based on the material and exchange mechanism, such as cation exchange membrane (CEM), anion exchange membrane (AEM), bipolar membrane (BPM), polymer composite porous membrane (PCPM), and ceramic membrane (CM). Among the various membranes, Nafion™ membranes are commonly used for laboratory-scale research.

Although the IEM is a crucial component of MFC, its high cost prevents MFC from being widely used. According to several previous reports, the IEM cost accounts for about 40–60% of the total cost of the MFC systems in field-scale applications (Aiken et al., 2019; Ge & He, 2016; Rozendal et al., 2008). Therefore, researchers are working hard to develop acceptable and cost-effective separators for biological fuel cells that can be replicated in practice. The IEM or separator studies in A-MFC systems are still challenging in the future.

13.2.1.3 Anolyte and Catholyte

In typical MFCs, anolyte and catholyte are known as the liquid present in anode and cathode compartments. These liquids play the role of stabilizing pH and increasing electrical conductivity (Gil et al., 2003). In anode chambers, the anolyte is usually wastewater (Mehrotra et al., 2021). For A-MFC systems, the catholyte is commonly the medium or wastewater for algae culture in cathode chambers (Gajda et al., 2013; Mohamed et al., 2020; Zhang et al., 2019). The pH has an impact on both the efficiency of microbial activity at the anode and the growth of algae at the cathode. As a result, pH control in anolyte and catholyte contributes to better A-MFC efficiency.

13.2.2 Designs of A-MFCs System

13.2.2.1 Single-chambered Algae-MFCs

Single-chambered A-MFCs are typically cubical in shape (Fu et al., 2010; Nishio et al., 2013) but can also be cylindrical (Abazarian et al., 2020). The basic construction consists of a membrane-less single-chamber with an anode and a cathode (Figure 13.1). Algae were employed to transport electrons to the electrode

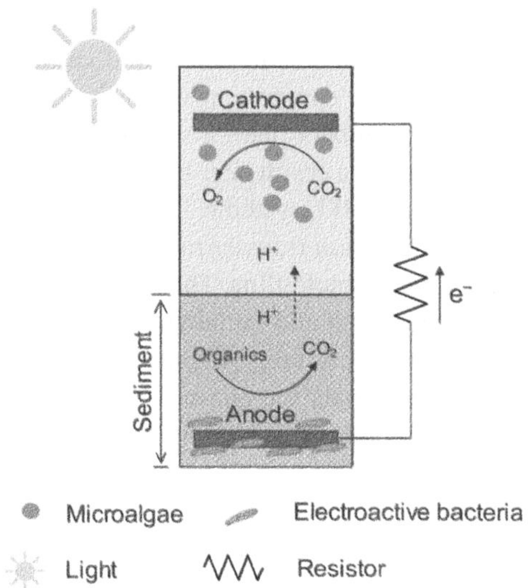

Figure 13.1 Single-chambered algae-MFCs.

surface without mediators. In this concept, pure algal culture can be employed, or algal–bacterial consortia to improve electron generation and A-MFCs efficiency.

In the case of pure algae culture, the A-MFC configuration can be described in terms of photosynthetic-sediment MFC. Microalgae can be used to provide oxygen as the final electron acceptor in the cathode compartment, while the anode is arranged in the sediment of A-MFCs (Song et al., 2020; Wang et al., 2014). The anodic bacterial activity generates CO_2, which is subsequently consumed by the microalgae cells in the upper cathode compartment. This design can overcome the limitations of the typical sediment MFC system which is the low oxygen reduction rate in the cathode. Furthermore, the combination of microalgae and sediment MFC-based technology represents a promising remediation strategy for organically contaminated sediments (Song et al., 2020).

Another type of single-chambered A-MFCs is known as a co-culture of algae and bacteria. It can be described in terms of the air cathode MFCs in which the cathode that is located outside the fuel cell is exposed to the air. The phototrophic bacteria will be chosen to mix with the microalgae (Sun et al., 2020a). This configuration can provide numerous advantages such as enhanced power generation, wastewater treatment, algal biomass cultivation for added-value production, CO_2 assimilation, and O_2 generation. Nevertheless, phototrophic bacteria can release oxygen during photosynthesis metabolism leading to the deterioration of the anaerobic condition. Besides, the interaction mechanism of consortium of photosynthetic bacteria and algae in A-MFCs system and the maintenance of appropriate ecological symbiotic partnerships are still poorly understood. Therefore, insights into the symbiotic relationships between microalgae and bacteria are needed for application in A-MFCs.

13.2.2.2 Dual-chambered Algae-MFCs

The dual-chambered A-MFCs are known to be the most common design using microalgae as oxygen generators in the cathode compartment of the MFC. This configuration consists of two different chambers (anode and cathode) separated by an IEM (Figure 13.2). Dual-chambered A-MFCs are typically rectangular in shape or H-shaped (Aiyer, 2021; Gonzalez Olias et al., 2019; Nayak & Ghosh, 2019). However, two drawbacks of an H-shaped are excessive fouling and small membrane area, both of which reduce power production (Kannan & Donnellan, 2021). In some cases, A-MFCs can also be roughly tubular (Khandelwal et al., 2020; Yadav et al., 2020).

Microalgae are typically cultivated in the anodic or cathodic chamber of dual-chambered A-MFCs. In the case of microalgae cultivated in the cathode compartment, it can serve as a bio-catholyte. Microalgae growth in the cathode supplies oxygen as a terminal electron acceptor to the cathode reaction; meanwhile, microalgae biomass can be used as a substrate for microorganisms in the anode chamber. Furthermore, cultivating microalgae in the cathode chamber can reduce operational costs since there is no requirement for external aeration. In the algae-assisted cathode compartment, according to previous studies, dissolved oxygen (DO) concentrations were produced as high as 6.6 mg/l, which is suitable for cathodic reactions (Kakarla & Min, 2014; Wang et al., 2010). According to Wu et al. (Wu et al., 2013), the concentration of DO has a clear positive correlation with the voltage output in A-MFCs. As a result, microalgae in suspended or attached form in a cathodic chamber may be suitable for A-MFC applications.

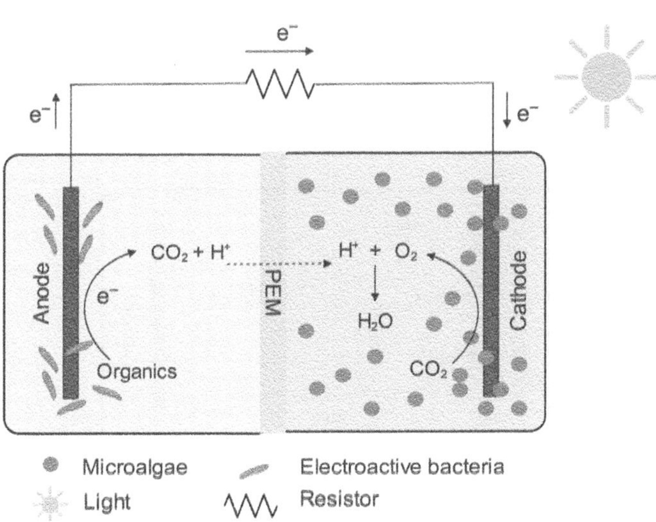

Figure 13.2 Dual-chambered algae-MFCs.

For the algae-assisted anode compartment, algae are used in both alive and non-alive biomass to generate electricity. In the case of dead algae, algal biomass is used as the substrate for the microorganism growth at the anode (Cui et al., 2014; Rashid et al., 2013). However, using algae as a substrate required pre-treatment to be degradable by bacteria (Kondaveeti et al., 2014; Rashid et al., 2013). In another way, alive algae can be cultivated in the anode chamber as electron donors to generate bioelectricity (Fu et al., 2010; Xu et al., 2015). Nevertheless, using algae in the anode compartment could significantly reduce the current because during photosynthesis the algae produced oxygen as an electron acceptor. Therefore, if algae are applied at the anode, it is necessary to control the oxygen content, light intensity, and algal cell density (Xu et al., 2015).

13.2.2.3 Triple-chambered Algae-MFCs

In recent years, an emerging design form of A-MFC is also proposed, which is the triple-chambered construction (Das et al., 2022; Yang et al., 2021). This concept is modified from microbial nutrient recovery cell (MNRC) configuration for simultaneous removal of heavy metals, nutrients, and organic matter in wastewater (Chen et al., 2015; Shahid et al., 2021). The triple-chambered A-MFC consisted of one algae compartment sandwiched between anode and cathode compartments (Pan et al., 2021). In this configuration, a CEM separated the anode and algae chambers, while an AEM separated the algae and cathode chambers (Figure 13.3). The triple-chambered A-MFC has great potential for treating wastewater

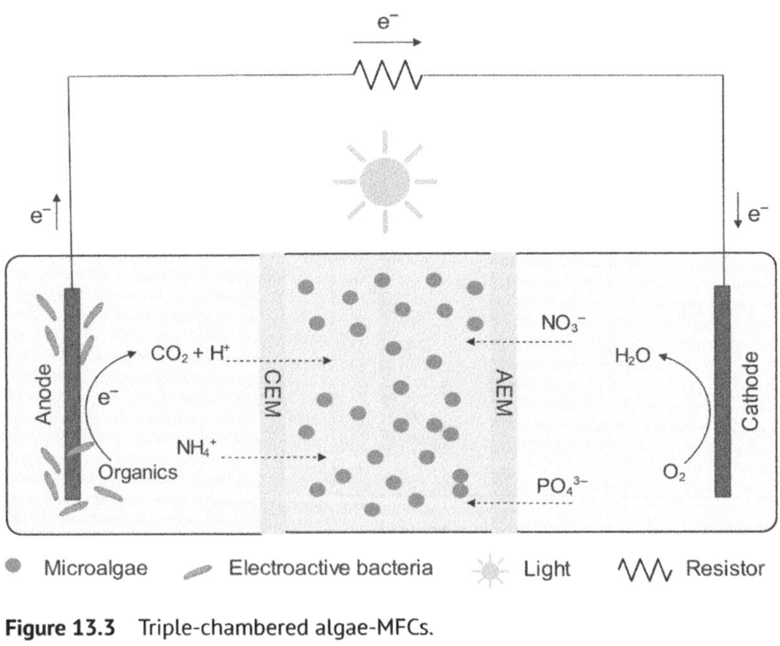

Figure 13.3 Triple-chambered algae-MFCs.

containing heavy metals, while also treating organic matter and generating electricity. However, a cost disadvantage of this configuration was the use of two distinct membranes (CEM and AEM) to separate the chambers. As a result, scaling up the application to wastewater treatment plants is problematic.

13.3 Factors Influencing the Performance of the Algal-assisted MFCs System

13.3.1 Algal Species

In addition to the operating parameters, the algae species also influenced the performance of A-MFCs. Table 13.1 shows the nutrient removal and power production capabilities of various algae species from various A-MFC designs. The choice of algae strain is crucial for the efficient removal of nutrients, as well as for generating maximum power in the A-MFCs. Furthermore, algae strains also have an effect on the biomass generated, which is an excellent resource for the recovery of value-added products. Previous studies have reported common algal genera used in MFC as *Chlorella*, *Scenedesmus*, *Spirulina*, *Desmodesmus,* and *Golenkinia* *(Apollon et al.,* 2021; *Nguyen & Min,* 2020; *Sharma et al.,* 2022*)*. The majority of these genera are capable of efficiently removing nutrients from wastewater. Green microalgae have a stronger ability to consume nitrogen and phosphate than many other species.

Some previous studies have shown that *Chlorella* and *Scenedesmus* are widely used in wastewater treatment (Cai et al., 2013; Jeong & Jang, 2020; Wang et al., 2022). Wang's research team cultured six freshwater microalgae in domestic wastewater to compare the wastewater treatment efficiency. As a result, *Scenedesmus* sp. demonstrated excellent removal of NH_4^+-N and TP, with removal rates of 93.8% and 87.2%, respectively (Wang et al., 2022). Likewise, Zhao et al. (Zhao et al., 2016) showed that *Scenedesmus obliquus* reached a removal rate of up to 82.5±2.2% COD, 61.2±2.1% TN, and 95.4±1.8% TP for high carbon loading wastewater and 80±2.2% COD, 60.1±2.2% TN, and 96.9±2% TP for high nitrogen loading wastewater. For the *Chlorella* genus, nitrogen and phosphorus removal efficiency is observed in the range of 8–100% (Cai et al., 2013). When growing *Chlorella* sp. mixed with activated sludge under light conditions, Wang et al. (Wang et al., 2016) found removal efficiencies of COD, NH_3-N, and TP of 87.3%, 99.2%, and 83.9%, respectively. Among the green algae, *Chlorella vulgaris* is regularly employed microalgae strain in MFCs since its high photosynthetic efficiency in producing oxygen (Colombo et al., 2017). Bazdar et al. (Bazdar et al., 2018) reported the DO concentration up to 9.5 mg/l in photosynthetic MFC using the microalgae *Chlorella vulgaris*. On the other hand, *Chlorella* sp. is one of the algae strains capable of using organic carbon as an energy source when grown

Table 13.1 Summary of the effect of algae species on the A-MFC performance.

Algae species	A-MFCs configuration	Results	References
Chlorella vulgaris (K-1801)	Triple-chambered	COD removal: 38.7–66.8% TKN removal: 49.8–69.0% PO_4^{3-} removal: 34.2–48.5%	**(Pan et al., 2021)**
Chlorella vulgaris (TISTR 8580)	Dual-chambered	COD removal: 74% NH_4^+-N removal: 79% Power generation: 0.0254 kWh/kg COD	**(Don & Babel, 2021)**
Chlorella regularis (FACHB-729)	Dual-chambered	COD removal: 65.2% Power density: 1.07 W/m**2**	**(Ndayisenga et al., 2018)**
Chlorella vulgaris	Dual-chambered	COD removal: 82% NH_4^+-N removal: 90.3% TN removal: 88.6%	**(Elmaadawy et al., 2022)**
Pseudokirchneriella subcapitata	Dual-chambered	TOC removal: 59.4% TN removal: 39.2% TP removal: 55.5%	**(Ribeiro et al., 2022)**
Scenedesmus dimorphus	Dual-chambered	COD removal: 72% TN removal: 75% PO_4^{3-} removal: 80.06% Power density: 0.38 W/m**3**	**(Mehrotra et al., 2021)**
Scenedesmus abundans	Dual-chambered	COD removal: 97.2% NO_3^--N removal: 97.1% PO_4^{3-} removal: 93.7% Power density: 0.84 W/m^2	**(Nayak & Ghosh, 2019)**
Chlamydomonas reinhardtii (cc-503)	Dual-chambered	COD removal: 73.3% Power density: 15.2 W/m**3**	**(Sharma & Chhabra, 2021)**
Spirulina platensis	Dual-chambered	COD removal: 67% Power density: 0.015 W/m**2**	**(Hadiyanto et al., 2019)**
Cladophora sp.	Single-chambered	Power density: 0.62 W/m**2**	**(Taşkan & Taşkan, 2022)**

under heterotrophic conditions (Mitra et al., 2012). On the whole, *Scenedesmus* sp. and *Chlorella* sp. are potential candidates for A-MFCs since the collected data emphasize that they have enhanced energy density with >80% pollutant removal.

13.3.2 Microbial Community

The A-MFCs were virtually identical to typical MFCs. In A-MFCs, anaerobic bacteria remove organic matter in wastewater and generate electricity. However, not all biodegradable microorganisms are electrochemically active, and the majority of them are not electrochemically active. Thus, the performance of A-MFCs is influenced by bacterial species. Exo-electrogens, electrochemically active bacteria (EAB), and anode-respiring bacteria (ARB) are some of the names given to the microorganisms that treat organics at the anode of A-MFCs. Most of the electrogenic bacteria belong to the Gram-negative group; only a few species belong to the Gram-positive group. Among the electroactive species, the most common electrochemical biodegradable bacteria found in the anode chamber include *Geobacter* sp., *Shewanella* sp., *Pseudomonas* sp., *Clostridium* sp., and *Enterobacter* sp. Some of the known electrogenic bacteria used for A-MFC applications are summarized in Figure 13.4.

Yang et al. (Yang et al., 2021) discovered that *Pseudomonas* and *Acinetobacter* were the dominant genera in both the anode and cathode of a triple-chamber A-MFC. These two genera could be crucial in the nitrogen elimination process. Furthermore, *Pseudomonas* is capable of creating mediators for electron transport, such as pyocyanin, which may improve system power output (Logan, 2009).

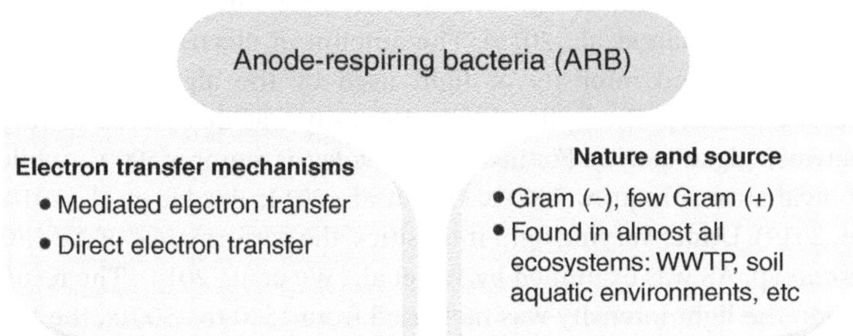

Figure 13.4 Features of anode-respiring bacteria.

In another research, a novel A-MFC system was developed with a nutrient removal efficiency of over 90% (Jiang et al., 2019). As a consequence of the microbial community study, it was discovered that the abundant denitrifying bacteria (*Thauera*, *Paracoccus*, *Stappia*, and *Azoarcus*) were responsible for the high nitrate removal. In terms of bacteria growing in the cathode compartment, *Flavobacteria*, *Thauera*, *Proteobacteria*, and *Arenimonas* were also found in the cathode compartment of the microalgae-assisted fixed-film activated sludge MFC (Elmaadawy et al., 2022). Furthermore, the culturing of powerful exoelectrogens (*Shewanella* and *Geobacter*) in the anode chamber and microalgae (*Chlorella* sp.) in the cathode chamber could increase power density in A-MFCs (Sharma et al., 2022).

In general, the bacteria used in A-MFCs mostly belong to the Proteobacteria phylum. They are Gram-negative bacteria, and the anode compartment environment in A-MFCs is conducive to bacterial growth (Prathiba et al., 2022). Moreover, the microorganisms that belong to the Proteobacteria phylum are quite beneficial in nitrification and denitrification processes (Wang et al., 2019). To increase the effectiveness of A-MFC, it is necessary to mix many species of bacteria with the ability to adapt to changing environmental conditions. On the other hand, there is a need for research on photosynthetic bacteria to apply to A-MFC in the future. This is a potential group of bacteria that provides a suitable process to generate power and remove pollutants.

13.3.3 Light Therapy

Light plays an essential role in the growth of algae in the cathode chamber, which affects the oxygen generation and uptake of nutrients (Bahman et al., 2020; Nzayisenga et al., 2020; Patel et al., 2019). Thus, light therapy influences the overall performance of the A-MFCs. Algae require photosynthetically active radiation (PAR) of at least 10–15 $\mu mol/m^2/s$ for glucose uptake; algae limit growth in complete darkness (Subramanian et al., 2016). The amount of electricity generated depends on the source and intensity of light used by the algae (Abazarian et al., 2020; Xu et al., 2021). The optimal light intensity was discovered to vary significantly between algae species. For instance, *C. vulgaris* requires 5000 lx while mixed culture algal strains require 7000 lx (He et al., 2014; Bazdar et al., 2018; Kakarla & Min, 2019). Under varying light intensities, the performance of A-MFC with *Desmodesmus* sp. A8 was examined by Wu et al. (Wu et al., 2014). The result showed that when the light intensity was increased from 1500 to 3500 lx, the DO content grew consistently from 7.5 to 13.2 mg/l. In another study conducted with the *Chlorella vulgaris* strain, the DO concentration in the cathode compartment increased from 7.8 to 9.5 mg/l corresponding to an increase in light intensity from 3500 to 10,000 lx (Bazdar et al., 2018).

Abazarian et al. (Abazarian et al., 2020) studied the effect of different light/dark cycles on the performance of the sediment MFCs with microalgae strain *Chlorella* sp. in the cathode. The results indicated the highest maximum power density of 19.6 mW/m^2 and open circuit voltage of 442.3 mV with a light/dark cycle of 12/12. The highest power density of 33.6 ± 0.5 mW/m^2 at 1600 lx for *Oscillatoria* sp. and the lowest power density of 23.2 ± 0.5 W/m^2 at 800 lx for *Scendesmus* sp. are shown in the report of Naina Mohamed et al. (Naina Mohamed et al., 2019). According to the research by Kakarla and Min (Kakarla & Min, 2019) with mixed microalgae culture (*Chlorealla* sp., *Desmodesmus* sp. and *Scenedismus* sp.), the greatest 0.35 V voltage and NH_4^+-N removal rate of 26.5 mg/l d achieved as photon flux density increased up to 92 μE/m^2/s. Likewise, the peak power generated of 62.7 mW/m^2 with a luminous intensity of 96 μE/m^2/s was also obtained in the study by Gouveia et al. when using *Chlorella vulgaris* in the cathode compartment (Gouveia et al., 2014).

Similar to the luminous intensity, the wavelength of light also directly affects the power density of the A-MFCs. Microalgae typically utilize light with wavelengths ranging from 400 to 700 nm for photosynthesis, though this varies depending on the species (Atta et al., 2013; Khan et al., 2022). According to some previous literature (Blair et al., 2014; Korbee et al., 2005; Wang et al., 2014b), microalgae growth is stimulated by red (600–700 nm) and blue (400–500 nm) lights. In the study by Lan et al. (Lan et al., 2013), red light is discovered to be more suited for *Chlamydomonas reinhardtii* transformation F5 to produce more electrons in PMFCs. Bahman et al. (Bahman et al., 2020) studied the effect of light intensity and wavelength on nitrogen and phosphate removal from urban wastewater by microalgae *Spirulina platensis (Bahman et al.,* 2020*)*. The results showed that the highest biomass and protein concentrations were recorded under blue light (400–500 nm) at a light intensity of 5800 lx, whereas the largest quantity of phosphate and ammonium removal was around 145 and 218 mg/l under violet light (380–450 nm) at a light intensity of 5800 lx, respectively. Nevertheless, Yan et al. (Yan et al., 2013) argued that the wavelength of red light was shown to be the most effective for treating synthetic domestic sewage with the microalgae *C. vulgaris (Yan et al.,* 2013*)*.

On the whole, the light conditions of microalgae culture directly affect the bioelectricity production in A-MFCs and also have an impact on the efficiency of nutrient treatment in wastewater. In addition to the above properties of light intensity, wavelength, and light/dark cycle, other factors such as light source and flashing light should be researched further in A-MFC applications. Finally, energy usage and prices should be factored into this procedure.

13.3.4 Temperature

Ambient temperature is also a factor affecting the overall performance of A-MFCs because this is the factor influencing the growth of both EAB and algae. In terms of algae, it has been found that algae species grow at temperatures ranging from 5 to 40°C (Singh & Singh, 2015). However, some strains of algae can tolerate different temperature thresholds. Gonçalves et al. (Gonçalves et al., 2016) reported that both microalgae and cyanobacteria demonstrated better biomass productivity and nutrient removal efficiency at 25°C (Gonçalves et al., 2016). *Chlorella vulgaris* can grow in temperatures ranging from 25 to 35 °C, while *Scenedesmus* species may grow in temperatures ranging from 10 to 40°C (Singh & Singh, 2015). In the temperature range of 19–35 °C, the generation of electricity and nitrogen removal were barely influenced (Kakarla & Min, 2019). As a result, the region and geographical location of the experimental sites should be considered to select suitable algae species for A-MFCs.

In terms of EAB, the suitable temperature for them to grow is observed in the ambient temperature range from 20 to 45 °C (Kakarla & Min, 2019; Anam et al., 2020; Mansoorian et al., 2016). Zhang et al. (Zhang et al., 2021) discovered a new EAB (*Alicyclobacillus hesperidum*) that generated a lot of electricity, with a maximum power density of 188.1 mW/m^2 at 50 °C (Zhang et al., 2021). According to a study on the effect of temperature on the catalytic activity of electrochemically active biofilms, the optimal working temperature of the biofilm is 45 °C, with a current of 1339 mA/cm^2 generated (Liu et al., 2011). It has been discovered that EABs are less susceptible to temperature changes than other bacteria and can adapt to both psychrophilic (<20 °C) and thermophilic (>50 °C) conditions (Michie et al., 2011). According to Dopson et al. (Dopson et al., 2016), MFCs working under thermophilic conditions have several advantages, including greater power generation, increased biodegradability, lesser restriction of mass transfer, and lower danger of contamination (Dopson et al., 2016). As a result, selecting the appropriate temperature range is critical for A-MFC applications. This temperature range must allow both algae and EAB to flourish.

13.3.5 pH

The pH of the culture medium or wastewater not only influences the metabolic and biochemical activities of microalgae but also electrogenic organisms. The majority of microalgae species thrive at neutral pH, and all microalgae strains appear to have a different optimal pH range (Gatamaneni et al., 2018; Reddy et al., 2019). Fagiri et al. (Fagiri et al., 2013) reported that maximum microalgal growth was observed at pH 8.0, suggesting that moderate alkalinity is required for optimal microalgae growth (Fagiri et al., 2013). However, Zhai et al. (Zhai et al., 2017) suggested that the range pH of 8.8–8.9 was optimal for nutrient removal by *Spirulina platensis* in synthetic municipal wastewater (Zhai et al., 2017). In another study, Liu et al. (Liu et al., 2015) discovered that utilizing *Chlorella vulgaris* increased the pH in the cathode compartment from 7.3 to 8.3 (Liu et al., 2015). Likewise, a tendency to increase pH also occurred in the study of Nguyen et al. (Nguyen et al., 2017) when using mixed algae culture to treat landfill leachate wastewater by A-MFCs (Nguyen et al., 2017). The elevated catholyte pH is explained to be caused by the H^+ ions consumed in the cathode compartment during carbon dioxide fixation in algae photosynthesis (Chi et al., 2011; Nguyen et al., 2019). In addition, the oxygen reduction reaction (ORR) at the cathode might produce $OH-$ ions, causing the pH catholyte to rise (Gajda et al., 2017; Popat et al., 2012). On the other hand, some earlier studies have shown that with an increase in the pH value of wastewater of more than 9, the microalgal cells tended to aggregate, which allowed auto-flocculation to occur (Li et al., 2021; Liang et al., 2022; Nguyen et al., 2014; Vandamme et al., 2013). Furthermore, increasing the catholyte pH also reduced the power density (Guerrini et al., 2013; Yuan et al., 2013). Thus, neutral pH is a suitable choice for the cathode compartment in A-MFCs.

For the anode compartment, the slightly alkaline anolyte pH~7.5 was shown to be beneficial for increased power generation and organic matter elimination in MFCs (Behera et al., 2010). This is because keeping the anolyte near neutral pH allows the electroactive biofilm to work better (Kumar & Mungray, 2017; Zhuang et al., 2012). According to earlier research, most anode-attached microorganisms prefer alkaline pH due to efficient extracellular electron transfer and proliferation of ARB (Behera & Ghangrekar, 2009; Yuan et al., 2011; Zhang et al., 2011). Zhang et al. (Zhang et al., 2011) studied the effect of pH on the performance and anodic microorganisms of MFCs (Zhang et al., 2011). The results showed that there were various anodic bacteria species at change pH conditions. Besides, low pH conditions ($pH \leq 5$) generated damaged biofilm, resulting in lower MFCs performance. In another previously reported study, the current density of *Thermincola ferriacetica* biofilm increased from 6.5 to $14.3\,A/m^2$ when the anolyte pH was raised from 5.2 to 8.0 (Lusk et al., 2018). Li et al. (Li et al., 2013) reported that maintaining a tight pH (6.3–7.6) proved critical for successful bioelectricity production and COD removal utilizing MFCs (Li et al., 2013). However, Luo et al. (Luo et al., 2011) reported the opposite result, stating that reducing pH from 7.0 to 2.0 increased maximum power densities from 9.5 to $16.8\,W/m$ (Luo et al., 2011). The inconsistencies in these results were due to the pH difference between the cathode and anode, the substrate used, and the organic matter decomposition process. Therefore, a pH of 6.5–7.0 is ideal for microbial growth in the anaerobic chamber.

13.3.6 Organic Loading Rate (OLR)

Similar to other wastewater treatment processes, the overall performance of A-MFCs is also influenced by wastewater characteristics and OLR. High OLR in the anodic chamber increases electrical output, but too high OLR concentration inhibits microbial metabolism. As a certain COD:N:P ratio is required for the growth of microorganisms in an anaerobic chamber, increasing the OLR can improve the elimination of nutrients in the anode chamber (Baby & Ahammed, 2022). In a study conducted by Subha et al. (Subha et al., 2019) using organic-rich chocolate wastewater, the trend of COD consumption rate and power generation increases linearly with an increase in OLR up to 1.6 gCOD/l d (Subha et al., 2019). Ye et al. (Ye et al., 2019) reported that MFC could remove more than 90% of organic matter in municipal wastewater with an OLR range of 435–725 mgCOD/l d (Ye et al., 2019). According to Prasertsung et al. (2014), COD removal efficiency was over 90% when applying single-MFC to cassava wastewater treatment with OLR of 0.56 kg COD/m^3 d (Prasertsung & Ratanatamskul, 2014). At present, there is no suitable recommendation on the optimal OLR range for A-MFC systems. The selection of the optimal OLR depends on the wastewater characteristics, the microflora in the anode compartment, and other operating modes. As mentioned, however, too high an OLR can disrupt the bioelectrochemical, causing membrane shutdown and thereby leading to a decrease in the overall efficiency of A-MFCs.

13.3.7 Hydraulic Retention Time (HRT)

Among the listed factors, HRT is also an important factor affecting the efficiency of organic matter treatment, nutrient removal as well as electricity generation of A-MFCs. In 2018, Yang et al. studied domestic wastewater treatment by A-MFC. The results reported that the treatment efficiency of COD, TN, and TP was 81.9%, 95.5%, and 96.4%, respectively, with an HRT of 12 days (Yang et al., 2018). Recently, a study was conducted by Sun et al. (2020b) on the use of MFC with microalgae-bacteria in the cathode compartment for aquaculture wastewater treatment and energy recovery. The results revealed that 100 mg/l of ammonia nitrogen was almost completely eliminated in the cathode within 90 hours (Sun et al., 2020a). Nguyen et al. (Nguyen & Min, 2020) conducted research on the effects of different HRTs on COD and nutrient removal from leachate using A-MFC. The study showed that the greatest removal efficiency for ammonium, phosphorus, and COD at 60-hours HRT was 76.4%, 86.3%, and 26%, respectively (Nguyen & Min, 2020). In dual-chambered A-MFC, HRT plays an important role in the removal of organic matter and nitrate (Nayak et al., 2022). Don and Babel (Don & Babel, 2021) studied synthetic wastewater treatment by circulating the anode effluent to the cathode chamber with HRT varying from 5 to 20 days (Don & Babel, 2021). This study demonstrated that COD removal efficiency is up to 74% in the anode with 10-days HRT. Ammonium removal efficiency reached 79% when combining anode and cathode with 20-days HRT. On the whole, HRTs depend on wastewater characteristics, initial contaminant concentration, and operating conditions of the A-MFCs system. However, so far, there have not been many studies on optimal HRT for MFC systems with algae in the cathode compartment. Therefore, current assessments are based solely on conventional MFC studies.

13.4 Limitations and Future Perspectives of A-MFCs

13.4.1 Limitations

The A-MFCs have the disadvantages of both conventional algae and MFC processes. Most of the barriers are related to the economic feasibility of A-MFCs. The first consideration is the design of the A-MFCs, which is directly related to cost. As mentioned in the structure, the proton exchange membrane (PEM) is the part that accounts for about 50% of the total cost required for the design. Thus, PEM cost, long-term stability, and sustainability are significant considerations when selecting membranes for scaling-up applications in A-MFCs. In practice, commercially available PEMs are frequently prohibitively expensive and are only appropriate for laboratory-scale studies. Table 13.2 summarizes the prices of some commercial PEMs from reputable manufacturers (reference information from the company's websites: https://ionexchangemembranes.com and https://www.fuelcellstore.com).

Because membranes are expensive, some research has suggested that MFC systems should be membrane-free, as mentioned in the design section. However, the MFCs using a separator have been shown to be more effective in wastewater treatment and nutrient recovery applications (Baby & Ahammed, 2022; Rossi & Logan, 2022). To overcome the cost disadvantage of PEMs, many scientists have constantly researched new materials suitable for creating low-cost membranes for MFCs. Among the potential materials, CMs are considered suitable for application to MFCs as well as A-MFCs on a practical scale (Jadhav et al., 2022; James, 2022).

Table 13.2 The reference price of some commercial PEMs.

Type of membranes	Price ($/m^2)	Company
AXM-100 (AMI-7001S)	350	Membranes International Inc.
CXM-200 (CMI-7000S)	400	Membranes International Inc.
Fumasep FKS-30 (CEM)	2000	Fuel Cell Store
Fumasep FAS-30 (AEM)	2100	Fuel Cell Store
Tion5-W PFSA (CEM)	3400	Fuel Cell Store
Fumasep FBM (BPM)	4733	Fuel Cell Store

Another limitation is related to cultivating algae in the cathode compartment. The environmental conditions must be favorable, with light being a key role in algae growth. Low light intensities can inhibit algal cell growth. On the other hand, the light penetration in the cathode can be reduced due to the high density of algae, which also leads to a decrease in the cell growth rate. In A-MFCs, biomass growth is directly connected to ammonium and phosphorus removal. The use of artificial light can improve A-MFC efficiency but also raises operating expenses (Baby & Ahammed, 2022). Using natural light is one of the solutions to reduce costs. Nevertheless, natural light sources have drawbacks such as fluctuations in radiation levels caused by day/night cycles, varying weather conditions, and seasonal changes. According to Bui et al. (Bui et al., 2022), the nutrient and COD removal rates of natural lighting were only 10% and 13% lower, respectively, when compared to artificial light (Bui et al., 2022). When using A-MFCs, it is vital to employ a combination of artificial and natural light sources to ensure a balance of cost and performance.

13.4.2 Future Perspectives

Although there are economic limitations for practical application, A-MFCs still have the potential for future development. Firstly, study various strains of algae in A-MFCs. In addition to microalgae strains, macroalgae should also be noted. Filamentous freshwater algae, also known as macroalgae, have several advantages over microalgae for wastewater treatment, including greater harvestability and resistance to predation (Cao et al., 2022a; Liu et al., 2020; Park et al., 2022). However, the use of filamentous algae in fuel cells is currently not investigated. Therefore, exploring the potential application of macroalgae in MFCs is a topic of future research.

Another research prospect comes from the group of photosynthetic bacteria. In fact, photosynthetic bacteria have shown high efficiency in wastewater treatment and the production of valuable products (protein and pigments) (Cao et al., 2020). So far, there have been only studies on the application of photosynthetic bacteria in conventional MFCs (Qi et al., 2018; Wong et al., 2016). Optimization of wastewater treatment efficiency and energy production from photosynthetic microorganisms, especially for a mixture of algae and photosynthetic bacteria, has not been fully studied. This has created a research gap, impeding a complete knowledge of the phototrophic MFC systems' unique combination of photosynthetic algae and bacteria. Thus, the study of integrating photosynthetic bacteria and algae in MFCs will be a future promise for sustainable energy and wastewater treatment applications.

Future research on algae-assisted MFC technology must solve current technological problems. This relates to various configurations and low-cost materials for the A-MFC components (membranes and electrodes). Furthermore, energy storage and stabilization methods for use in A-MFCs must be researched further. The comprehensive investigation of the aforementioned challenges can ensure that the field progresses from concept to implementation.

13.5 Conclusion

Algae-assisted MFCs are a revolutionary biotechnology that addresses two major issues: wastewater treatment and bioenergy recovery. Nonetheless, research on A-MFCs in the literature is relatively restricted, particularly for large-scale applications. When using A-MFCs technology, it is vital to achieve high nutrient removal/recovery capacity without interfering with the energy generation process, while also maintaining cheap construction costs. In fact, this technology has a lot of potentials, but it is still being developed. Some of the challenges that need to be addressed include reducing the cost of electrodes and membranes, developing more efficient biocathodes, and designing simpler systems.

References

Abazarian, E., Gheshlaghi, R., and Mahdavi, M.A. (2020). Impact of light/dark cycle on electrical and electrochemical characteristics of algal cathode sediment microbial fuel cells. *Journal of Power Sources* 475: 228686.

Aiken, D.C., Curtis, T.P., and Heidrich, E.S. (2019). Avenues to the financial viability of microbial electrolysis cells [MEC] for domestic wastewater treatment and hydrogen production. *International Journal of Hydrogen Energy* 44 (5): 2426–2434.

Aiyer, K.S. (2021). Synergistic effects in a microbial fuel cell between co-cultures and a photosynthetic alga *Chlorella vulgaris* improve performance. *Heliyon* 7 (1): e05935.

Anam, M., Yousaf, S., and Ali, N. (2020). Comparative assessment of microbial-fuel-cell operating parameters. *Journal of Environmental Engineering* 146 (9): 04020093.

Apollon, W., Luna-Maldonado, A.I., Kamaraj, S.-K. et al. (2021). Progress and recent trends in photosynthetic assisted microbial fuel cells: a review. *Biomass and Bioenergy* 148: 106028.

Atta, M., Idris, A., Bukhari, A. et al. (2013). Intensity of blue LED light: a potential stimulus for biomass and lipid content in fresh water microalgae *Chlorella vulgaris*. *Bioresource Technology* 148: 373–378.

Baby, M.G. and Ahammed, M.M. (2022). Nutrient removal and recovery from wastewater by microbial fuel cell-based systems–a review. *Water Science and Technology* 86 (1): 29–55.

Bahman, M., Aghanoori, M., Jalili, H. et al. (2020). Effect of light intensity and wavelength on nitrogen and phosphate removal from municipal wastewater by microalgae under semi-batch cultivation. *Environmental Technology* 43 (9): 1352–1358.

Bazdar, E., Roshandel, R., Yaghmaei, S. et al. (2018). The effect of different light intensities and light/dark regimes on the performance of photosynthetic microalgae microbial fuel cell. *Bioresource Technology* 261: 350–360.

Behera, M. and Ghangrekar, M.Á. (2009). Performance of microbial fuel cell in response to change in sludge loading rate at different anodic feed pH. *Bioresource Technology* 100 (21): 5114–5121.

Behera, M., Jana, P.S., More, T.T. et al. (2010). Rice mill wastewater treatment in microbial fuel cells fabricated using proton exchange membrane and earthen pot at different pH. *Bioelectrochemistry* 79 (2): 228–233.

Blair, M.F., Kokabian, B., and Gude, V.G. (2014). Light and growth medium effect on *Chlorella vulgaris* biomass production. *Journal of Environmental Chemical Engineering* 2 (1): 665–674.

Bui, X.-T., Nguyen, T.-T., Ngo, H.H. et al. (2022). Co-culture of microalgae-activated sludge in sequencing batch photobioreactor systems: effects of natural and artificial lighting on wastewater treatment. *Bioresource Technology* 343: 126091.

Cai, T., Park, S.Y., and Li, Y. (2013). Nutrient recovery from wastewater streams by microalgae: status and prospects. *Renewable and Sustainable Energy Reviews* 19: 360–369.

Cao, K., Zhi, R., and Zhang, G. (2020). Photosynthetic bacteria wastewater treatment with the production of value-added products: a review. *Bioresource Technology* 299: 122648.

Cao, J., Chen, F., Fang, Z. et al. (2022a). Effect of filamentous algae in a microalgal-bacterial granular sludge system treating saline wastewater: assessing stability, lipid production and nutrients removal. *Bioresource Technology* 354: 127182.

Cao, T.N.-D., Mukhtar, H., Yu, C.-P. et al. (2022b). Agricultural waste-derived biochar in microbial fuel cells towards a carbon-negative circular economy. *Renewable and Sustainable Energy Reviews* 170: 112965.

Chen, X., Sun, D., Zhang, X. et al. (2015). Novel self-driven microbial nutrient recovery cell with simultaneous wastewater purification. *Scientific Reports* 5 (1): 1–10.

Chi, Z., O'fallon, J.V., and Chen, S. (2011). Bicarbonate produced from carbon capture for algae culture. *Trends in Biotechnology* 29 (11): 537–541.

Colombo, A., Marzorati, S., Lucchini, G. et al. (2017). Assisting cultivation of photosynthetic microorganisms by microbial fuel cells to enhance nutrients recovery from wastewater. *Bioresource Technology* 237: 240–248.

Cui, Y., Rashid, N., Hu, N. et al. (2014). Electricity generation and microalgae cultivation in microbial fuel cell using microalgae-enriched anode and bio-cathode. *Energy Conversion and Management* 79: 674–680.

Das, S., and Ghangrekar, M. (2022). Enzymatic cell disruption followed by application of imposed potential for enhanced lipid extraction from wet algal biomass employing photosynthetic microbial fuel cell. *Bioresource Technology* 127924.

Del Campo, A.G., Perez, J.F., Cañizares, P. et al. (2015). Characterization of light/dark cycle and long-term performance test in a photosynthetic microbial fuel cell. *Fuel* 140: 209–216.

Don, C.D.Y.A. and Babel, S. (2021). Circulation of anodic effluent to the cathode chamber for subsequent treatment of wastewater in photosynthetic microbial fuel cell with generation of bioelectricity and algal biomass. *Chemosphere* 278: 130455.

Dopson, M., Ni, G., and Sleutels, T.H. (2016). Possibilities for extremophilic microorganisms in microbial electrochemical systems. *FEMS Microbiology Reviews* 40 (2): 164–181.

Elmaadawy, K., Liu, B., Hassan, G. et al. (2022). Microalgae-assisted fixed-film activated sludge MFC for landfill leachate treatment and energy recovery. *Process Safety and Environmental Protection* 160: 221–231.

Fagiri, Y.M.A., Salleh, A., and El-Nagerabi, S.A. (2013). Influence of chemical and environmental factors on the growth performance of *Spirulina platensis* strain SZ100. *Journal of Algal Biomass Utilization* 4 (2): 7–15.

Fu, C.-C., Hung, T.-C., Wu, W.-T. et al. (2010). Current and voltage responses in instant photosynthetic microbial cells with *Spirulina platensis*. *Biochemical Engineering Journal* 52 (2-3): 175–180.

Gajda, I., Greenman, J., Melhuish, C. et al. (2013). Photosynthetic cathodes for microbial fuel cells. *International Journal of Hydrogen Energy* 38 (26): 11559–11564.

Gajda, I., Stinchcombe, A., Greenman, J. et al. (2017). Microbial fuel cell–a novel self-powered wastewater electrolyser for electrocoagulation of heavy metals. *International Journal of Hydrogen Energy* 42 (3): 1813–1819.

Gatamaneni, B.L., Orsat, V., and Lefsrud, M. (2018). Factors affecting growth of various microalgal species. *Environmental Engineering Science* 35 (10): 1037–1048.

Ge, Z. and He, Z. (2016). Long-term performance of a 200 liter modularized microbial fuel cell system treating municipal wastewater: treatment, energy, and cost. *Environmental Science: Water Research & Technology* 2 (2): 274–281.

Gil, G.-C., Chang, I.-S., Kim, B.H. et al. (2003). Operational parameters affecting the performannce of a mediator-less microbial fuel cell. *Biosensors and Bioelectronics* 18 (4): 327–334.

Gonçalves, A.L., Pires, J.C., and Simoes, M. (2016). The effects of light and temperature on microalgal growth and nutrient removal: an experimental and mathematical approach. *RSC Advances* 6 (27): 22896–22907.

Gonzalez Olias, L., Cameron, P.J., and Di Lorenzo, M. (2019). Effect of electrode properties on the performance of a photosynthetic microbial fuel cell for atrazine detection. *Frontiers in Energy Research* 7: 105.

Gouveia, L., Neves, C., Sebastião, D. et al. (2014). Effect of light on the production of bioelectricity and added-value microalgae biomass in a photosynthetic alga microbial fuel cell. *Bioresource Technology* 154: 171–177.

Guerrini, E., Cristiani, P., and Trasatti, S.P.M. (2013). Relation of anodic and cathodic performance to pH variations in membraneless microbial fuel cells. *International Journal of Hydrogen Energy* 38 (1): 345–353.

Hadiyanto, H., Christwardana, M., and Da Costa, C. (2019). Electrogenic and biomass production capabilities of a Microalgae–Microbial fuel cell (MMFC) system using tapioca wastewater and Spirulina platensis for COD reduction. *Energy Sources, Part A: Recovery, Utilization, and Environmental Effects* 1–12.

He, H., Zhou, M., Yang, J. et al. (2014). Simultaneous wastewater treatment, electricity generation and biomass production by an immobilized photosynthetic algal microbial fuel cell. *Bioprocess and Biosystems Engineering* 37 (5): 873–880.

Jadhav, D.A., Park, S.-G., Eisa, T. et al. (2022). Current outlook towards feasibility and sustainability of ceramic membranes for practical scalable applications of microbial fuel cells. *Renewable and Sustainable Energy Reviews* 167: 112769.

James, A. (2022). Ceramic-microbial fuel cell (C-MFC) for waste water treatment: a mini review. *Environmental Research* 210: 112963.

Jeong, D. and Jang, A. (2020). Exploration of microalgal species for simultaneous wastewater treatment and biofuel production. *Environmental Research* 188: 109772.

Jiang, Q., Song, X., Liu, J. et al. (2019). Enhanced nutrients enrichment and removal from eutrophic water using a self-sustaining in situ photomicrobial nutrients recovery cell (PNRC). *Water Research* 167: 115097.

Kakarla, R. and Min, B. (2014). Photoautotrophic microalgae *Scenedesmus obliquus* attached on a cathode as oxygen producers for microbial fuel cell (MFC) operation. *International Journal of Hydrogen Energy* 39 (19): 10275–10283.

Kakarla, R. and Min, B. (2019). Sustainable electricity generation and ammonium removal by microbial fuel cell with a microalgae assisted cathode at various environmental conditions. *Bioresource Technology* 284: 161–167.

Kannan, N. and Donnellan, P. (2021). Algae-assisted microbial fuel cells: a practical overview. *Bioresource Technology Reports* 15: 100747.

Khan, M.J., Singh, N., Mishra, S. et al. (2022). Impact of light on microalgal photosynthetic microbial fuel cells and removal of pollutants by nanoadsorbent biopolymers: updates, challenges and innovations. *Chemosphere* 288: 132589.

Khandelwal, A., Chhabra, M., and Yadav, P. (2020). Performance evaluation of algae assisted microbial fuel cell under outdoor conditions. *Bioresource Technology* 310: 123418.

Kondaveeti, S., Choi, K.S., Kakarla, R. et al. (2014). Microalgae Scenedesmus obliquus as renewable biomass feedstock for electricity generation in microbial fuel cells (MFCs). *Frontiers of Environmental Science & Engineering* 8 (5): 784–791.

Korbee, N., Figueroa, F.L., and Aguilera, J. (2005). Effect of light quality on the accumulation of photosynthetic pigments, proteins and mycosporine-like amino acids in the red alga *Porphyra leucosticta* (Bangiales, Rhodophyta). *Journal of Photochemistry and Photobiology B: Biology* 80 (2): 71–78.

Kumar, P. and Mungray, A.K. (2017). Microbial fuel cell: optimizing pH of anolyte and catholyte by using taguchi method. *Environmental Progress & Sustainable Energy* 36 (1): 120–128.

Lan, J.C.-W., Raman, K., Huang, C.-M. et al. (2013). The impact of monochromatic blue and red LED light upon performance of photo microbial fuel cells (PMFCs) using Chlamydomonas reinhardtii transformation F5 as biocatalyst. *Biochemical Engineering Journal* 78: 39–43.

Li, X.M., Cheng, K.Y., and Wong, J.W. (2013). Bioelectricity production from food waste leachate using microbial fuel cells: Effect of NaCl and pH. *Bioresource Technology* 149: 452–458.

Li, Y., Ma, Q., Pan, Y. et al. (2021). Development of an effective flocculation method by utilizing the auto-flocculation capability of Phaeodactylum tricornutum. *Algal Research* 58: 102413.

Liang, C., Yang, Y., Xia, Y. et al. (2022). The optimization of chlorella vulgaris flocculation harvesting by chitosan and calcium hydroxide. *Indian Journal of Microbiology* 62 (2): 266–272.

Liu, Y., Climent, V., Berná, A. et al. (2011). Effect of temperature on the catalytic ability of electrochemically active biofilm as anode catalyst in microbial fuel cells. *Electroanalysis* 23 (2): 387–394.

Liu, T., Rao, L., Yuan, Y. et al. (2015). Bioelectricity generation in a microbial fuel cell with a self-sustainable photocathode. *The Scientific World Journal* 2015.

Liu, J., Pemberton, B., Lewis, J. et al. (2020). Wastewater treatment using filamentous algae–a review. *Bioresource Technology* 298: 122556.

Logan, B.E. (2009). Exoelectrogenic bacteria that power microbial fuel cells. *Nature Reviews Microbiology* 7 (5): 375–381.

Luo, Y., Zhang, R., Liu, G. et al. (2011). Simultaneous degradation of refractory contaminants in both the anode and cathode chambers of the microbial fuel cell. *Bioresource Technology* 102 (4): 3827–3832.

Lusk, B.G., Peraza, I., Albal, G. et al. (2018). pH dependency in anode biofilms of *Thermincola ferriacetica* suggests a proton-dependent electrochemical response. *Journal of the American Chemical Society* 140 (16): 5527–5534.

Mansoorian, H.J., Mahvi, A.H., Jafari, A.J. et al. (2016). Evaluation of dairy industry wastewater treatment and simultaneous bioelectricity generation in a catalyst-less and mediator-less membrane microbial fuel cell. *Journal of Saudi Chemical Society* 20 (1): 88–100.

Mehrotra, S., Kumar, V.K., Gajalakshmi, S. et al. (2021). Bioelectrogenesis from ceramic membrane-based algal-microbial fuel cells treating dairy industry wastewater. *Sustainable Energy Technologies and Assessments* 48: 101653.

Michie, I.S., Kim, J.R., Dinsdale, R.M. et al. (2011). The influence of psychrophilic and mesophilic start-up temperature on microbial fuel cell system performance. *Energy & Environmental Science* 4 (3): 1011–1019.

Mitra, D., Van Leeuwen, J.H., and Lamsal, B. (2012). Heterotrophic/mixotrophic cultivation of oleaginous *Chlorella vulgaris* on industrial co-products. *Algal Research* 1 (1): 40–48.

Mohamed, S.N., Hiraman, P.A., Muthukumar, K. et al. (2020). Bioelectricity production from kitchen wastewater using microbial fuel cell with photosynthetic algal cathode. *Bioresource Technology* 295: 122226.

Naina Mohamed, S., Jayabalan, T., and Muthukumar, K. (2019). Simultaneous bioenergy generation and carbon dioxide sequestration from food wastewater using algae microbial fuel cell. *Energy Sources, Part A: Recovery, Utilization, and Environmental Effects* 1–9.

Nayak, J.K. and Ghosh, U.K. (2019). Post treatment of microalgae treated pharmaceutical wastewater in photosynthetic microbial fuel cell (PMFC) and biodiesel production. *Biomass and Bioenergy* 131: 105415.

Nayak, J.K., Mishra, S., and Ghosh, U.K. (2022). Simultaneous production of bio-energy and bio-treatment of wastewater using photosynthetic microbial fuel cell: optimization and kinetic modeling approach. *Waste and Biomass Valorization* 1–16.

Ndayisenga, F., Yu, Z., Yu, Y. et al. (2018). Bioelectricity generation using microalgal biomass as electron donor in a bio-anode microbial fuel cell. *Bioresource Technology* 270: 286–293.

Nguyen, H.T. and Min, B. (2020). Leachate treatment and electricity generation using an algae-cathode microbial fuel cell with continuous flow through the chambers in series. *Science of the Total Environment* 723: 138054.

Nguyen, T.D.P., Frappart, M., Jaouen, P. et al. (2014). Harvesting *Chlorella vulgaris* by natural increase in pH: effect of medium composition. *Environmental Technology* 35 (11): 1378–1388.

Nguyen, H.T., Kakarla, R., and Min, B. (2017). Algae cathode microbial fuel cells for electricity generation and nutrient removal from landfill leachate wastewater. *International Journal of Hydrogen Energy* 42 (49): 29433–29442.

Nguyen, T.D.P., Tran, T.N.T., Le, T.V.A. et al. (2019). Auto-flocculation through cultivation of Chlorella vulgaris in seafood wastewater discharge: influence of culture conditions on microalgae growth and nutrient removal. *Journal of Bioscience and Bioengineering* 127 (4): 492–498.

Nishio, K., Hashimoto, K., and Watanabe, K. (2013). Light/electricity conversion by defined cocultures of Chlamydomonas and Geobacter. *Journal of Bioscience and Bioengineering* 115 (4): 412–417.

Nzayisenga, J.C., Farge, X., Groll, S.L. et al. (2020). Effects of light intensity on growth and lipid production in microalgae grown in wastewater. *Biotechnology for Biofuels* 13 (1): 1–8.

Pan, M., Su, Y., Zhu, X. et al. (2021). Bioelectrochemically assisted sustainable conversion of industrial organic wastewater and clean production of microalgal protein. *Resources, Conservation and Recycling* 168: 105441.

Park, J.B., Montemezzani, V., Picken, C. et al. (2022). Effect of algal contact time and horizontal water velocity on the performance of Filamentous Algal Nutrient Scrubbers (FANS). *Journal of Environmental Management* 312: 114882.

Patel, A.K., Joun, J.M., Hong, M.E. et al. (2019). Effect of light conditions on mixotrophic cultivation of green microalgae. *Bioresource Technology* 282: 245–253.

Popat, S.C., Ki, D., Rittmann, B.E. et al. (2012). Importance of OH−transport from cathodes in microbial fuel cells. *ChemSusChem* 5 (6): 1071–1079.

Prasertsung, N. and Ratanatamskul, C. (2014). Effects of organic loading rate and operating temperature on power generation from cassava wastewater by a single-chamber microbial fuel cell. *Desalination and Water Treatment* 52 (4–6): 937–946.

Prathiba, S., Kumar, P.S., and Vo, D.-V.N. (2022). Recent advancements in microbial fuel cells: a review on its electron transfer mechanisms, microbial community, types of substrates and design for bio-electrochemical treatment. *Chemosphere* 286: 131856.

Qi, X., Ren, Y., Liang, P. et al. (2018). New insights in photosynthetic microbial fuel cell using anoxygenic phototrophic bacteria. *Bioresource Technology* 258: 310–317.

Rashid, N., Cui, Y.-F., Rehman, M.S.U. et al. (2013). Enhanced electricity generation by using algae biomass and activated sludge in microbial fuel cell. *Science of the Total Environment* 456: 91–94.

Reddy, C.N., Nguyen, H.T., Noori, M.T. et al. (2019). Potential applications of algae in the cathode of microbial fuel cells for enhanced electricity generation with simultaneous nutrient removal and algae biorefinery: current status and future perspectives. *Bioresource Technology* 292: 122010.

Ribeiro, V.R., Osório, H.D.D., Ulrich, A.C. et al. (2022). The use of microalgae-microbial fuel cells in wastewater bioremediation and bioelectricity generation. *Journal of Water Process Engineering* 48: 102882.

Rossi, R. and Logan, B.E. (2022). Impact of reactor configuration on pilot-scale microbial fuel cell performance. *Water Research* 119179.

Rozendal, R.A., Hamelers, H.V., Rabaey, K. et al. (2008). Towards practical implementation of bioelectrochemical wastewater treatment. *Trends in Biotechnology* 26 (8): 450–459.

Shahid, K., Ramasamy, D.L., Kaur, P. et al. (2021). Effect of modified anode on bioenergy harvesting and nutrients removal in a microbial nutrient recovery cell. *Bioresource Technology* 332: 125077.

Sharma, A. and Chhabra, M. (2021). Performance evaluation of a photosynthetic microbial fuel cell (PMFC) using *Chlamydomonas reinhardtii* at cathode. *Bioresource Technology* 338: 125499.

Sharma, M., Salama, E.-S., Zhang, P. et al. (2022). Microalgae-assisted microbial fuel cells for electricity generation coupled with wastewater treatment: Biotechnological perspective. *Journal of Water Process Engineering* 49: 102966.

Singh, S. and Singh, P. (2015). Effect of temperature and light on the growth of algae species: a review. *Renewable and Sustainable Energy Reviews* 50: 431–444.

Song, X., Wang, W., Cao, X. et al. (2020). Chlorella vulgaris on the cathode promoted the performance of sediment microbial fuel cells for electrogenesis and pollutant removal. *Science of the Total Environment* 728: 138011.

Subha, C., Kavitha, S., Abisheka, S. et al. (2019). Bioelectricity generation and effect studies from organic rich chocolaterie wastewater using continuous upflow anaerobic microbial fuel cell. *Fuel* 251: 224–232.

Subramanian, G., Yadav, G., and Sen, R. (2016). Rationally leveraging mixotrophic growth of microalgae in different photobioreactor configurations for reducing the carbon footprint of an algal biorefinery: a techno-economic perspective. *RSC Advances* 6 (77): 72897–72904.

Sun, J., Yang, P., Li, N. et al. (2020a). Extraction of photosynthetic electron from mixed photosynthetic consortium of bacteria and algae towards sustainable bioelectrical energy harvesting. *Electrochimica Acta* 336: 135710.

Sun, J., Li, N., Yang, P. et al. (2020b). Simultaneous antibiotic degradation, nitrogen removal and power generation in a microalgae-bacteria powered biofuel cell designed for aquaculture wastewater treatment and energy recovery. *International Journal of Hydrogen Energy* 45 (18): 10871–10881.

Taşkan, B. and Taşkan, E. (2022). Sustainable bioelectricity generation using Cladophora sp. as a biocathode in membrane-less microbial fuel cell. *Bioresource Technology* 347: 126704.

Vandamme, D., Foubert, I., and Muylaert, K. (2013). Flocculation as a low-cost method for harvesting microalgae for bulk biomass production. *Trends in Biotechnology* 31 (4): 233–239.

Wang, X., Feng, Y., Liu, J. et al. (2010). Sequestration of CO_2 discharged from anode by algal cathode in microbial carbon capture cells (MCCs). *Biosensors and Bioelectronics* 25 (12): 2639–2643.

Wang, D.-B., Song, T.-S., Guo, T. et al. (2014). Electricity generation from sediment microbial fuel cells with algae-assisted cathodes. *International Journal of Hydrogen Energy* 39 (25): 13224–13230.

Wang, S.K., Stiles, A.R., Guo, C. et al. (2014b). Microalgae cultivation in photobioreactors: An overview of light characteristics. *Engineering in Life Sciences* 14 (6): 550–559.

Wang, L., Liu, J., Zhao, Q. et al. (2016). Comparative study of wastewater treatment and nutrient recycle via activated sludge, microalgae and combination systems. *Bioresource Technology* 211: 1–5.

Wang, Y., Lin, Z., Wang, Y. et al. (2019). Sulfur and iron cycles promoted nitrogen and phosphorus removal in electrochemically assisted vertical flow constructed wetland treating wastewater treatment plant effluent with high S/N ratio. *Water Research* 151: 20–30.

Wang, Q., Wang, X., Hong, Y. et al. (2022). Microalgae cultivation in domestic wastewater for wastewater treatment and high value-added production: species selection and comparison. *Biochemical Engineering Journal* 108493.

Wong, M.T., Cheng, D., Wang, R. et al. (2016). Modifying the endogenous electron fluxes of Rhodobacter sphaeroides 2.4. 1 for improved electricity generation. *Enzyme and Microbial Technology* 86: 45–51.

Wu, X.-Y., Song, T.-S., Zhu, X.-J. et al. (2013). Construction and operation of microbial fuel cell with *Chlorella vulgaris* biocathode for electricity generation. *Applied Biochemistry and Biotechnology* 171 (8): 2082–2092.

Wu, Y.-C., Wang, Z.-J., Zheng, Y. et al. (2014). Light intensity affects the performance of photo microbial fuel cells with Desmodesmus sp. A8 as cathodic microorganism. *Applied Energy* 116: 86–90.

Xu, C., Poon, K., Choi, M.M. et al. (2015). Using live algae at the anode of a microbial fuel cell to generate electricity. *Environmental Science and Pollution Research* 22 (20): 15621–15635.

Xu, Z., Chen, S., Guo, S. et al. (2021). New insights in light-assisted microbial fuel cells for wastewater treatment and power generation: a win-win cooperation. *Journal of Power Sources* 501: 230000.

Yadav, G., Sharma, I., Ghangrekar, M. et al. (2020). A live bio-cathode to enhance power output steered by bacteria-microalgae synergistic metabolism in microbial fuel cell. *Journal of Power Sources* 449: 227560.

Yan, C., Zhang, L., Luo, X. et al. (2013). Effects of various LED light wavelengths and intensities on the performance of purifying synthetic domestic sewage by microalgae at different influent C/N ratios. *Ecological Engineering* 51: 24–32.

Yang, Z., Pei, H., Hou, Q. et al. (2018). Algal biofilm-assisted microbial fuel cell to enhance domestic wastewater treatment: nutrient, organics removal and bioenergy production. *Chemical Engineering Journal* 332: 277–285.

Yang, Z., Li, J., Chen, F. et al. (2021). Bioelectrochemical process for simultaneous removal of copper, ammonium and organic matter using an algae-assisted triple-chamber microbial fuel cell. *Science of the Total Environment* 798: 149327.

Ye, Y., Ngo, H.H., Guo, W. et al. (2019). Effect of organic loading rate on the recovery of nutrients and energy in a dual-chamber microbial fuel cell. *Bioresource Technology* 281: 367–373.

Yi, G., Wang, B., Feng, Y. et al. (2022). The ins and outs of photo-assisted microbial electrochemical systems for synchronous wastewater treatment and bioenergy recovery. *Resources, Conservation and Recycling* 181: 106230.

Yuan, Y., Zhao, B., Zhou, S. et al. (2011). Electrocatalytic activity of anodic biofilm responses to pH changes in microbial fuel cells. *Bioresource Technology* 102 (13): 6887–6891.

Yuan, Y., Zhou, S., and Tang, J. (2013). In situ investigation of cathode and local biofilm microenvironments reveals important roles of OH–and oxygen transport in microbial fuel cells. *Environmental Science & Technology* 47 (9): 4911–4917.

Zhai, J., Li, X., Li, W. et al. (2017). Optimization of biomass production and nutrients removal by Spirulina platensis from municipal wastewater. *Ecological Engineering* 108: 83–92.

Zhang, L., Li, C., Ding, L. et al. (2011). Influences of initial pH on performance and anodic microbes of fed-batch microbial fuel cells. *Journal of Chemical Technology & Biotechnology* 86 (9): 1226–1232.

Zhang, Y., Zhao, Y., and Zhou, M. (2019). A photosynthetic algal microbial fuel cell for treating swine wastewater. *Environmental Science and Pollution Research* 26 (6): 6182–6190.

Zhang, X., Wang, H., Xia, T. et al. (2021). Characterization of a new electrochemically active bacterium phylogenetically related to *Alicyclobacillus hesperidum* and its electrochemical performance in microbial fuel cell. *Biosensors and Bioelectronics* 175: 112865.

Zhao, Y., Ge, Z., Lui, H. et al. (2016). Ability of different microalgae species in synthetic high-strength wastewater treatment and potential lipid production. *Journal of Chemical Technology & Biotechnology* 91 (11): 2888–2895.

Zhou, M., Chi, M., Luo, J. et al. (2011). An overview of electrode materials in microbial fuel cells. *Journal of Power Sources* 196 (10): 4427–4435.

Zhuang, L., Zheng, Y., Zhou, S. et al. (2012). Scalable microbial fuel cell (MFC) stack for continuous real wastewater treatment. *Bioresource Technology* 106: 82–88.

14

Development and Achievements of Photo-bioelectrochemical Fuel Cell (PBFC) in Metal, Antibiotics, and Dyes Removal

Anwesha Mukherjee

Department of Microbiology, Institute of Science, Nirma University, Ahmedabad, Gujarat, India

14.1 Introduction

Contamination of soil and water with various organic and inorganic pollutants impacts environment profoundly and is the preliminary cause of threat for human health and other life forms. The presence of toxic nonbiodegradable contaminants or heavy metals in surface or groundwater is a severe concern recently (Kurwadkar, 2019). Partially treated wastewater discharge is the preliminary reason for freshwater contamination by toxic compounds. Recovering bioenergy from various contaminants such as metal, antibiotics, or dyes present in wastewater facilitates new technologies such as microbial fuel cell (MFC) which could be integrated with photo bioelectrochemical fuel cells (PBFC) to achieve higher energy (Apollon et al., 2022a; Sun et al., 2019a). Recently major efforts are given in research to understand the application of native photosynthesis to power photoelectrochemical cells. Photosynthetic bacteria, algae used in PBFC can fix CO_2 which eventually lower down deadly greenhouse gas in environment and generate O_2 and biomass. Oxygen, generated during photosynthesis in the PBFC, actually acts as an electron acceptor in the cathodes. Like regular MFC, PBFC could be applied for pollutant removal in the anodic as well as cathodic chamber. The anodic biofilm plays the same role as regular MFC as it collects electrons from the pollutants using them as substrate and degrades those pollutants (Wu et al., 2014).

Photosynthesis-Assisted Energy Generation: From Fundamentals to Lab Scale and In-Field Applications, First Edition. Edited by Sathish-Kumar Kamaraj and Iryna Rusyn.

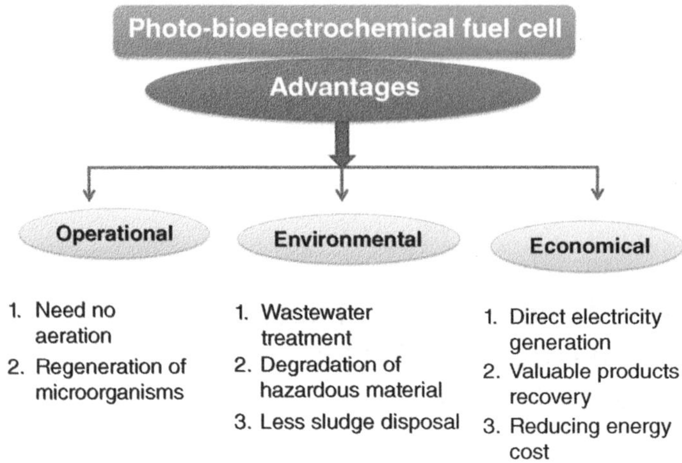

Figure 14.1 Advantages of photo-bioelectrochemical fuel cell in terms of operational, environmental, and economical.

Though the use of solar energy for bioelectricity generation is usually helpful in lowering down the cost of mechanical aeration in PBFC by generating oxygen in cathodic compartment, the system could also be used for various pollutant removals. The primary reason of integration of solar energy with MFC is to generate metabolic oxygen which is released into the aqueous media of the cathode and generates water to complete the circuit for the generation of electricity. In a recent study, a plant-based MFC, when inoculated with livestock's urine, was able to generate a maximum power density of 132 mW/m^2, and 95% improvement in plant root growth was observed (Apollon et al., 2022b). Moreover, from an integrative perspective, the microalgae-MFC is also able to produce biomass and chemical products (Fischer, 2018). Phyto-power system is a hybrid system of PBFC and phytoremediation for simultaneous bioenergy generation and contaminants removal (organic and/or inorganic) from the soil and water. Though there are several reports of removal of various pollutants through PBFC (He et al., 2018), there are very few reports on antibiotic, dyes, and metal removal through PBFC (Habibul et al., 2016). Figure 14.1 depicts various advantages of PBFC in terms of operational, environmental, and economic.

The carcinogenic aromatic amines and the intermediate compounds, generated during the process of the degradation of azo dye, possess a hazardous effect on human health. Recently, the operation of MFCs for azo dye degradation has increased severely. There are also studies on photocatalysis for the degradation of azo dyes and intermediate products effectively (Akpan & Hameed, 2009). Integration of MFC with photoelectrocatalysis (PEC) is now an emerging technology to achieve higher dye degradation along with electricity generation (Logroño et al., 2017). In a study, p-nitrophenol degradation was accelerated by the PEC–MFC system which offers various advantages for dyes and intermediate products removal (Long et al., 2017; Yuan et al., 2010).

Moreover, the amount of pharmaceuticals has risen exponentially due to the progress in medicine industry, most of the drugs are xenobiotic compounds and very difficult in biodegradation. There are several reports stated the presence of higher concentration of antibiotics in different wastewater streams, including livestock wastewater, pharmaceutical industry wastewater, and hospital effluents which have raised significant public health concerns (Guo et al., 2018; Zakaria & Dhar, 2022; Zhao et al., 2019). Pharmaceuticals, abuse of antibiotics, steroidal drugs, and personal care products are some examples of high source of contamination in wastewater that are not treated efficiently through conventional wastewater treatment and needs an additional treatment system where PBFC can play an emerging role (Sun et al., 2019b). Oxytetracyclin removal through PBFC was increased up to 61% in the presence of 50 Ω external resistance instead of in an open circuit system (Sun et al., 2019c). There are very few reports regarding the biodegradation of various antibiotics and other drugs through photobioreactors. Thus, it is essential to summarize the studies related to biodegradation of pharmaceutical compounds through photobioreactor, use of microorganisms during degradation process, and recent developments in this area.

Thus, this study aims to provide a comprehensive review on PBFC as an emerging technology for metal removal and antibiotic and dye degradation. Various designs are also discussed in terms of improving the removal efficiency of photobioreactor for metal, dyes, and antibiotics. It also focuses on the recent developments and advantages of PBFC for pollutant removal. Microorganisms involved in pollutant degradation through PBFC are also discussed. Various environmental conditions to enhance electricity generation and use and control of microbial community are discussed in this chapter. Some coupling systems, including bioelectrochemical systems coupled with the constructed wetlands, and photocatalysis, are also demonstrated. Finally, the challenges and outlooks on PBFCs for metal removal and pharmaceuticals pollutants and dye degradation are discussed, which will help to understand the degradation mechanisms and ways to improve the degradation performance of PBFCs.

14.2 Microorganisms Involved in Metal, Antibiotic and Dye Removal

There are various interesting combinations of MFC with photobioreactor or photosynthetic systems which are collectively described as PBFC. Though the use of photosynthetic organisms/plants in the fuel cell was not much explored, there are some recent studies demonstrate that the photosynthetic organisms/plants could be used in anodic as well as in cathodic compartment, possessing different functional activities in different chambers (Rafieenia et al., 2022; Rusyn, 2021). Photosynthesis in the anodic chamber generates various organic and volatile compounds which are utilized by the organisms present in the anodic chamber. There are various cyanobacteria such as *Synechosystis* sp., *Anabaena variabilis,* and other photosynthetic organisms including *Spirulina platensis*, *Rhodospirillum* sp., and *Rhodobacter* sp. were already reported as inoculum used in anodic chamber of photo-MFC (Greenman et al., 2019). The major problem of using photosynthetic organisms in the anodic compartment is the generation of oxygen which leads to high redox potential at anodic chamber causing low performance in comparison to anodic reaction conducted by the anaerobic organisms. On the other hand, the use of photosynthetic organisms in the cathodic chamber is more beneficial as the generation of oxygen can act as electron acceptor in that chamber and accelerate electricity generation. Even it is reported that the presence of Cyanobacteria or microalgae in the cathodic chamber can improve the efficiency of MFC up to 20% (Gajda et al., 2013). This system is also advantageous as it helps in CO_2 sequestration and generation of cell biomass in cathodic chamber due to cathodic reaction. New biomass grown in the cathodic chamber could be used as feedstock for the anodic chamber of its own system and would be advantageous for bioelectricity production (Greenman et al., 2019). Mechanism and function of photosynthetic organisms in PBFC are demonstrated in Figure 14.2.

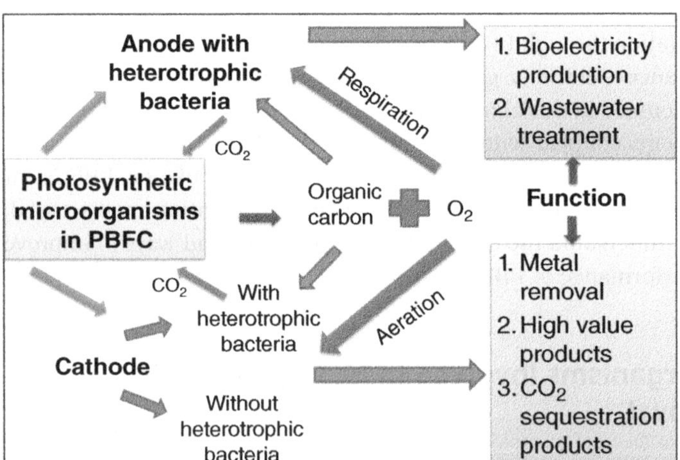

Figure 14.2 Schematic representation of mechanism and function of photosynthetic organisms in PBFC.

In a study with *Desmodesmus* sp. A8 as cathodic microorganism showed that with increasing light intensity from 0 to 1500 lx, cathodic and anodic resistance was decreased and fuel cell performance was increased up to threefold (Wu et al., 2014). In an integrated photo bioelectrochemical system, a single-chambered MFC was placed in a glass beaker which operated as cathode as well as algal bioreactor. The cathode compartment was initially inoculated with green algae *Pseudokirchneriella subcapitata,* while the anodic compartment was inoculated with anaerobic sludge. The system was able to remove 92% chemical oxygen demand (COD), 98% ammonium nitrogen, and 82% phosphate and was able to generate 2.2 W/m^3 power density during one-year operation (Xiao et al., 2012). Even genetically modified cyanobacteria were also experimented for the generation of higher current (Sekar et al., 2016). In many cases, the photosynthetic microorganisms are used as co-culture with heterotrophic bacteria (Fischer, 2018; Aiyer, 2021; He et al., 2009; Liu & Choi, 2017). An indirect synergistic relationship was maintained in the presence of both types of microorganism which enhance fuel cell performance.

Photosynthetic organisms have a unique ability to adapt environmental conditions and can biodegrade various hazardous organic compounds, including pharmaceuticals. Microalgae and other photosynthetic bacteria proficiently remove various organic pollutants, heavy metals, and other contaminants such as dye from wastewater. Various genera of microalgae such as Anabaena, Chlorella, and Oscillatoria are known for their high absorption capability for heavy metals. For example, 65% of cadmium (Cd) could be accumulated by a *Chlorella* sp. of its organic mass from a Cd-rich suspension. Another species of Chlorella (*Chlorella pyrenoidosa*) was also reported for arsenic accumulation. Some microalgae are also reported for copper (Cu) removal through metabolism (approximately 90%) or through adsorption (rest 10%). A marine microalga, *Nannochloropsis oculata*, was reported for copper removal from acid mine water (Tanvir et al., 2021). Generally, metal removal occurs in the cathode compartment where the photosynthetic organisms accumulate metal from wastewater during metabolism and generate O_2 which acts as electron acceptor. A plant-based (Rye grass *Lolium perenne*) MFC was capable for almost 99% Cr (VI) removal with 55 mA/m^2 current density generation. Toxic and carcinogenic Cr (VI) often co-existed with dye. It is also reported that due to photosynthetic activity various chemicals secreted in the surroundings help to increase the current density. Even, sands became polluted by antibiotics, dyes, and heavy metals are common in coastal areas, seashore wetlands, and riversides due to industrial production and wastewater discharge. In a study, Rhodamin B containing sand was used in the anodic chamber of an integrated photo-electro-catalytic fuel cell with *Shewanella* sp., whereas photocathode was used for the removal of Cr (VI) (He et al., 2018). Rhodamin B removal was achieved up to 95.6% and even Cr (VI) reduction became faster. Microalgae

Chlorella sp. QB-102 was used as biocathode for Cd (II) removal by the method of hydroxide precipitation and biosorption as reported by Zhang et al. (Zhang et al., 2018). Microalgae could be applied in the anodic compartment for bioremediation of heavy metal containing dye present in industrial effluent in combination with MFC (Deka et al., 2022).

Specifically, the detailed cellular responses of *Rhodobacter sphaeroides* to cobalt, chromium, mercury, nickel, and gold were already reported. A most interesting fact is that the organism was able to survive in the presence of these heavy metals not only due to the selection processes. It was clear from the study by the same group that *R. sphaeroides* is able to adapt to high concentrations of heavy metals during their metabolic process. These studies provide an excellent knowledge base for the future development of sunlight-powered photo-bioelectrochemical systems for heavy metals detection and removal based on purple bacteria (Grattieri, 2020).

14.3 Mechanism of Toxic Compounds Removal Through Photo-Bioelectrochemical Fuel Cell (PBFC)

Photo-bioelectrochemical fuel cells (PBFCs) are solar/light-assisted fuel cell where photosynthetic microorganisms are used in the anodic/cathodic chamber. Microorganisms absorb light and fix CO_2 as carbohydrate that could be consumed by other heterotrophic bacteria simultaneously present in the anodic chamber and produce electricity similar to other bio-electrochemical fuel cells. This system is self-sustained where electricity is generated in the presence of light without the addition of extra nutrients. An integrated/hybrid system of photoelectrochemical cell with MFC (PEC-MFC) contains two individual systems where both were connected by series connection (Yang et al., 2020).

14.3.1 Mechanism of PBFC

Photocatalytic degradation of pollutants occurs due to the generation of electrons which are transferred from the valence band to the conduction band of photocatalyst because of light irradiation. The reaction between pollutants and electrons occurs on the surface of photocatalysts. Thus, the selection of photocatalyst should be cost-effective, and on the basis of chemical stability, nontoxic nature, and biocompatibility. The photocatalysts can be induced by either visible light or ultraviolet (UV) light on the basis of their band gap energy. However, the use of UV light is not recommended for pollutant removal in PBFC because of the high operational cost associated with the UV light source and adverse effect of UV light on the microorganisms involved in biofilm formation.

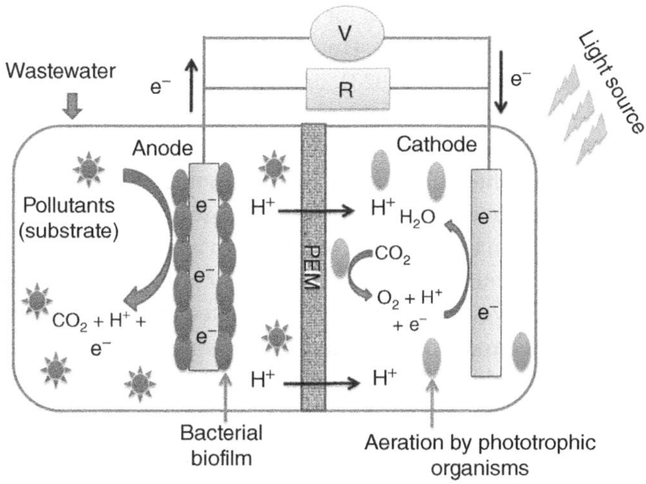

Figure 14.3 Representation of the generalized mechanism of PBFC.

A generalized functional mechanism of PBFC is shown in Figure 14.3 where a simple bioanode–photocathode system is represented which is briefly described in the next section.

Various pollutants present in wastewater are generally degraded in the anodic compartment due to metabolism of the heterotrophic organisms whereas algae, present in the cathodic compartment are mainly responsible for the generation of oxygen which completes the electrical circuit. Reduction of various toxic metals could also be possible in the cathodic compartment. There are some other configurations such as photoanode–biocathode system and biophotoanodic system which are also described in the next section with their mechanism.

14.3.2 Various Configurations of PBFC for Toxic Compound Removal

14.3.2.1 Bioanode–Photocathode System

This type of system consists of a bioanode connected with a photocathode via an external circuit. The electrons generated by the microorganisms in the anodic compartment migrated toward the cathode due to potential difference between anode and cathode. The electrons are combined with the photo-generated holes (h^+) in the cathodic compartment (Rafieenia et al., 2022). In this system, pollutants might be degraded in the anodic compartment due to oxidation reaction by the microorganisms, as well as they could be reduced at the cathode by accepting electrons generated in the anode. There are various reports where bioanode–photocathode systems had been used for degradation of various pollutants such as azo dyes (Ahmadpour et al., 2022; Hou et al., 2017), pharmaceutical chemicals (Chen et al., 2022; Wang et al., 2023), and various metal ions (Tong et al., 2022) present in the wastewater with a degradation efficiency higher than 50%. In a recent study, cathodic degradation of reactive black 5 (RB5) dye is reported in a dual-chambered photo-electric bioanode–photocathode system where AgBr/CuO hybrid photocathode was used. The system was able to degrade about 55.56% of RB5 over 72 hour in the presence of light with a maximum power density of 61.11 mW/m^2 (Ahmadpour et al., 2020). In another study, degradation of 2,4-dichlorophenol (2,4-DCP) was reported under simulated solar irradiation in a single chamber bio-photoelectrochemical system comprising $BiVO_4$/reduced graphene oxide (RGO) photocathode. The system achieved 68.5% degradation of 2,4-DCP with 60.31 A/m^3 current density (Tu et al., 2020). Nitrofurazone (NFZ), a broad-spectrum antibacterial drug, was degraded in a g-C_3N_4/CdS photocathode containing photo-bioelectrochemical system and was able to achieve 83.14% removal of NFZ within 4 hour of light illumination with approximately 9 mA current generation.

14.3.2.2 Photoanode–Biocathode System

Photoanode–biocathode systems have been designed and used to achieve higher removal of pollutants with the generation of energy simultaneously. A good alternative of high-cost metal-based cathodes is biocathodes having microbial catalysts. They are cost-effective and have long-term stability. Biocathode mainly works by adhering to microorganisms on the cathode surface by performing the cathodic reactions using the electrons transferred from the anode. There are several electron acceptors such as O_2, Fe_3^+, NO_3^-, SO_4^{2-} are reported in biocathodes. The first study on the removal of refractory pollutants using biocathode-based photo-electrochemical system was reported by (Du et al., 2014) for degradation of methyl orange (MO). An MFC system consisting of a TiO_2 based photoanode and a biocathode was able to degrade the dye with electricity generation. A total degradation of 65.95% MO was achieved through this system though the current generation was low in comparison to other carbon-based cathode used in the system. In a recent study, degradation of chlortetracycline (CTC) was reported where oxygen-defect-rich $BiVO_4/Ni_9S_8$ photoanode was used (Tu et al., 2022). The current density was about 0.68 A/m^3 (7.6 times higher) and CTC removal efficiency was 1.3 times higher (82.3%) than unilluminated system.

14.3.2.3 Biophotoanodic System

A biophotoanode is a two-sided electrode where one side is illuminated and the other side is colonized with electroactive bacteria (EAB) (Fischer, 2018). EAB accelerates extracellular electron migration generated from the degradation of pollutants in the biophotoanodic chamber, resulting in a synergistic effect by increasing the photocatalytic rate (Rafieenia et al., 2022). The synergistic degradation of pollutants is mainly conducted by three steps (i) pollutant adsorption on the electrode; (ii) Photocatalysis on the illuminating side of the electrode, and (iii) biodegradation of the pollutants and the intermediates by the biofilm generated by the EAB. Qian et al. (Qian et al., 2014) were the first to report a biophotoanode fabricated by a hematite nanowire photoanode and *Shewanella oneidensis* MR-1 as an EAB. The result showed a 150% higher current generation when compared with its abiotic system. There are very few studies about the degradation of pollutants in biophotoanodic system. A biophotoanode, comprising of nano-Ag_3PO_4 and *Shewanella oneidensis* MR-1, was applied for degradation of p-chlorophenol and 21% degradation efficiency was achieved.

A novel technology, intimate coupling photocataysis and biodegradation (ICPB), which combines biodegradation and photocatalysis in the same reactor, is gaining interest recently to increase the potential of pollutant degradation. This system was first reported by Marsolek et al. (Marsolek et al., 2008) to degrade 2,4,5-trichlorophenol in a photocatalytic circulating bed biofilm reactor. The mechanism involved in ICPBs is similar to biophotoanodes; however, the biofilm is formed inside the porous carrier instead of anode surface. It prevents microorganisms from being destroyed by the sheer force of the anodic solution and direct contact between the photocatalyst and biofilm. Furthermore, this mechanism protects microorganisms from UV light (if used) as they are hidden in the interior of the porous carrier (Rafieenia et al., 2022). In a study, use of visible light instead UV light was described as more advantageous for degradation of toxic components as UV irradiation became responsible for biofilm detachment during phenolic wastewater treatment using ICPB technology (Zhou et al., 2015). The system is able to achieve 99.8% removal of phenolic compounds from wastewater in presence of visible light. There are also various reports of the use of ICPB system for degradation of dyes (Xiong et al., 2020), atrazine (Zhang et al., 2021), and tetracycline (Xiong et al., 2017a). Despite of various promising results regarding ICPB, the interaction between biofilm and photocatalyst, the effect of operational parameters on degradation efficiency, and the development of a novel photocatalysts are still to be explored.

14.3.3 Various Materials Involved in PBFC Structure

The influence of various materials for higher electricity production and degradation potential is crucial in terms of metal removal and antibiotic and dye degradation. The main components of PBFC are electrode material, membrane, microorganisms, and light source. There are various factors that influence the function of electrode material such as transparency of the electrode material, electrical conductivity, biocompatibility, and surface area of electrode (Anam et al., 2021). The fundamental property of an electrode of PBFC is that light should be able to penetrate through the electrode surface so that light could be illuminated through the photosynthetic microorganisms embedded on electrode surface. High electrical conductivity amplifies the electron transfer rate in the anodic chamber which eventually enhances power output. Even biocompatibility of the microorganisms with electrode surface is a key factor for good amount of electricity generation with pollutant removal efficiency. Carbon, graphite, and some metal oxides are the most commonly used anode material due to their nontoxic nature toward biofilm formation. Increasing the surface area of electrode and its porous nature enhance biofilm formation on electrode surface by good amount of microbial attachment. The most common approach applied for increasing the surface area and conductivity of the electrode surface is to change the physical chemistry of electrode material by varying the number and nature of the electrode component. Various materials such as stainless steel or gold mesh, platinum strip, carbonaceous compounds (graphite, carbon paper, carbon cloth, and carbon brush) coated with osmium, multiwalled nanotube, or PTFE (polytetrafluoroethylene) are most commonly used to modify the electrode of PBFC for better function (Bazdar et al., 2018; Fu et al., 2010). The maximum power densities achieved $100\,mW/m^2$ after electrode modification.

The proton exchange membrane used as separator needs to be compatible for the microorganisms and should have good ionic conductivity and support good transport of ions across the membrane (Kannan & Donnellan, 2021). Usually, the membranes are made up of hydrogels (nafion membrane) or cross-linked polymers (Ultrex material). There were various modifications applied in separator membranes such as the combination of polyvinyl alcohol hydrogel and clay which were able to increase power density up to fourfold (Angioni et al., 2018; Li et al., 2013). There are some reports of earthenware-based membrane or using porous ceramics as the membrane in MFCs but very few studies were available on the effect of various PEMs for the production of electricity and to enhance treatment efficiency of a photo-bioelectrochemical fuel cell.

14.3.4 Photoelectrode-Microbe Interaction

PBFCs could be characterized into two types according to the function of microorganisms in the system. For the simple type of PBFCs, the system requires sunlight and an external source of organic matter to generate electricity due to metabolism of anoxygenic microorganism (Figure 14.3). However, for complex PBFCs, oxygenic photosynthetic organisms harvest solar energy and generate organic biomass, which is subsequently metabolized by heterotrophic bacteria to generate electricity (Figure 14.4a and b) (Anam et al., 2021). Co-culture of

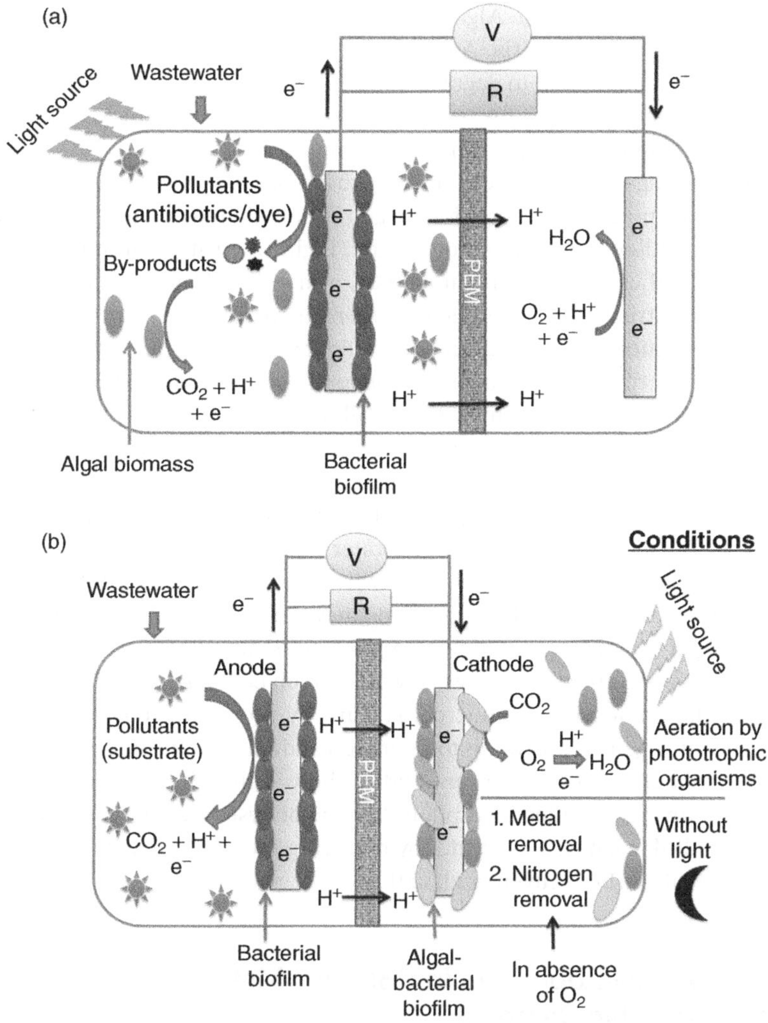

Figure 14.4 Synergistic effect of bacterial and algal biomass for pollutant degradation. (a) Pollutants (such as antibiotic/dye) degradation through photo-bio anodic system, (b) activity of algal-bacterial biofilm in the cathodic chamber in presence/absence of light source.

heterotrophic bacteria and photosynthetic organisms has proven to be more beneficial in terms of pollutant degradation. Algal-bacterial biofilm could be active both in light and dark reaction. During light reaction, algae is able to generate oxygen which could be utilized for metabolism of heterotrophic bacteria whereas generation of CO_2 during that metabolism could be utilized by photosynthetic organisms for their light reaction. Algal biomass also could be utilized as organic substrate for heterotrophic bacteria. A potential synergistic relationship between of algal-bacterial biofilm is able to degrade toxic by-products also.

Various studies showed that use of photosynthetic microorganisms at the anodic chamber could enhance PBFC performance (Cui et al., 2014; Xu et al., 2015). Microalgae at cathodic chamber are primarily used as oxygen supplier. It is most advantageous strategy for oxygen supply than using reducing salt or catalyst at the cathode or sparging of oxygen. Light intensity is an important factor for photosynthetic organisms. In a study, a modified MFC with a photobioreactor was constructed by introducing *Chlorella vulgaris* for *in-situ* oxygen generation. The reaction was carried out both in light and dark cycle. Catalyst (Pt) coated cathode material (carbon paper) was able to achieve a maximum power density of 24.4 mW/m^2 under intermittent illumination (Wu et al., 2013). Biocathode of photosynthetic organisms could also be utilized for capturing the CO_2 generated due to oxidation of carbon material at the anodic compartment. The captured CO_2 is reduced by the photosynthetic bacteria/microalgae during photosynthesis. The supply of treated wastewater from anodic to cathodic compartment provides carbon source to the cathodic photosynthetic organisms but continuous supply of heterotrophic organisms may act as electron donor which eventually hamper electricity generation. A possible strategy to overcome this problem is to transfer limited treated wastewater which could actually accelerate the growth of the photosynthetic microorganisms by providing good source of phosphate and nitrogen and helping to degrade residual carbon present in the wastewater (Sevda et al., 2019). Various studies showed that the algal-microbial community at cathodes of integrated photo-bioreactor system could be used for treatment of landfill leachate wastewater by removing the COD along with other nutrients (Luo et al., 2017; Nguyen et al., 2017).

14.4 Recent Developments in PBFC for Metal, Antibiotics, and Dye Removal

Pollutant degradation due to sunlight irradiation was mostly studied and recent developments were carried out with various nanoparticles (Venkidusamy et al., 2023). However, acclimatization of photosynthetic organisms and toleration of severe conditions such as limitation of food and the presence of toxic compounds are some important factors for biodegradation and bioremediation of pollutants present in wastewater through PBFC. Several reports showed that adaptation of organisms to grow in harsh and limiting conditions could improve the removal efficiency of contaminants. An experiment showed that if *Chlorella vulgaris* was pre-exposed with 200 mg/l of levofloxacin for 11 days, levofloxacin removal efficiency was increased from 16% to 28% (Xiong et al., 2017b). Co-metabolism or the addition of another substrate is an alternative strategy to improve the biodegradation efficiency of the photosynthetic organisms in PBFC. In a study by Xiong et al. (Xiong et al., 2017c), by addition of sodium acetate in the bioreactor, removal efficiency of ciprofloxacin increased from 13% to 56% by *Chlamydomonas Mexicana*. There are various factors that influence co-metabolism such as microorganism used in the study and concentration of substrate and pollutant. However, the presence of some organic substrates or the generation of by-products and secondary metabolites during oxidation–reduction process can lower removal efficiency of the contaminants, possibly due to catabolic repression. Thus, it is necessary to evaluate the effects of various carbon sources used as substrate for biodegradation and wastewater treatment (Xiong et al., 2021).

Recent studies have reported that the use of consortium could support biomass production which facilitates wastewater treatment and becomes cost-effective. Among the different consortium that could be used in the PBFC, microalgae–microalgae consortium and microalgae-bacteria consortium are the most commonly used for wastewater treatment. It has been already reported that the use of consortium for bioremediation of wastewater is more advantageous than using individual microorganism as inoculum. Symbiotic relationships occurred between the co-cultures could accelerate uptake of nutrients, biodegradation of pollutants, and be more resistant in harsh environment. According to various reports, it has been shown that the combination of phototrophic algae and heterotrophic bacteria as consortium in the PBFC became advantageous for wastewater treatment due to its cost-effectiveness and potential pollutant removal efficiency. Furthermore, beside the cooperative relationship between bacteria and algae, exhibition of antagonistic effect cannot be ignored. Such as production of toxic metabolite could inhibit algal growth whereas the release of exotoxin by algae inhibits bacterial growth (Fuentes et al., 2016). Therefore, a detailed study on cooperative interaction mechanisms between microalgae and bacteria is necessary to upscale the applications of PBFC with microalgae–bacteria consortium. Pollutant (such as antibiotic, dye, and metal) degradation mechanism and electricity generation through PBFC in the presence of various substrate and inoculum is summarized in Table 14.1.

Emerging photo-bioelectrochemical technology is advantageous over traditional metal recovery process because of energy saving and chemical-free mechanism (Huang et al., 2021). In a study by Ren et al. (Ren et al., 2018), improvement in TiO_2/Fe_2O_3 photoanode influence performance of PBFC in presence of visible light and achieved 90.9% Cr (VI) removal. The system was able to enhance treatment of metal-containing wastewater by forming a heterogeneous structure on photoaode which was favorable for the attachment of bacterial communities. In another study, 95% of cadmium [Cd (II)] removal was reported by using a

Table 14.1 Mechanism of biodegradation and electricity generation efficiency of photo bio-electrochemical fuel cell for various antibiotics, dye, and metals.

Sr no.	Pollutant type	Mechanism	Substrate	Inoculum	Anode and cathode	Light source	Removal efficiency (%)	Power density (mW/m^2)	References
1	Oxytetracycline	Photocatalysis at biocathode (*Chlorella vulgaris*)	Glucose	Anaerobic sludge	Carbon felt	Visible	86	54	Sun et al. (2019a)
2	Oxytetracycline	Biodegradation at bioanode	Glucose	Anaerobic sludge	Carbon felt	Visible	61	54	Sun et al. (2019c)
3	Dye Zn Cr	Biodegradation	Dye wastewater	Anode-Sludge Cathode-*Chlorella vulgaris*	Carbon fiber	—	93 98 80	6.46	Logroño et al. (2017)
4	Chromium [Cr(VI)]	Phytoremediation	Sodium acetate	Inoculated from other lab scale MFC	Carbon felt	—	99	Current density-$55\,mA/m^2$	Habibul et al. (2016)
5	Cadmium [Cd(II)]	Photocatalysis at biocathode (*Chlorella*)	Sodium acetate	Activated sludge	Nickel foam/graphene	—	95	36.4	Zhang et al. (2018)
6	Rhodamin B	Photocatalysis at photoanode	Acetate	Domestic wastewater	Ag-TiO_2 electrode/carbon fiber brush	UV	99.5	318.19	Sui et al. (2017)
7	Methyl orange	Photo catalytic decolorization at anode	—	—	Graphene/carbon paper	UV-visible	90	—	Su et al. (2020)
8	Metronidazole	Photo assisted MFC	Acetate	Effluent from acetate-fed MFCs	Anode-graphite felt, Cathode-Mo/W coated graphite felt	Visible	94.5 ± 1.4	—	Wang et al. (2019)

9	Brilliant red	Photo-MFC	Acetate	Effluent from MFC anode	Graphite plate	UV-Visible	82	Current-6 mA	Long et al. (2019)
10	Ofloxacin	Photocatalysis	Glucose	Consortium from MFC anode	Anode-PANi@CNTs/SS Cathode-$LiNbO_3$/carbon felt	UV	86.5	0.546 W/m^2	Xu et al. (2020)
11	Methyl orange	Biodegradation + photocatalysis	Acetate	Domestic wastewater	Anode-graphite brush, Cathode-g-C_3N_4/BiOBr photocathode	UV-Visible	97.8	Current-11.85 mA	Hou et al. (2017)
12	Nitrofurazone	Biodegradation + photocatalysis	Acetate	Domestic wastewater	Anode-Carbon brush, Cathode-g-nickel foam loaded C_3N_4/CdS photo cathode	Solar	83.5	Current-9 mA	Hou et al. (2020)
13	4-Chlorophenol	Photocatalysis	Acetate	Biofilm dominated by *Geobacter sulfurreducens*	Anode-carbon foam, Cathode- Stainless steel mess	Visible	41	Current density-0.4 A/m^2	Zhou et al. (2017)

photosynthetic algae *Chlorella* sp. QB-102 in a photosynthetic bacteria/algae MES (Zhang et al., 2018). As there are few reports on metal recovery by using PBFC, further research is required to achieve application level of up-scaling with actual metal ion-containing wastewater (Yang et al., 2022).

Increased light intensity enhances potential of PBFC and improves biomass (biopolymer) generation. In the anodic and cathodic compartment, bacteria and algae produce biopolymer which could be used in wastewater treatment alone or in combination with nanoparticles to increase pollutant removal efficiency of PBFC (Khan et al., 2022). Though various photocatalyst have toxic effect on the growth of microorganisms, it has been also reported that they are also able to reduce the toxic effect of recalcitrant substrate which eventually becomes advantageous for biofilm growth. In a study, the biotoxicity and mutagenicity of diclofenec were reduced by photocatalysis (Liu et al., 2019). Thus, choosing of photocatalysts, which have minimum biotoxicity, is a crucial factor for designing the combined system of PBFC.

14.5 Challenges and Future Outlook

Recent studies have already proved that PBFCs are an attractive technology for power generation along with value-added chemical production and wastewater treatment. However, still now there are many challenges to become independent technology and commercially successful. Despite of high efficiency of light-to-energy conversion, there are still some problems in the application of photocatalytic process for the removal of various recalcitrant pollutants. One of the most important reasons is that the light reaction is fast; therefore, identification of intermediate products becomes difficult. However, foremost limitation of implementing this technology is algal cell separation, evaporation, excess amount of sludge, low-efficiency of organic matter removal, and release of particulate matter in the treated effluent during metal removal or pharmaceutical and dye-containing wastewater treatment (Tawfik et al., 2022). One of the most commonly used photocatalyst, TiO_2, are highly reactive under UV light which is not recommended for bioelectrochemical system. Furthermore, maintaining biocathodes, in terms of biofilm formation on the electrode surface, is a challenge for its efficient application. One of the main problems of using biocathodes in bioelectrochemical cells is to maintain the pH of the compartment for proper utilization of the buffer system. Since to maintain high rate of operation, photoelectrochemical cells integrated with MFCs needs high chemical cost for buffering agent to maintain the pH (Jaiswal et al., 2020). Formation of CO_2 is an additional main concern, as CO_2 adversely affects on biofilm formation and thus power generation (Tawfik et al., 2022). Oxygen diffusion is also a key factor in achieving high performance of biocathodes for proper functioning of biofilm on electrode surface. Though the use of bacteria and algae as consortium is a well-established method for pollutant removal and generation of high power from PBFC, but the effect of their quorum sensing crosstalk on system performance remains unknown (Wang et al., 2020).

Employing multidisciplinary approaches has provided unique tools to critically expand our understanding of photobioelectrocatalysis. Specifically, integration of quantum mechanism with electrochemical experiments helped to understand the reaction mechanism of the biohybrid systems (Zhang et al., 2019). Recent studies and advance technologies facilitate the understanding of the photoelectrochemical mechanisms in bacterial cells, biofilm formation on electrode surface, and selection of exogenous redox mediators if needed, which are otherwise particularly challenging to be studied. More precisely combining biology with engineering and implementing artificial redox mediating systems with artificial light-harvesting complexes could influence the efficiency of the light conversion process and promote its practical applications (Xiao & He, 2014).

Economic viability and scalability are two other major issues that need to be considered to build a practical photo/bioelectrochemical system (Zhang et al., 2019). Finding and development of cost effective materials are in progress for PBFC development Instead of using various costly semiconductors materials and various co-catalyst, like platinum, TiO_2. Non-precious metal or even nonmetal-based catalysts are in high demand for future applications. In addition to these components, the membrane separator also contributes to the high cost of the overall system and limits performance due to resistance and the substance crossover rate (Yang et al., 2020).

Degradation of azo dye leads to the formation of various toxic metabolites such as phenylenediamine and 5-aminosalicylic acid during bioelectrochemical degradation which requires posttreatment of the effluent to prevent the environment from disposal of toxic compounds. During bioelectrochemical treatment of antibiotic-containing wastewaters, horizontal transfer of antibiotic resistance gene could be possible from antibiotic-degrading microorganisms to electroactive microorganisms which lead to an increase in antibiotic tolerance of the electroactive microorganisms. Release of antibiotic resistance gene in the environment should be prohibited as it poses a significant health risk (Yan et al., 2019). Transfer/spreading of antibiotic resistance genes to the microorganisms and finally in the environment could be prevented by integration of bioelectrochemical system with photocatalysis. Photocatalysis readily degrades the antibiotics and reduces the chance of gene transfer to the environment. In a recent study, the removal of oxytetracycline and a decrease in toxicity were achieved in MES combined with photocatalysis (Rafieenia et al., 2022).

Finally, in spite of various challenges regarding electrode materials optimization, screening, and culture of functional microorganisms, large-scale engineering applications of photo-assisted bioelectrochemical fuel cells, there are glorious future possibilities. New achievements and developments in the areas of the architecture, electrode materials, membrane, and solution chemistry of PBFCs, all demonstrate a significant impact on the development of new technologies for pollutant removal as well as power generation. However, as MFCs were not able to generate electricity for large-scale applications, thereby the concept of using algae and photosynthetic bacteria became advantageous to enhance the output of bioelectrochemical fuel cells (Saratale et al., 2017). Therefore, improving the PBFC performances could be enhanced with improvement in the internal structure of the system (electrode materials, microbial species, membrane, etc.) and a proper understanding of operational parameters (temperature, pH, hydraulic retention time, etc). In addition, engineering application based on photo-assisted bioelectrochemical fuel cells is worth investigating (Yi et al., 2022).

14.6 Conclusion

This chapter summarized and critically assessed the recent developments of various photo-assisted bioelectrochemical fuel cells, specifically their fundamental mechanism, and practical application in terms of metal removal, antibiotic, and dye degradation. In the last few decades, the progress of PBFC systems for wastewater treatment has been limited to the optimization and standardization of the experimental setup. Finally, it can be concluded that the combination of the multidisciplinary approaches could provide a better effect for pollutant removal with bioelectricity generation instead of single implementation of photo-bioelectrocatalysis. However, to achieve the full potential of the PBFC system for successful commercialization, it will be necessary to improve the underlying process mechanisms, optimization of operational efficiency and manufacturing parameters, as well as further investigate of suitable microorganisms for metal removal and antibiotic and dye degradation. Development of biohybrid photosynthesis systems with extraordinary proficiencies could be possible with strong collaboration of scientists working on different fields such as bio-electrochemists, engineers, organic chemists, computational scientists, material scientists, and biophysicists to explore the required tools for the development and advancement of photo-bioelectrocatalysis process. Furthermore, the role of bacterial–algal interaction in photo-bioelectrochemical systems should be explored more in terms of genetical and functional analysis of microbial communities to manage their optimum performance in PBFCs.

References

Ahmadpour, T., Aber, S., and Hosseini, M.G. (2020). Improved dye degradation and simultaneous electricity generation in a photoelectrocatalytic microbial fuel cell equipped with AgBr/CuO hybrid photocathode. *Journal of Power Sources* 474: 228589. https://doi.org/10.1016/j.jpowsour.2020.228589.

Ahmadpour, T., Aber, S., and Hosseini, M.G. (2022). Visible-light enhanced azo dye degradation and power generation in a microbial photoelectrochemical cell using AgBr/ZnO composite photocathode. *Bioelectrochemistry* 146: 108139. https://doi.org/10.1016/j.bioelechem.2022.108139.

Aiyer, K.S. (2021). Synergistic effects in a microbial fuel cell between co-cultures and a photosynthetic alga *Chlorella vulgaris* improve performance. *Heliyon* 7 (1): e05935. https://doi.org/10.1016/j.heliyon.2021.e05935.

Akpan, U.G. and Hameed, B.H. (2009). Parameters affecting the photocatalytic degradation of dyes using TiO_2-based photocatalysts: a review. *Journal of Hazardous Materials* 170 (2–3): 520–529. https://doi.org/10.1016/j.jhazmat.2009.05.039.

Anam, M., Gomes, H.I., Rivers, G. et al. (2021). Evaluation of photoanode materials used in biophotovoltaic systems for renewable energy generation. *Sustainable Energy & Fuels* 5 (17): 4209–4232. https://doi.org/10.1039/D1SE00396H.

Angioni, S., Millia, L., Mustarelli, P. et al. (2018). Photosynthetic microbial fuel cell with polybenzimidazole membrane: synergy between bacteria and algae for wastewater removal and biorefinery. *Heliyon* 4 (3): e00560. https://doi.org/10.1016/j.heliyon.2018.e00560.

Apollon, W., Rusyn, I., González-Gamboa, N. et al. (2022a). Improvement of zero waste sustainable recovery using microbial energy generation systems: a comprehensive review. *Science of the Total Environment* 817: 153055. https://doi.org/10.1016/j.scitotenv.2022.153055.

Apollon, W., Vidales-Contreras, J.A., Rodríguez-Fuentes, H. et al. (2022b). Livestock's urine-based plant microbial fuel cells improve plant growth and power generation. *Energies* 15 (19): 6985. https://doi.org/10.3390/en15196985.

Bazdar, E., Roshandel, R., Yaghmaei, S., and Mardanpour, M.M. (2018). The effect of different light intensities and light/dark regimes on the performance of photosynthetic microalgae microbial fuel cell. *Bioresource Technology* 261: 350–360. https://doi.org/10.1016/j.biortech.2018.04.026.

Chen, S., Hou, Y., Rong, Y. et al. (2022). Hydroxyl radical and carbonate radical facilitate chlortetracycline degradation in the bio-photoelectrochemical system with a bioanode and a Bi_2O_3/CuO photocathode using bicarbonate buffer. *Chemosphere* 296: 134040. https://doi.org/10.1016/j.chemosphere.2022.134040.

Cui, Y., Rashid, N., Hu, N. et al. (2014). Electricity generation and microalgae cultivation in microbial fuel cell using microalgae-enriched anode and bio-cathode. *Energy Conversion and Management* 79: 674–680. https://doi.org/10.1016/j.enconman.2013.12.032.

Deka, R., Shreya, S., Mourya, M. et al. (2022). A techno-economic approach for eliminating dye pollutants from industrial effluent employing microalgae through microbial fuel cells: Barriers and perspectives. *Environmental Research* 212: 113454. https://doi.org/10.1016/j.envres.2022.113454.

Du, Y., Feng, Y., Qu, Y. et al. (2014). Electricity generation and pollutant degradation using a novel biocathode coupled photoelectrochemical cell. *Environmental Science & Technology* 48 (13): 7634–7641. https://doi.org/10.1021/es5011994.

Fischer, F. (2018). Photoelectrode, photovoltaic and photosynthetic microbial fuel cells. *Renewable and Sustainable Energy Reviews* 90: 16–27. https://doi.org/10.1016/j.rser.2018.03.053.

Fu, C.C., Hung, T.C., Wu, W.T. et al. (2010). Current and voltage responses in instant photosynthetic microbial cells with *Spirulina platensis*. *Biochemical Engineering Journal* 52 (2-3): 175–180. https://doi.org/10.1016/j.bej.2010.08.004.

Fuentes, J.L., Garbayo, I., Cuaresma, M. et al. (2016). Impact of microalgae-bacteria interactions on the production of algal biomass and associated compounds. *Marine Drugs 14* (5): 100. https://doi.org/10.3390/md14050100.

Gajda, I., Greenman, J., Melhuish, C., and Ieropoulos, I. (2013). Photosynthetic cathodes formicrobial fuel cells. *International Journal of Hydrogen Energy* 38: 11559–11564. https://doi.org/10.1016/j.ijhydene.2013.02.111.

Grattieri, M. (2020). Purple bacteria photo-bioelectrochemistry: enthralling challenges and opportunities. *Photochemical & Photobiological Sciences* 19 (4): 424–435. https://doi.org/10.1039/C9PP00470J.

Greenman, J., Gajda, I., and Ieropoulos, I. (2019). Microbial fuel cells (MFC) and microalgae; photo microbial fuel cell (PMFC) as complete recycling machines. *Sustainable Energy & Fuels* 3 (10): 2546–2560. https://doi.org/10.1039/C9SE00354A.

Guo, X., Yan, Z., Zhang, Y. et al. (2018). Behavior of antibiotic resistance genes under extremely high-level antibiotic selection pressures in pharmaceutical wastewater treatment plants. *Science of the Total Environment* 612: 119–128. https://doi.org/10.1016/j.scitotenv.2017.08.229.

Habibul, N., Hu, Y., Wang, Y.K. et al. (2016). Bioelectrochemical chromium (VI) removal in plant-microbial fuel cells. *Environmental Science & Technology* 50 (7): 3882–3889. https://doi.org/10.1021/acs.est.5b06376.

He, Z., Kan, J., Mansfeld, F. et al. (2009). Self-sustained phototrophic microbial fuel cells based on the synergistic cooperation between photosynthetic microorganisms and heterotrophic bacteria. *Environmental Science & Technology 43* (5): 1648–1654. https://doi.org/10.1021/es803084a.

He, R., Liu, L., Shi, P., and Nie, C. (2018). Environmental decontamination using photocatalytic fuel cells and photoelectrocatalysis-microbial fuel cells. *Journal of Chemical Technology & Biotechnology* 93 (11): 3336–3346. https://doi.org/10.1002/jctb.5729.

Hou, Y., Gan, Y., Yu, Z. et al. (2017). Solar promoted azo dye degradation and energy production in the bio-photoelectrochemical system with a g-C_3N_4/BiOBr heterojunction photocathode. *Journal of Power Sources* 371: 26–34. https://doi.org/10.1016/j.jpowsour.2017.10.033.

Hou, Y., Yuan, G., Qin, S. et al. (2020). Photocathode optimization and microbial community in the solar-illuminated bio-photoelectrochemical system for nitrofurazone degradation. *Bioresource Technology* 302: 122761. https://doi.org/10.1016/j.biortech.2020.122761.

Huang, L., Song, S., Cai, Z. et al. (2021). Efficient conversion of bicarbonate (HCO_3−) to acetate and simultaneous heavy metal Cr (VI) removal in photo-assisted microbial electrosynthesis systems combining WO_3/MoO_3/g-C_3N_4 heterojunctions and *Serratia marcescens* electrotroph. *Chemical Engineering Journal 406*: 126786. https://doi.org/10.1016/j.cej.2020.12678.

Jaiswal, K.K., Kumar, V., Vlaskin, M.S. et al. (2020). Microalgae fuel cell for wastewater treatment: recent advances and challenges. *Journal of Water Process Engineering* 38: 101549. https://doi.org/10.1016/j.jwpe.2020.101549.

Kannan, N. and Donnellan, P. (2021). Algae-assisted microbial fuel cells: a practical overview. *Bioresource Technology Reports* 15: 100747. https://doi.org/10.1016/j.biteb.2021.100747.

Khan, M.J., Singh, N., Mishra, S. et al. (2022). Impact of light on microalgal photosynthetic microbial fuel cells and removal of pollutants by nanoadsorbent biopolymers: updates, challenges and innovations. *Chemosphere* 288: 132589. https://doi.org/10.1016/j.chemosphere.2021.132589.

Kurwadkar, S. (2019). Occurrence and distribution of organic and inorganic pollutants in groundwater. *Water Environment Research* 91 (10): 1001–1008. https://doi.org/10.1002/wer.1166.

Li, J., Li, M., Zhang, J. et al. (2013). A microbial fuel cell capable of converting gaseous toluene to electricity. *Biochemical Engineering Journal* 75: 39–46. https://doi.org/10.1016/j.bej.2013.03.015.

Liu, L. and Choi, S. (2017). Self-sustaining, solar-driven bioelectricity generation in micro-sized microbial fuel cell using co-culture of heterotrophic and photosynthetic bacteria. *Journal of Power Sources* 348: 138–144. https://doi.org/10.1016/j.jpowsour.2017.03.014.

Liu, W., Li, Y., Liu, F. et al. (2019). Visible-light-driven photocatalytic degradation of diclofenac by carbon quantum dots modified porous g-C_3N_4: mechanisms, degradation pathway and DFT calculation. *Water Research* 151: 8–19. https://doi.org/10.1016/j.watres.2018.11.084.

Logroño, W., Pérez, M., Urquizo, G. et al. (2017). Single chamber microbial fuel cell (SCMFC) with a cathodic microalgal biofilm: a preliminary assessment of the generation of bioelectricity and biodegradation of real dye textile wastewater. *Chemosphere* 176: 378–388. https://doi.org/10.1016/j.chemosphere.2017.02.099.

Long, X., Pan, Q., Wang, C. et al. (2017). Microbial fuel cell-photoelectrocatalytic cell combined system for the removal of azo dye wastewater. *Bioresource Technology* 244: 182–191. https://doi.org/10.1016/j.biortech.2017.07.088.

Long, X., Wang, H., Wang, C. et al. (2019). Enhancement of azo dye degradation and power generation in a photoelectrocatalytic microbial fuel cell by simple cathodic reduction on titanium nanotube arrays electrode. *Journal of Power Sources 415*: 145–153. https://doi.org/10.1016/j.jpowsour.2019.01.069.

Luo, S., Berges, J.A., He, Z., and Young, E.B. (2017). Algal-microbial community collaboration for energy recovery and nutrient remediation from wastewater in integrated photobioelectrochemical systems. *Algal Research* 24: 527–539. https://doi.org/10.1016/j.algal.2016.10.006.

Marsolek, M.D., Torres, C.I., Hausner, M., and Rittmann, B.E. (2008). Intimate coupling of photocatalysis and biodegradation in a photocatalytic circulating-bed biofilm reactor. *Biotechnology and Bioengineering* 101 (1): 83–92. https://doi.org/10.1002/bit.21889.

Nguyen, H.T., Kakarla, R., and Min, B. (2017). Algae cathode microbial fuel cells for electricity generation and nutrient removal from landfill leachate wastewater. *International Journal of Hydrogen Energy* 42 (49): 29433–29442. https://doi.org/10.1016/j.ijhydene.2017.10.011.

Qian, F., Wang, H., Ling, Y. et al. (2014). Photo enhanced electrochemical interaction between *Shewanella* and a hematite nanowire photoanode. *Nano Letters* 14 (6): 3688–3693. https://doi.org/10.1021/nl501664n.

Rafieenia, R., Sulonen, M., Mahmoud, M. et al. (2022). Integration of microbial electrochemical systems and photocatalysis for sustainable treatment of organic recalcitrant wastewaters: main mechanisms, recent advances, and present prospects. *Science of the Total Environment* 824: 153923. https://doi.org/10.1016/j.scitotenv.2022.153923.

Ren, G., Sun, Y., Lu, A. et al. (2018). Boosting electricity generation and Cr (VI) reduction based on a novel silicon solar cell coupled double-anode (photoanode/bioanode) microbial fuel cell. *Journal of Power Sources* 408: 46–50. https://doi.org/10.1016/j.jpowsour.2018.10.081.

Rusyn, I. (2021). Role of microbial community and plant species in performance of plant microbial fuel cells. *Renewable and Sustainable Energy Reviews* 152: 111697. https://doi.org/10.1016/j.rser.2021.111697.

Saratale, R.G., Kuppam, C., Mudhoo, A. et al. (2017). Bioelectrochemical systems using microalgae—a concise research update. *Chemosphere* 177: 35–43. https://doi.org/10.1016/j.chemosphere.2017.02.132.

Sekar, N., Jain, R., Yan, Y., and Ramasamy, R.P. (2016). Enhanced photo-bioelectrochemical energy conversion by genetically engineered cyanobacteria. *Biotechnology and Bioengineering 113* (3): 675–679. https://doi.org/10.1002/bit.25829.

Sevda, S., Garlapati, V.K., Sharma, S. et al. (2019). Microalgae at niches of bioelectrochemical systems: a new platform for sustainable energy production coupled industrial effluent treatment. *Bioresource Technology Reports* 7: 100290. https://doi.org/10.1016/j.biteb.2019.100290.

Su, S., Zhang, Y., Hu, W. et al. (2020). Efficient and synergistic decolourization and nitrate removal using a single-chamber with a coupled biocathode-photoanode system. *Bioelectrochemistry* 132: 107439. https://doi.org/10.1016/j.bioelechem.2019.107439.

Sui, M., Dong, Y., Wang, Z. et al. (2017). A biocathode-driven photocatalytic fuel cell using an Ag-doped TiO_2/Ti mesh photoanode for electricity generation and pollutant degradation. *Journal of Photochemistry and Photobiology A: Chemistry* 348: 238–245. https://doi.org/10.1016/j.jphotochem.2017.08.047.

Sun, J., Xu, W., Yuan, Y. et al. (2019a). Bioelectrical power generation coupled with high-strength nitrogen removal using a photo-bioelectrochemical fuel cell under oxytetracycline stress. *Electrochimica Acta* 299: 500–508. https://doi.org/10.1016/j.electacta.2019.01.036.

Sun, J., Xu, W., Yang, P. et al. (2019b). Enhancing the performance of photo-bioelectrochemical fuel cell using graphene oxide/cobalt/polypyrrole composite modified photo-biocathode in the presence of antibiotic. *International Journal of Hydrogen Energy* 44 (3): 1919–1929. https://doi.org/10.1016/j.ijhydene.2018.11.111.

Sun, J., Xu, W., Yang, P. et al. (2019c). Enhanced oxytetracycline removal coupling with increased power generation using a self-sustained photo-bioelectrochemical fuel cell. *Chemosphere* 221: 21–29. https://doi.org/10.1016/j.chemosphere.2018.12.152.

Tanvir, R.U., Zhang, J., Canter, T. et al. (2021). Harnessing solar energy using phototrophic microorganisms: a sustainable pathway to bioenergy, biomaterials, and environmental solutions. *Renewable and Sustainable Energy Reviews* 146: 111181. https://doi.org/10.1016/j.rser.2021.111181.

Tawfik, A., Niaz, H., Qadeer, K. et al. (2022). Valorization of algal cells for biomass and bioenergy production from wastewater: sustainable strategies, challenges, and techno-economic limitations. *Renewable and Sustainable Energy Reviews* 157: 112024. https://doi.org/10.1016/j.rser.2021.112024.

Tong, Y., Wei, J., Mo, R. et al. (2022). Photocatalytic microbial fuel cells and performance applications: a review. *Frontiers in Chemistry* 10: 953434. https://doi.org/10.3389/fchem.2022.953434.

Tu, L., Hou, Y., Yuan, G. et al. (2020). Bio-photoelectrochemcial system constructed with $BiVO_4$/RGO photocathode for 2, 4-dichlorophenol degradation: $BiVO_4$/RGO optimization, degradation performance and mechanism. *Journal of Hazardous Materials* 389: 121917. https://doi.org/10.1016/j.jhazmat.2019.121917.

Tu, L., Rong, Y., Yu, Z. et al. (2022). Chlortetracycline degradation performance and mechanism in the self-biased bio-photoelectrochemical system constructed with an oxygen-defect-rich BiVO4/Ni9S8 photoanode. *Chemosphere* 295: 133787. https://doi.org/10.1016/j.chemosphere.2022.133787.

Venkidusamy, V., Nallusamy, S., Nammalvar, G. et al. (2023). ZnO/graphene composite from solvent-exfoliated few-layer graphene nanosheets for photocatalytic dye degradation under sunlight irradiation. *Micromachines 14* (1): 189. https://doi.org/10.3390/mi14010189.

Wang, Q., Cai, Z., Huang, L. et al. (2019). Intensified degradation and mineralization of antibiotic metronidazole in photo-assisted microbial fuel cells with Mo-W catalytic cathodes under anaerobic or aerobic conditions in the presence of Fe (III). *Chemical Engineering Journal* 376: 119566. https://doi.org/10.1016/j.cej.2018.07.168.

Wang, Z., He, Z., and Young, E.B. (2020). Toward enhanced performance of integrated photo-bioelectrochemical systems: Taxa and functions in bacteria–algae communities. *Current Opinion in Chemical Biology* 59: 130–139. https://doi.org/10.1016/j.cbpa.2020.05.010.

Wang, C., Xing, Y., Zhang, K. et al. (2023). Evaluation of photocathode coupling-mediated hydroxychloroquine degradation in a single-chamber microbial fuel cell based on electron transfer mechanism and power generation. *Journal of Power Sources* 559: 232625. https://doi.org/10.1016/j.jpowsour.2022.232625.

Wu, X.Y., Song, T.S., Zhu, X.J. et al. (2013). Construction and operation of microbial fuel cell with *Chlorella vulgaris* biocathode for electricity generation. *Applied Biochemistry and Biotechnology* 171: 2082–2092. https://doi.org/10.1007/s12010-013-0476-8.

Wu, Y.C., Wang, Z.J., Zheng, Y. et al. (2014). Light intensity affects the performance of photo microbial fuel cells with *Desmodesmus* sp. A8 as cathodic microorganism. *Applied Energy 116*: 86–90. https://doi.org/10.1016/j.apenergy.2013.11.066.

Xiao, L. and He, Z. (2014). Applications and perspectives of phototrophic microorganisms for electricity generation from organic compounds in microbial fuel cells. *Renewable and Sustainable Energy Reviews* 37: 550–559. https://doi.org/10.1016/j.rser.2014.05.066.

Xiao, L., Young, E.B., Berges, J.A., and He, Z. (2012). Integrated photo-bioelectrochemical system for contaminants removal and bioenergy production. *Environmental Science & Technology* 46 (20): 11459–11466. https://doi.org/10.1021/es303144n.

Xiong, H., Zou, D., Zhou, D. et al. (2017a). Enhancing degradation and mineralization of tetracycline using intimately coupled photocatalysis and biodegradation (ICPB). *Chemical Engineering Journal* 316: 7–14. https://doi.org/10.1016/j.cej.2017.01.083.

Xiong, J.Q., Kurade, M.B., and Jeon, B.H. (2017b). Biodegradation of levofloxacin by an acclimated freshwater microalga, *Chlorella vulgaris*. *Chemical Engineering Journal* 313: 1251–1257. https://doi.org/10.1016/j.cej.2016.11.017.

Xiong, J.Q., Kurade, M.B., and Jeon, B.H. (2017c). Ecotoxicological effects of enrofloxacin and its removal by monoculture of microalgal species and their consortium. *Environmental Pollution* 226: 486–493. https://doi.org/10.1016/j.envpol.2017.04.044.

Xiong, J., Guo, S., Zhao, T. et al. (2020). Degradation of methylene blue by intimate coupling photocatalysis and biodegradation with bagasse cellulose composite carrier. *Cellulose* 27 (6): 3391–3404. https://doi.org/10.1007/s10570-020-02995-0.

Xiong, Q., Hu, L.X., Liu, Y.S. et al. (2021). Microalgae-based technology for antibiotics removal: from mechanisms to application of innovational hybrid systems. *Environment International* 155: 106594. https://doi.org/10.1016/j.envint.2021.106594.

Xu, C., Poon, K., Choi, M.M., and Wang, R. (2015). Using live algae at the anode of a microbial fuel cell to generate electricity. *Environmental Science and Pollution Research* 22: 15621–15635. https://doi.org/10.1007/s11356-015-4744-8.

Xu, P., Zheng, D., He, Q., and Yu, J. (2020). The feasibility of ofloxacin degradation and electricity generation in photo-assisted microbial fuel cells with $LiNbO_3$/CF photocatalytic cathode. *Separation and Purification Technology* 250: 117106. https://doi.org/10.1016/j.seppur.2020.117106.

Yan, W., Xiao, Y., Yan, W. et al. (2019). The effect of bioelectrochemical systems on antibiotics removal and antibiotic resistance genes: a review. *Chemical Engineering Journal* 358: 1421–1437. https://doi.org/10.1016/j.cej.2018.10.128.

Yang, F., Ke, Z., Li, Z. et al. (2020). Photo/bio-electrochemical systems for environmental remediation and energy harvesting. *ChemSusChem* 13 (13): 3391–3403. https://doi.org/10.1002/cssc.202000203.

Yang, L., Yi, G., Wang, B. et al. (2022). Atomic H* enhanced electrochemical recovery towards high-value-added metallic Sb from complex mine flotation wastewater. *Resources, Conservation and Recycling* 178: 106020. https://doi.org/10.1016/j.resconrec.2021.106020.

Yi, G., Wang, B., Feng, Y. et al. (2022). The ins and outs of photo-assisted microbial electrochemical systems for synchronous wastewater treatment and bioenergy recovery. *Resources, Conservation and Recycling* 181: 106230. https://doi.org/10.1016/j.resconrec.2022.106230.

Yuan, S.J., Sheng, G.P., Li, W.W. et al. (2010). Degradation of organic pollutants in a photoelectrocatalytic system enhanced by a microbial fuel cell. *Environmental Science & Technology* 44 (14): 5575–5580. https://doi.org/10.1021/es101317z.

Zakaria, B.S. and Dhar, B.R. (2022). A review of stand-alone and hybrid microbial electrochemical systems for antibiotics removal from wastewater. *Processes* 10 (4): 714. https://doi.org/10.3390/pr10040714.

Zhang, Y., He, Q., Xia, L. et al. (2018). Algae cathode microbial fuel cells for cadmium removal with simultaneous electricity production using nickel foam/graphene electrode. *Biochemical Engineering Journal* 138: 179–187. https://doi.org/10.1016/j.bej.2018.07.021.

Zhang, Y., Liu, M., Zhou, M. et al. (2019). Microbial fuel cell hybrid systems for wastewater treatment and bioenergy production: synergistic effects, mechanisms and challenges. *Renewable and Sustainable Energy Reviews* 103: 13–29. https://doi.org/10.1016/j.rser.2018.12.027.

Zhang, L., Zhang, Y., Xiao, K. et al. (2021). The intimate coupling of photocatalysis and biodegradation for the degradation and mineralization of atrazine in water. *New Journal of Chemistry* 45 (29): 13029–13039. https://doi.org/10.1039/D1NJ00134E.

Zhao, R., Feng, J., Liu, J. et al. (2019). Deciphering of microbial community and antibiotic resistance genes in activated sludge reactors under high selective pressure of different antibiotics. *Water Research* 151: 388–402. https://doi.org/10.1016/j.watres.2018.12.034.

Zhou, D., Xu, Z., Dong, S. et al. (2015). Intimate coupling of photocatalysis and biodegradation for degrading phenol using different light types: visible light vs. UV light. *Environmental Science & Technology* 49: 7776–7783. https://doi.org/10.1021/acs.est.5b00989.

Zhou, D., Dong, S., Shi, J. et al. (2017). Intimate coupling of an N-doped TiO_2 photocatalyst and anode respiring bacteria for enhancing 4-chlorophenol degradation and current generation. *Chemical Engineering Journal* 317: 882–889. https://doi.org/10.1016/j.cej.2017.02.128.

15

Agriculture-based Crop in PMFCs for the Futuristic Sustainable Protected Agriculture

Divya Shanmugavel[1], Omar Solorza-Feria[2], and Sathish-Kumar Kamaraj[3]

[1] *Programme of Nanoscience and Nanotechnology, CINVESTAV del IPN, Hydrogen and Fuel Cells Group, México, México*
[2] *Department of Chemistry, CINVESTAV del IPN, Hydrogen and Fuel Cells Group, México, México*
[2] *Instituto Politécnico Nacional (IPN)-Centro de Investigación en Ciencia Aplicada y Tecnología Avanzada, Unidad Altamira (CICATA-Altamira), Altamira, Tamps., México*

15.1 Introduction

As the population increases day by day, so does the consumption of food and energy resources. This poses major demanding situations to agriculture and electricity manufacturing exacerbated with the aid of the impacts of weather exchange, land degradation, and depletion of resources which include freshwater, fossil fuels, and fertile topsoil. According to Alexandratos and Bruinsma (2012), annual yields will increase by less than 1% over the next few years, leaving very limited scope for the expansion of cultivated land. The world's food demand for agricultural crops is increasing and is likely to continue to rise over the next few decades with a growing world population of 2.3 billion and projected increases in per capita income by mid-century (Godfray et al., 2010). Asia, especially China, India, and Southeast Asia, will see the highest population growth, with Central America accounting for about 60%, and by 2050, the world's population will grow. Africa will grow the fastest, more than 70% faster than Asia. Asia's annual growth rate is 2.5%, while the global average is 2.3%, and only 1.3% in many developed countries (UNDP 2020). In addition, the total area available for agricultural production is

Photosynthesis-Assisted Energy Generation: From Fundamentals to Lab Scale and In-Field Applications, First Edition. Edited by Sathish-Kumar Kamaraj and Iryna Rusyn.

driven by the need for land for other purposes such as infrastructure development, urbanization, bioenergy production, biodiversity conservation (Sands and Leimbach, 2003), and land degradation increasingly restricted. Agriculture is inherently sensitive to climate variability and change, both from natural reasons and from human pastime.

Agricultural production is another factor affected by global climate change in the second half of the 21st century. The impact of climate change on agricultural productivity will lead to a decrease in agricultural land, and land scarcity is expected by 2050 (Nelson et al. 2010). The global challenge for agriculture and energy consumption is to provide food and energy for a growing world population. This is because the population is projected to reach nine billion by 2050. Sustainable and increasing agriculture includes the adoption of economically and environmentally sustainable farming practices. This is a useful approach to addressing the food security challenges facing the world due to population growth and environmental degradation that are having long-term impacts on agriculture around the world (Simon et al. 2013). Sustainable agriculture is typically categorized into three major targets: economic viability, environmental sustainability, and ethical acceptability. The meaning of the term "sustainability" has long been debated. Originally, the term was used to describe agricultural and industrial techniques that mitigate or prevent environmental degradation often associated with economic activity. Sustainable agriculture is defined as agriculture that can economically sustain constant consumption or productivity by alternately using natural resources and man-made capital in production. This was noted by Hartwick (1978) and Solow (1974). Define it ecologically, noting that natural assets and man-made capital are complementary within the manufacturing system and that herbal sources are the restricting elements of production and consequently have to be conserved. Applying this concept to agriculture, the Batie (1989) describes "sustainable agriculture as improving the long-term environmental quality and resource base on which agriculture depends and providing basic human resources for food and fiber." This satisfies requirements, is affordable, and sustainable, and enhances farmers' and society's quality of life. Worldwide strength call for is developing at a price of 7% every 12 months as the sector populace grows. Plant microbial fuel cells (PMFC) have received a super deal of attention in recent years as a rising energy-green era for wastewater treatment, desalination, sustainable power generation, price-brought energy technology, and chemical manufacturing. On this machine, electricity era is carried out by using combining the era of electricity with the aid of microorganism through the oxidation of organic compounds and the deposition of natural compounds by using plant life within the rhizosphere (Sophia and Sreeja, 2017). Essentially, living vegetation performs photosynthetic approaches and brings natural compounds consisting of glucose and sucrose for boom and replica. Most of these organic compounds are released into the soil through their roots. Microorganisms surrounding the roots certainly convert natural compounds into electricity. The excess from this process are electrons released into the environment (Effendi et al., 2022).

PMFC can be used not only for waste treatment but also for energy and value-added product generation. For instance, electrical power can be received at once from the oxidation of natural compounds in PMFC. Due to the fact that they are the most simple and great efforts were made to broaden them. Power conversion with the aid of microorganisms has numerous blessings. It is environmentally friendly, sustainable and may be operated with unique substrates over an extensive concentration variety. PMFC has the capacity for power-efficient wastewater treatment, renewable strength era, water reuse, and bioremediation. PMFCs seem like a super hobby for solving some of the maximum essential water, power, and nutrient troubles for sustainable agriculture. The purpose of this chapter is to introduce plants used in agriculture and to discuss key factors important for the development of PMFCs for practical applications in sustainable agriculture, including energy production (Dhanarajan., 2017).

15.2 Challenges for Agriculture

Over the centuries, the growth of agriculture has contributed to the upward thrust of human civilization. Earlier than agriculture advanced, human beings spent maximum of their time collecting meals, hunting wild animals, and collecting wild vegetation. Approximately 11,500 years in the past, human beings found out the way to develop grains and root vegetation and tailored their way of life to agriculture. 2000 years ago, humans on Earth depended on agriculture. When people first started growing crops, rice and corn were probably the first crops to be domesticated. Agriculture allowed people to produce surplus food (PreventionWeb, n.d.; Kwon and Park, 2021). Food surpluses allowed people to pursue other jobs not related to agriculture. Agriculture developed independently in every region of the world, but always began with the domestication of domesticated plants and animals. Some of the plants are cereals and legumes that serve as the main source of carbohydrates and proteins for mankind, complementing and enriching the human diet by providing the energy needed for agricultural activities such as tillage supplemented by a variety of animals that carry water, carry products, grind grain, etc. (Diamond, 1997). With the onset of the Industrial Revolution, many important scientific discoveries were made, and technological changes led to a better understanding and development of agricultural activities. The use of fossil fuels has replaced animal and human power in agriculture with self-propelled machines. Agriculture's current challenge is to ensure global food security as the population is projected to grow to 9 billion people in the next 30 years. Agriculture provides enough food to feed all mankind, but food shortages are occurring in some parts of the world, with nearly 1 billion people undernourished. But by 2050, agriculture will need to produce 60–110% more than it did in 2006 to feed everyone on the planet (Ray et al., 2013). This means that production of crops such as corn, rice, wheat, and soybeans will need to increase by 2.5% annually. The next challenge is that agricultural production must be based on sustainable standards. Currently, only 38% of the land area on earth is used for growing agricultural products. This shows the impact of agriculture on the environment. Cultivated acreage may increase but is unlikely in the future, as can be inferred from the fact that it has increased by only 3% over the past 20 years (Foley et al., 2011). Even the cost of new land may be too high to generate meaningful benefits from an environmental and economic standpoint, and some of the new farmland may be used as primary farmland or nonedible land for bioenergy production. It is used for urban growth of crops. Only 62% of agricultural production is used for human consumption, 35% for livestock, and 3% for biofuel production. Water and soil quality, as well as biodiversity conservation, are of great importance for agro-environmental sustainability. One of the most important factors for plant growth is water, which is ubiquitous on Earth. More than 70% of water is used to water crops and reused, but misuse of this water can lead to major problems such as water pollution (Chel and Koushik, 2011). This pollution not only destroys soil but also contaminates important areas such as estuaries and can run off into the sea, severely limiting biodiversity.

Activities may become difficult or even impossible to carry out. The products used to improve crop yields are the same which cause some of the water and soil pollutants. The former controls pests, diseases, and weeds that cause significant losses to agricultural production. However, many of them are toxic to other plant and animal species and threaten agricultural biodiversity. These fertilizers are readily available and create an optimal environment for root development, resulting in vigorous growth and high yields. Instead, it is estimated to either remain in the soil or migrate to other ecosystems (Foley et al., 2011; Rusyn et al., 2020). Soil chemistry varies spatially with topography and various interacting factors. The challenge is to quantify the soil differences. In-field biological variability is as great as soil variability, including soil microbial populations, weed populations, insect populations, disease incidence, crop growth, and yields, and farmers are aware of all biological variations that result from the fluctuation.

Climate change, caused by recent changes in atmospheric composition, is the final and major problem. Because of the growing concentrations of greenhouse gases along with carbon dioxide, methane, and nitrous oxide, the results of greenhouse gases increase temperatures. Furthermore, one of the main reasons for the production of these gases is the use of fossil fuels. However, agriculture and forestry directly produce a significant portion (21%) of these gases, which could rise to 30% when all contributing factors are considered (e.g. food industry and transport; FAO 2016).

15.3 Development of Plant Microbial Fuel Cells

Over the past decades, many researchers have worked on bioelectricity generation by the rhizosphere, now known as PMFCs. This technology has attracted the attention of many researchers and scientists. Because PMFCs can convert chemical energy stored in organic matter through microbial metabolism into electrical energy without leaving an ecological footprint (Syed et al., 2021).

PMFCs have received a remarkable deal of interest in recent years as a brand-new technology for strength-green wastewater remedy, desalination, sustainable strength technology, and fee-brought chemical manufacturing. A standard PMFC includes an anode, a cathode, and a membrane keeping apart those chambers. At the anode, natural dependencies are oxidized by microbes, releasing electrons and protons. Electrons are carried away by using wires that act as modern-day transporters, and protons journey across the cathode through an optional membrane. For the PMFC gadget, the soil acts as a membrane and acts as a single sedimentary formation with awesome anaerobic and cardio zones developing. A reduction reaction happens on the cathode, wherein electrons, protons, and oxygen react to form water (Figure 15.1).

Paddy fields are used for harvesting bioenergy in the form of biomass and bioelectricity. Over time, PMFC generation has visible inexperienced roofs, floating bodies of water, wetlands, and paddy fields (Rusyn & Hamkalo, 2020). From lighting your lawn to working your radio, this technology is a robust contender for future inexperienced power generation. Researchers are actively working on PMFCs with a focal point on energy generation under natural conditions. Figure 15.2 shows the historical progress of PMFC research (Regmi et al., 2018a).

Alternatives together with membrane bioreactor technology restoration, wetlands, membrane filtration, and lagoon pools require excessive capital prices and land. PMFC could be a way to counter this technology. PMFCs have the potential to generate clean energy and also help sustainable and protected agriculture, which is the main theme of this chapter. In summary, PMFCs can be an effective approach for generating electricity in agriculture. Optimization of monitoring system design for agricultural drainage characteristics such as irrigation, fertilization, pesticide management, pH, dissolved oxygen, turbidity, conductivity, and nutrients. The decision to use PMFCs as an energy source for power generation

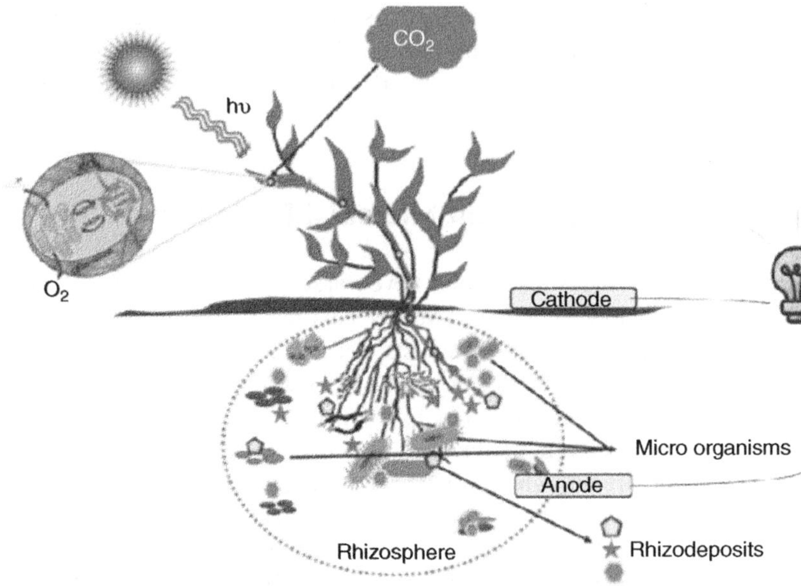

Figure 15.1 Mechanism of power generation by PMFC and formation of root exudates during photosynthetic process by plants.

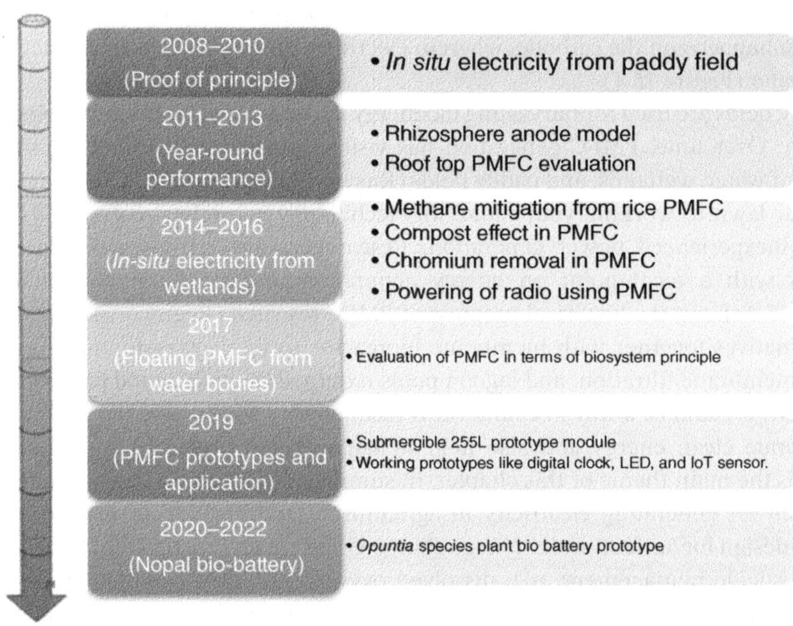

Figure 15.2 Timeline of PMFC evolution.

from agricultural crops depends on accessibility to water sources, water levels, sediment or soil properties (organic matter/moisture content, permeability, etc.), and richness and diversity of microbial communities. It also helps monitor many factors and installation space availability (Li et al., 2015).

15.4 Agriculture-Based Crops in PMFCs

Bioelectricity can be generated without harvesting plants by using PMFCs. By converting atmospheric CO_2 into carbohydrates, plants use chlorophyll for photosynthesis. In the process of photosynthesis, terrestrial flora repair 67.5 billion heaps of carbon annually between the land and the ecosystem. In rhizome deposition, about 47.25 billion tonnes of constant carbon are transferred to the roots, in which they're transported into the rhizosphere by means of exudates (sugars, natural acids, and so forth), polymeric secretions containing carbohydrates and enzymes, gases inclusive of ethylene, CO_2, and useless cells are emitted to material. Root exudate not only provides carbon but also stimulates chemotaxis by soil microbes to reach the rhizosphere. Rhizobia are attracted to root hairs inside the rhizosphere via chemoattractants including aromatic acids, dicarboxylic acids, flavonoids, and amino acids. Figure 15.1 shows how root exudate and microbial metabolites such as formate, succinate, and biotin function as mediators of electron transfer between cells (Chiranjeevi et al., 2019).

It is found that microorganisms can donate electrons from nodule deposits to solid surfaces such as graphite electrodes under anaerobic conditions. At the cathode, electrons are used together with protons to reduce oxygen or some other compound, and the protons are transported throughout the membrane to the cathode compartment. Vegetation developing within the anode chamber is submerged in water together with their roots, creating an anaerobic environment within the anode. PMFC was first described by Helder et al. (2012) and aims to generate nondestructive and sustainable bioelectricity. This requirement stems from the fact that only rhizome deposits, not biomass, need to be harvested for bioenergy production in PMFC. This avoids biomass transport and nutrient degradation in the ecosystem. Selection of plants for PMFCs is typically based on local availability rather than knowledge of plant physiology or morphology. Plant selection is based on salt tolerance, widespread occurrence in the region, and the fact that they do not compete with the land used for food production. Vegetation is labeled in step with their photosynthetic pathway into C3, C4, and CAM. C4 plant life is the high-quality type to transform CO_2 into four-carbon sugar compounds and inputs the Calvin cycle. This type of plant could be very green in a dry, warm ecosystem and produces copious amounts of bioenergy. C_4 plant life introduces more nodules as microbial substrates near the roots, growing strength recuperation in PMFCs. The photosynthetic rate is in the range of 2.5–5%. Photosynthesis depends on CO_2 attention and light intensity PAR (photosynthetically active radiation) at the plant level. Different plant species, inclusive of grasses and monocotyledons, have adventitious root systems that provide a bigger rhizosphere surface area for microbial propagation. Plants can grow in humid conditions to prevent atmospheric oxygen from reaching the anode. Therefore, plant species selection helps increase energy production in PMFC (Apollon et al., 2022).

Plant subsidies for MFCs using flowering species plants such as *Ipomoea aquatica*, *Typha latifolia*, and *Canna indica*, and grass species plants such as *Oryza sativa*, *Glyceria maxima*, *Spartina anglica*, and *Arundinella anomala*. Numerous studies have demonstrated material-related bioenergy production. Among these plants, *Spartina anglica*, *Arundinella anomala*, and *Arundo donax* are wetland plants that can survive and grow under low-water conditions such as those found in PMFC (Awerbuch et al., 2002; Wijte et al., 2005). However, some of these plants used in PMFC are not agricultural crops. For this reason, many researchers have started working on agricultural plants that can enhance crop fertility and biopower generation (Helder, 2012). These include rice, green beans, pear cacti, tomatoes, corn, sweet leaves, soft wheat, red rice, and bulrush. Some medicinal plants such as green amaranth are also used in PMFCs.

15.4.1 C3 Photosynthetic Plant

A normal plant that does not have photosynthetic variations to reduce photorespiration is called a C3 plant. The first step of the Calvin cycle is the fixation of carbon dioxide by way of rubisco, and plant life that use only this "widespread" mechanism of carbon fixation is known as C3 vegetation plants, for the 3-carbon compound (3-PGA) the reaction produces. About 85% of the plant species on the earth are C3 plants, inclusive of rice, wheat, soybeans, and all timber trees.

15.4.1.1 *Oryza sativa*

Oryza sativa was developed to utilize bioelectricity and simultaneously purify wastewater with PMFCs. Between 2008 and 2018, 47 plant species have been integrated into the PMFC and *Oryza sativa* (Asian rice) become the best studied for PMFC operation. Japonica types being studied for PMFC development are Bomba, Coral, Koshihikari, Samgwang, Sasanishiki, and Satojiman. In many Asian and African countries, the demand for rice is big because of population increase and economic growth, making rice manufacturing one of the most important economic activities. PMFC using *Oryza sativa* have maximum electricity density which is increased from 2.81 to 70.14 mW/m^2. The use of PMFC during energy harvesting does not appear to limit the expansion of rice plant life. Indeed, rice PMFCs operating in a closed cycle contained more soil organic matter and produced more electrical energy and plant biomass than those operating in open cycle (Khudzari et al., 2019).

Srivastava et al. (2015) designed and developed a PMFC in paddy fields with an integrated algal (*Cyanobacterium phormidium*) cathode. Electrochemically active bacteria in the anodic compartment were supported by way of paddy vegetation to fulfill their carbon necessities. Voltage generation with the aid of the plant microbial fuel mobile of the algae cathode device changed into dynamic throughout its operational lifestyles. Although all the performances of the algal cathodic PMFCs are noteworthy among the studies listed, the reported values for the algal cathodic microbial fuel cells are normalized by the volume of the anode chamber. Their examination correctly tested low-price algae-assisted oxygen generation for cathodic reactions and root sediments and root exudates from paddy vegetation as carbon assets for anodic reactions. Their experiments performed a power density of 29.78 mW/m^3 and a current density of 610 mA/m^3. This was the first design of a paddy field-based microbial fuel cell with a cyanobacteria cathode that can operate in a self-sufficient mode in a completely natural process, without an external supply of organic or oxidant (Srivastava et al., 2015).

15.4.1.2 *Vigna radiata*

Construction of a plant-microbial fuel cell using graphite rods and stainless steel mesh as electrode material with mung beans (*Vigna radiata Wilzeck*). Comparison of potentials should determine the ability to generate electricity in fuel cell chamber designs using different electrode materials. Both systems using graphite and stainless steel electrodes with *V. radiata* have confirmed the ability to simultaneously produce food and generate electricity in the designed single-chamber PMFC. However, the low power density produced is still a challenge as it is insufficient for applicable applications (Pamintuan & Sanchez, 2019). Plants in the PMFC system containing *Vigna radiata* grew faster, were taller, and had higher pod yields than those in the control system. These results demonstrate that PMFC technology can be deployed in farmland to continuously generate passive power while growing food, eliminating competition between power generation and agriculture (Rusyn, 2021).

15.4.1.3 *Solanum lycopersicum*

Tomatoes are a popular economic crop and can be grown anywhere. This type of plant has been little studied in the PMFC field. The bacterial colonies produced by each plant root system are thought to play a role in biofilm formation and contribution to electricity generation. These plants help us better understand the relationship between photosynthesis and the co-generation of bioelectricity. The average power density produced was 24 mW/m^2. Crops such as tomatoes can also generate electricity, but these plants are usually growth-stage dependent and require long start-up times to generate electricity (Nitisoravut and Regmi, 2017).

15.4.1.4 *Stevia rebaudiana*

Stevia rebaudiana, a plant species, was used to generate bioelectricity in PMFC. Due to its special properties (e.g. sugar substitute), it is a plant of particular importance worldwide. Thus, Apollon et al. (2022) used this plant species in a PMFC system. This represents an exceptionally good opportunity to increase the biomass needed for sugar production. Moreover, embedding PMFC in this plant species is a new challenge for them around the world. This may provide further avenues for future research to improve *Stevia* biomass. Similarly, their work could be carried out at a huge scale and at a low price with the usage of revolutionary natural materials. It aims to develop a PMFC system or technique with high efficiency that can be achieved. Such a technique to convert organic matter from the PMFC inoculated with domesticated animals' urine into stored energy could be an excellent option for improving the efficiency of commercial use of PMFC (Apollon et al., 2022).

15.4.2 C4 Photosynthetic Plant

In C4 plants, the light response and the Calvin cycle are physically separated, with the mildly structured response occurring within the mesophyll cells (the spongy tissue in the center of the leaf) and the Calvin cycle in specialized cells around the leaf veins occurs in cells. These cells are called packet-trading envelope cells. For example, in C4 photosynthesis, CO_2 is first fixed in mesophyll cells to form a 4-carbon organic acid (oxaloacetate). This step is performed using a non-Rubisco enzyme, PEP carboxylase, which has no tendency to bind O_2. Oxaloacetate is then converted to a similar molecule, malate, and transported to the bundle sheath cell. Inside the shell, malic acid breaks down, releasing molecules of CO_2. The CO_2 is then held constant using Rubisco and processed into sugars via the Calvin cycle, similar to C3 photosynthesis.

This system is not without an energy cost: ATP must be consumed to bring a three-carbon "ferry" molecule back out of the envelope cell, ready to pick up another molecule of CO_2 in the atmosphere it won't work. However, mesophyll cells are constantly pumping CO_2 in the form of malic acid into the adjacent bundle sheath cells, so there is usually more CO_2 to O_2 around Rubisco. This strategy minimizes photorespiration. The C4 pathway is used in about 3% of all vascular flowers. Some examples are crabgrass, sugar cane, and corn. C4 plants are common in potentially hot habitats, but less noticeable in cooler regions. Under hot conditions, the benefit of reduced photorespiration is the ATP rate of CO_2 transport from mesophyll cells to bundle-envelope cells may exceed.

15.4.2.1 *Zea mays*

Perspectives on maize (Zea mays), a modern-day agricultural leader, as a bioelectricity generation target. Maize is one of the most productive crops that produces more organic biomass than other crops in a fleeting period of time, providing benefits for electrobiological systems (Rusyn 2021). Corn stalks are usually processed through energy-intensive or expensive processes to obtain sugars for bioenergy production. It is possible to generate electricity directly from corn stalks (Wang et al., 2009).

15.4.3 CAM Photosynthetic Plant

A few vegetational plants, which can be adapted to dry environments, which include cacti and pineapples, use the crassulacean acid metabolism (CAM) pathway to limit photorespiration. This name comes from its own family of vegetation, the *Crassulaceae,* in which scientists first found the pathway. In place of isolating the mild-established reactions and the use of CO_2 within the Calvin cycle in area, CAM plants separate those techniques in time. At night, CAM plant life opens their stomata, allowing CO_2 to diffuse into the leaves. This CO_2 is fixed into oxaloacetate by way of PEP carboxylase (the same step utilized by C4 plant), then transformed to malate or every other form of organic acid. The organic acid is saved inside vacuoles till the following day. In the sunlight hours, the CAM plants do not open their stomata; however, they are able to still photosynthesize. That is due to the fact the organic acids are transported out of the vacuole and broken all the way down to launch CO_2, which enters the Calvin cycle. This controlled release continues an excessive awareness of CO_2 spherical rubisco. The CAM pathway requires ATP at multiple steps, so like C4 photosynthesis, it isn't always a lively "freebie". However, plant species that use CAM photosynthesis not only keep away from photorespiration but are also very water green. Their stomata open at night, while humidity has a tendency to be higher and temperatures are cooler, both factors that reduce water loss from leaves. CAM vegetation plants are typically dominant in very hot, dry areas, like deserts.

15.4.4 *Opuntia* species

Opuntia is commercially used in prickly pear production (for human consumption), animal feed production (as animal feed), and biogas production as a renewable energy source with great possibility. The 2020 work of Wilgince Apollon, Sathish-Kumar Kamaraj et al. aims to develop a novel vertically integrated PMFC based on ceramic rods to reduce the cathode space requirement on the soil surface using four different *Opuntia* species (*Opuntia ficus-indica*, *O. albicarpa*, *O. robusta*, and *O. joconostle*) under unsaturated and open environmental conditions. Furthermore, the authors observed sustained energy production and morphological changes in Opuntia species. Finally, they aim to demonstrate the feasibility of new PMFC designs that can power electronic devices such as LEDs and digital clocks without the need for additional power amplifiers (Apollon et al., 2020).

The performance of various PMFCs investigated in recent years was compiled and is shown in Table 15.1.

Table 15.1 Performance of PMFCs using different plant species.

Plant species	Mode of respiration	Reactor type	Anode	Cathode	Current density (mA/m^2)		Power density (mW/m^2)		References
					Average	Maximum	Average	Maximum	
Oryza sativa	C3	Membrane less	Graphite felt	Graphite felt	—	52	—	5.75	Kaku et al. (2008)
Glyceria maxima	C3	—	Graphite granule	Graphite felt	32	153	4	67	Strik et al. (2008)
Oryza sativa	C3	Membrane less	Graphite felt	Graphite felt	—	163	—	14	Takanezawa et al. (2010)
Arundinella anomala	C4	Cation exchange membrane	Graphite rod in graphite grains	Graphite felt	—	—	10	22	Helder et al. (2010)
Spartina anglica	C4	Cation exchange membrane	Graphite rod in graphite grains	Graphite felt	—	—	2	222	Helder et al. (2010)
Spartina anglica	C4	Proton exchange membrane	Graphite granule	Graphite felt	214	—	50	100	Timmers et al. (2010)
Glyceria maxima	C3	Cation exchange membrane	Graphite granule	Graphite felt	67	—	32	80	Timmers et al. (2012)
Echinochloa glabrescens	C4	Proton exchange membrane	Carbon fiber	Stainless steel	—	—	—	0.088 GJ/ha/Year	Bombelli et al. (2013)
Oryza sativa	C3	Proton exchange membrane	Carbon fiber	Stainless steel	—	—	—	0.980GJ/ha/Year	Bombelli et al. (2013)
Oryza sativa	C3	-	Carbon fiber	Carbon fiber	—	—	—	23	Moqsud et al. (2015)
Lolium perenne	C3	Membrane less	Graphite granule	Carbon felt	—	55	—	—	Habibul et al. (2016)
Canna Stuttgart	C3	Proton exchange membrane	Carbon brush	carbon cloth	—	—	—	222.54	Sophia and Sreeja (2017)
Trigonella foenum graecum	C3					—	80.26		

Brassica juncea	C3					—	69.32		
Phragmites australis	C3	Ultrafiltration membrane	Graphite felt	Graphite felt	—	122	—	22	Wetser et al. (2017)
Spartina anglica	C3				—	132	—	22.76	
Vetiveria zizaniodes	C4	Proton exchange membrane	Steel mesh with graphite fiber	Graphite fiber	—	242	—	68	Regmi et al. (2018a)
Puccinellia distans	C4	Proton exchange membrane	Carbon felt	Nickel mesh	5.28	—	—	83.7	Khudzari et al. (2018a)
Chlorophytum comosum	C3 and CAM	Cation exchange membrane	Graphite sheet	Graphite sheet	—	25	—	18	Azri et al. (2018)
Epipremnum aureum	C3	Bentonite clay-based membrane	Carbon fiber brush	Carbon fiber cloth	—	20.76	—	14.05	Sarma and Mohanty (2018)
Oryza sativa	C3	Membrane less	Carbon felt	Nickel mesh	—	—	—	41.41	Regmi et al. (2018b)
Spartina anglica	C4	Cation exchange membrane	Graphite granule	Graphite felt	74	384	47	155	Sudirjo et al. (2019)
Opuntia sp	CAM	-	Graphite felt	Zinc sheet	332.1	— (mA/m^3)	—	157.93 (mW/m^3)	Apollon et al. (2020)
Oryza sativa	C3	-	Graphite felt	Graphite felt	—	—	—	16.8	Kumar et al. (2020)
Canna indica	C3	-	Graphite felt	Graphite felt	—	—	—	22.76	Sarma and Mohanty (2018)
Ipomoea aquatic	C3	Membrane less	Activated carbon with stainless steel mesh	Stainless steel mesh	—	41.08	—	10.60	Liu et al. (2021)
Stevia rebaudiana	C3	Clay cup membrane	Graphite felt	Stainless steel mesh	9.37	—	—	43.68	Apollon et al. (2022)

15.5 Development of Green Energy System to Promote Sustainable Agriculture

Agriculture remains the backbone of the worldwide economic system and the number one career for most rural people. Restricted water and land assets make it tough to feed a rapidly developing international population with the usage of cutting-edge agricultural practices. Science and technology will play a fair greater great role inside the subsequent technology of sustainable agriculture. Continuous monitoring of crop health and requirements (water, soil, nutrients, pH, temperature) is crucial for smart farming. In this regard, the PMFC can serve two purposes: acting as a biosensor and generating electricity. Plant photosynthesis and power generation by PMFCs using root exudates are carefully associated as they produce root exudates with a healthful excessive photosynthetic price. A study by Brunelli et al. (2016) showed that wireless sensors powered by PMFCs serve the dual purpose of monitoring environmental parameters (light intensity and soil moisture) and plant health.

The electricity generated via the PMFC becomes used to rate the supercapacitor. The strength stored in the supercapacitor becomes measured and used to wirelessly broadcast a signal with a frequency of 15 minutes. In this way, the monitored system (device) acts as a generator, providing energy equivalent to a neutral sensor (Arulmani et al., 2021). The dynamics of changes in the electric properties of PMFCs can also be used in plant monitoring structures. For example, differences in biopotentials in the rhizosphere under various water environments, including water scarcity, have been demonstrated. PMFCs have been successfully used to reveal soil moisture content in inexperienced roof systems containing sedum species. A fantastic correlation was found between the soil moisture content and the amount of power generation. Precision agriculture, based on precise crop needs and proper monitoring, is essential to sustainably feed a growing world population.

The two most principal factors are plant species selection and reactor component optimization. Historically, PMFC plants have performed well with high root exudation rates and high rates of photosynthetic activity. Reactor components such as electrode materials, anode placement, design, and internal resistance must be matched to obtain the desired PMFC performance. Given the first-rate research efforts of interdisciplinary studies professionals to set up green PMFC designs and resolve scale-up hurdles, the potential for plants to meet all of our food and energy needs is highly promising. Not a far-fetched idea (Kuleshova et al., 2022).

Depending on the installation and other factors, a PMFC can be used as an alternative power source for these wireless sensors. In addition to soil/sediment organic matter, plants can excrete organic matter as nodule sediments that can be used as substrates for PMFCs. During the rice harvest season, the PMFC was put in a paddy field, with graphite-felt anodic and cathodic electrodes placed over the rice rhizosphere zone, respectively. The study found that PMFCs require sunlight to generate electricity and that acetate (one of the major organic molecules produced by roots) increases electricity production in low-light conditions. This PMFC changed into capable of generating the most power density of $6\,mW/m^2$ (anode vicinity) and the anode-dominant species found to be a particular bacterial community composed of *Natronocella acetinitrilica, Beijerinckiaceae* micro-organism, and *Rhizobiales* micro-organism. In summary, PMFC is a successful strategy to improve agriculture via fertilization, pesticide management, and monitoring and improvement of agricultural surplus characteristics such as pH, DO, turbidity, conductivity, nutrients (NO_3), and so forth. Similar studies are wanted from different perspectives, consisting of investigation of distinctly efficient electrodes and optimization of the device structure. Applications of PMFCs in the areas of waste minimization, resource recovery, water supply, and agricultural monitoring are promising (Li et al., 2015).

15.6 Conclusion

Agronomically important crops provide PMFC with both high-quality food and green, clean energy. Over the past decade, there have been many advances in PMFC research. With regard to producing power, the achievement is substantially greater. The immense potential use of PMFCs lies in waste reduction, resource recovery, water supply, and agricultural monitoring (King, 2017). However, there are other important uses for PMFCs, such as creating PMFCs on green roofs, converting crops into power sources, and collecting energy from natural bodies of water. Interestingly, electrical signals from PMFCs can also be used to monitor soil, microbial, and plant health. As a result, PMFCs may one day offer a cost-effective and energy-efficient option for wetland wastewater treatment and sustainable agriculture. This chapter describes and concludes the technological development of PMFCs to address energy issues and agricultural expansion. PMFCs can convert organic compounds directly into electricity, improving agriculture in the process.

References

Alexandratos, N. and J. Bruinsma. 2012. World agriculture towards 2030/2050: the 2012 revision. ESA Working paper No. 12-03. Rome, Food and Agricultural Organization, FAO.

Apollon, W., Kamaraj, S.-K., Silos-Espino, H. et al. (2020). Impact of Opuntia species plant bio-battery in a semi-arid environment: demonstration of their applications. *Applied Energy* 279: https://doi.org/10.1016/j.apenergy.2020.115788.

Apollon, W., Vidales-Contreras, J.A., Rodríguez-Fuentes, H. et al. (2022). Livestock's urine-based plant microbial fuel cells improve plant growth and power generation. *Energies* 15 (19): 6985.

Arulmani, S.R.B., Gnanamuthu, H.L., Kandasamy, S. et al. (2021). Sustainable bioelectricity production from *Amaranthus viridis* and *Triticum ostium* mediated plant microbial fuel cells with efficient electrogenic bacteria selections. *Process Biochemistry* 107: 1359–5113. Elsevier.

Awerbuch, B.Holmer, D., Nita-Rotaru, C., and Rubens, H. (2002) An on-demand secure routing protocol resilient to byzantine failures. *Proceedings of the Workshop on Wireless Security*, pp. 21–30.

Azri, Y.M., Tou, I., Sadi, M., and Benhabyles, L. (2018). Bioelectricity generation from three ornamental plants: *Chlorophytum comosum*, *Chasmanthe floribunda* and *Papyrus diffuses*. *International Journal of Green Energy* 15: 254–263.

Batie, S.S. (1989). Sustainable Development: Challenges to Profession of Agricultural Economics. *American Journal of Agricultural Economics* **71** (5): 1083–1101.

Bombelli, P., Iyer, D.M.R., Covshoff, S. et al. (2013). Comparison of power output by rice (Oryza sativa) and an associated weed (Echinochloa glabrescens) in vascular plant bio-photovoltaic (VP-BPV) systems. *Applied Microbiology and Biotechnology* **97** (1): 429–438.

Brunelli, D., Tosato, P., and Rossi, M. (2016). Flora health wireless monitoring with plant-microbial fuel cell. *Procedia Engineering* **168**: 1646–1650.

Chel, A. and Koushik, G. (2011). Renewable energy for sustainable agriculture. *Agronomy for Sustainable Development* 31: 91–118.

Chiranjeevi, P., Kumar Yeruva, D., Kiran Kumar, A. et al. (2019). Plant-microbial fuel cell technology. In: *Microbial Electrochemical Technology*, 549–564. Elsevier.

Dhanarajan, A. (2017). *Sustainable Agriculture towards Food Security*, 3–15. Springer.

Diamond, J. (1997). Guns, Germs and Steel. A Short History of Everybody for the Last 13,000 Years. 480 pp. London: Jonathan Cape. Price £18.99 (hard covers). ISBN 0 224 03809 5. *Geological Magazine* **136** (1): 83–108.

Effendi, A.J., Ramadan, B.S., and Helmy, Q. (2022). Enhanced remediation of hydrocarbons contaminated soil using electrokinetic soil flushing – Landfarming processes. *Bioresource Technology Reports* **17**: 100959.

Foley, J.A., Ramankutty, N., Brauman, K.A. et al. (2011). Solutions for a cultivated planet. *Nature* 478 (7369): 337–342.

Food and Agriculture Organization (2016). *The State of Food and Agriculture. Climate Change, Agriculture and Food Security*. Food and Agriculture Organization of the United Nations.

Food and Agriculture Organization of the United Nations (FAO) (2016). The agriculture sectors in the Intended Nationally Determined Contributions: Analysis. ISBN 978-92-5-109244-6, Rome

Godfray, H.C.J., Beddington, J.R., Crute, I.R. et al. (2010). Food security: the challenge of feeding 9 billion people. *Science* **327** (5967): 812–818.

Habibul, N., Hu, Y., Wang, Y.K. et al. (2016). Bioelectrochemical chromium(VI) removal in plant-microbial fuel cells. *Environmental Science & Technology* **50** (7): 3882–3889.

Hartwick, J.M. (1978). Substitution among exhaustible resources and intergenerational equity. *Review of Economic Studies* 45 (2): 347–354.

Helder, M. (2012). Design criteria for the Plant-Microbial Fuel Cell Electricity generation with living plants – from lab to application ISBN: 978-94-6173-351-1, pp. 3–40.

Helder, M., Strik, D.P.B.T.B., Hamelers, H.V.M. et al. (2010). Concurrent bio-electricity and biomass production in three Plant-Microbial Fuel Cells using *Spartina anglica*, *Arundinella anomala* and *Arundo donax*. *Bioresource Technology* **101** (10): 3541–3547.

Helder, M., Strik, D.P., Hamelers, H.V., and Buisman, C.J. (2012). The flat-plate plant-microbial fuel cell: The effect of a new design on internal resistances. *Biotechnology for Biofuels* **5** (1): 1–11.

Kaku, N., Yonezawa, N., Kodama, Y., and Watanabe, K. (2008). Plant/microbe cooperation for electricity generation in a rice paddy field. *Applied Microbiology and Biotechnology* **79** (1): 43–49.

Khudzari, J.M., Kurian, J., Gariépy, Y. et al. (2018a). Effects of salinity, growing media, and photoperiod on bioelectricity production in plant microbial fuel cells with weeping alkali grass. *Biomass and Bioenergy* 109: 1–9.

Khudzari, J.M., Kurian, J., Tartakovsky, B., and Raghavan, G.S.V. (2018b). Bibliometric analysis of global research trends on microbial fuel cells using Scopus database. *Biochemical Engineering Journal* **136**: 51–60.

Khudzari, J.M., Gariépy, Y., Kurian, J. et al. (2019). Effects of biochar anodes in rice plant microbial fuel cells on the production of bioelectricity, biomass, and methane. *Biochemical Engineering Journal* 141: 190–199.

King, A. (2017). Technology: the future of agriculture. *Nature* 544: 21–23.

Kuleshova, T., Rao, A., Bhadra, S. et al. (2022). Plant microbial fuel cells as an innovative, versatile agro-technology for green energy generation combined with wastewater treatment and food production. *Biomass and Bioenergy* **167**: 106629.

Kumar, V.K., Man Mohan, K., Manangath, S.P. et al. (2020). Resource recovery from paddy field using plant microbial fuel cell. *Process Biochemistry* 99: 270–281.

Kwon, K.J. and Park, B.J. (2021). Efficiency of Spathiphyllum spp. as a plant-microbial fuel cell. *Ornamental Horticulture* 27 (2): 173–182.

Li, X., Abu-Reesh, I.M., and He, Z. (2015). Review-development of bioelectrochemical systems to promote sustainable agriculture. *Agriculture* 5: 367–388.

Liu, S., Feng, X., Xue, H. et al. (2021). Bioenergy generation and nitrogen removal in a novel ecological-microbial fuel cell. *Chemosphere* 278: 130450.

Moqsud, M.A., Yoshitake, J., Bushra, Q.S. et al. (2015). Compost in plant microbial fuel cell for bioelectricity generation. *Waste Management* 36: 63–69.

Nelson, GC; Rosegrant M.W.; Palazzo A, Gray I, Ingersol C, Robertson R, Tokgoz S, Zhu T, Sulser T; Ringler C, Msangi S, You L. (2010) Food security, farming, and climate change to 2050. *Food Security, Farming, and Climate Change to 2050: Scenarios, Results, Policy Options*.

Nitisoravut, R. and Regmi, R. (2017). Plant microbial fuel cells: a promising biosystems engineering. *Renewable and Sustainable Energy Reviews* 76: 81–89.

Pamintuan, K.R.S. and Sanchez, K.M. (2019). Power generation in a plant-microbial fuel cell assembly with graphite and stainless steel electrodes growing Vigna Radiata. *Materials Science and Engineering* 703: 012037.

PreventionWeb (n.d.). The state of food and agriculture: Climate change, agriculture and food security.

Ray, D.K., Mueller, N.D., West, P.C., and Foley, J.A. (2013). Yield trends are insufficient to double global crop production by 2050. *PLoS One* 8: e66428.

Regmi, R., Nitisoravut, R., Charoenroongtavee, S. et al. (2018a). Earthen pot–plant microbial fuel cell powered by Vetiver for bioelectricity production and wastewater treatment. *Clean: Soil, Air, Water* 46: 1700193.

Regmi, R., Nitisoravut, R., and Ketchaimongkol, J. (2018b). A decade of plant-assisted microbial fuel cells: looking back and moving forward. *Biofuels* 9 (5): 605–612.

Rusyn, I. (2021). Role of microbial community and plant species in performance of plant microbial fuel cells. *Renewable and Sustainable Energy Reviews* **152**: 111697.

Rusyn, I.B. and Hamkalo, K.R. (2020). Electro-biosystems with mosses on green roofs. *Environmental Research, Engineering and Management* 76 (1): 20–31. https://doi.org/10.5755/j01.erem.76.1.22212.

Rusyn, I., Malovanyy, M., Tymchuk, I., and Synelnikov, S. (2020). Effect of mineral fertilizer encapsulated with zeolite and polyethylene terephthalate on the soil microbiota, pH and plant germination. *Ecological Questions* 32 (1): 1–12. https://doi.org/10.12775/EQ.2021.007.

Sands, R.D. and Leimbach, M. (2003). Modeling agriculture and land use in an integrated assessment framework. *Climatic Change* **56** (1–2): 185–210.

Sarma, P.J. and Mohanty, K. (2018). Epipremnum aureum and Dracaena braunii as indoor plants for enhanced bio-electricity generation in a plant microbial fuel cell with electrochemically modified carbon fiber brush anode. *Journal of Bioscience and Bioengineering* 126: 404–410.

Sayed, E.T., Ali Abdelkareem, M., Obaideen, K. et al. (2021). Progress in plant-based bio electrochemical systems and their connection with sustainable development goals. *Carbon Resources Conversion* 4: 169–183.

Simon, M.M., Greenaway, S., White, J.K. et al. (2013). A comparative phenotypic and genomic analysis of C57BL/6J and C57BL/6N mouse strains. *Genome Biology* **14** (7): 1–22.

Solow, M.R. (1974). The economic resource or the resource of economics. *The American Economic Review* 64 (2): 1–14.

Sophia, A.C. and Sreeja, S. (2017). Green energy generation from plant microbial fuel cells (PMFC) using compost and a novel clay separator. *Sustainable Energy Technologies and Assessments* 21: 59–66.

Srivastava, P., Yadav, A.K., and Mishra, B.K. (2015). The effects of microbial fuel cell integration into constructed wetland on the performance of constructed wetland. *Bioresource Technology* **195**: 223–230.

Strik, D.P.B.T.B., Hamelers, H.V.M., Snel, J.F.H., and Buisman, C.J.N. (2008). Green electricity production with living plants and bacteria in a fuel cell. *International Journal of Energy Research* **32** (9): 870–876.

Sudirjo, E., Buisman, C.J.N., and Strik (2019). Activated carbon mixed with marine sediment is suitable as bioanode material for Spartina anglica sediment/plant microbial fuel cell: plant growth, electricity generation, and spatial microbial community diversity. *Water (Switzerland)* 11: 1810.

Syed, E.T., Abdelkareem, M.A., Obaideen, K. et al. (2021). Progress in plant-based bioelectrochemical systems and their connection with sustainable development goals. *Carbon Resources Conversion* **4**: 169–183.

Takanezawa, K., Nishio, K., Kato, S. et al. (2010). Factors affecting electric output from rice-paddy microbial fuel cells. *Bioscience, Biotechnology, and Biochemistry* **74** (6): 1271–1273.

Timmers, R.A., Strik, D.P., Hamelers, H.V.M., and Buisman, C.J.N. (2010). Long-term performance of a plant microbial fuel cell with Spartina anglica. *Applied Microbiology and Biotechnology* 86 (3): 973–981.

Timmers, R.A., Rothballer, M., Strik, D.P. et al. (2012). Microbial community structure elucidates performance of Glyceria maxima plant microbial fuel cell. *Applied Microbiology and Biotechnology* 94 (2): 537–548.

United Nations Development Programme (UNDP) (2020). Annual Report 2020 https://annualreport.undp.org/2020/.

Wang, X., Feng, Y., Wang, H. et al. (2009). Bioaugmentation for electricity generation from corn stover biomass using microbial fuel cells. *Environmental Science & Technology* **43** (15): 6088–6093.

Wetser, K., Dieleman, K., Buisman, C., and Strik, D. (2017). Electricity from wetlands: tubular plant microbial fuels with silicone gas-diffusion biocathodes. *Applied Energy* 185: 642–649.

Wijte, A.H.B.M., Mizutani, T., Motamed, E.R. et al. (2005). Temperature and endogenous factors cause seasonal patterns in rooting by stem fragments of the invasive giant reed, Arundo donax (Poaceae). *International Journal of Plant Sciences* **166** (3): 507–517.).

Part IV

Sustainable Issues Associated with Photosynthesis-Assisted Power Generation

The main crucial elements of sustainability of photosynthesis-assisted power generation are considered. Issues of renewable energy sources, reducing, recycling, and reusing liquid effluents and solid residues associated with photosynthesis-assisted power generation, the environmental benefits and impacts of bioelectrochemical systems on the environment, cost problems and scaling up to a practical level, and social issues have been clarified in the section.

16

An Overview of Sustainable Issues Associated with Bio-Assisted Power Generation Systems

Lakshmipathy Muthukrishnan[1], Sathish-Kumar Kamaraj[2], Manuel Sánchez-Cárdenas[3], and Luis Antonio Sánchez-Olmos[3]

[1]*Department of Conservative Dentistry & Endodontics, Saveetha Dental College and Hospitals, Saveetha Institute of Medical and Technical Sciences (SIMATS), Saveetha University, Chennai, Tamil Nadu, India*
[2]*Instituto Politécnico Nacional (IPN)-Centro de Investigación en Ciencia Aplicada y Tecnología Avanzada, Unidad Altamira (CICATA-Altamira), Altamira, Tamps., México*
[3]*Direccion de Postgrado e Investigación, Universidad Politécnica de Aguascalientes, Aguascalientes, México*

16.1 Introduction – Paradigm Shift toward Sustainability

Sustainability does not stand alone but entails dynamic and reliable actions to re-create and maintain balance among society, environment, and economy. It is predicted that in the forthcoming decades, there will be a steady surge in the population worldwide, incessant competition for resources, and the deteriorating effects brought about by climatic change – the three foremost determinants of science policy drafting. Innovations in agriculture undoubtedly would have a greater impact on sustainability provided they would address the environmental defacements and sustainable resources such as water, good-quality soil, green energy, the importance of biodiversity, and the security and safety of food and feed. There must be a holistic approach for materializing the measures pertaining to build a sustainable environment (Church et al. 2014).

There is a growing need and widespread demand in the utilization of renewable energy (RE) such as wind, solar power, energy storage, nuclear power, and

Photosynthesis-Assisted Energy Generation: From Fundamentals to Lab Scale and In-Field Applications, First Edition. Edited by Sathish-Kumar Kamaraj and Iryna Rusyn.

energy-saving advanced materials such as carbon composites and biofuels have been enlisted under future sustainable energy infrastructure. Although the cost of energy storage remains high, the durability remains low. Furthermore, the storage of spent nuclear fuel remains technically and politically contentious, there is an ever-increasing demand (for the expansion of biofuels) for land for agriculture, wilderness, and energy crops. These challenges pertaining to the increased energy needs have increased the global energy needs of the developed and developing worlds (Lovins 2005; Deutch et al. 2009; Tilman et al. 2009; Dunn et al. 2011; Peralta-Yahya et al. 2012; Woodward et al. 2014).

However, there exists a significant difference between medicinal and energy products. For instance, a few grams of any antibiotic could bring about potential therapeutic effects, whereas a few kilograms of hydrocarbons are required for running automobiles. Although the antibiotic produced from low conversion efficiency might cost high in the market, it is not so with the hydrocarbons. This difference in efficiency in the biological applications toward energy conversion varies totally with the biological application toward health. Some of the biological capabilities in the event of sustainable energy applications have been illustrated. Further biological capabilities include catalytic competencies capturing energy and storage, fuel synthesis, metal reduction and oxidation, and material fabrication. But it provides the capabilities of self-replication, -assembly and –repair augmenting low cost technology development (Brenner et al. 2006; Emerson et al. 2010; Shi et al. 2012).

16.2 Sustainable Systems

Immense potential of biology provides capabilities to harvest sustainable energy, but the biological applications toward building sustainable energy infrastructure remain the need of the hour. The sustainable applications were not only limited to biofuels but also enabled non-biological sustainable energy technologies. Such applications were graded by the prominent role of biology in supporting an abiotic technology developing into techno-biological applications (Figure 16.1).

16.2.1 Exploring the Bio-catalytic and Self-Assembly Characteristics

One of the notable challenges in sustainable systems is not only to provide energy from fossil fuels but the global distribution of the same at an affordable cost. The two most fundamental features of this system viz. enzyme catalysis and self-assembly make them unique in resolving the problem encountered in sustainable energy production.

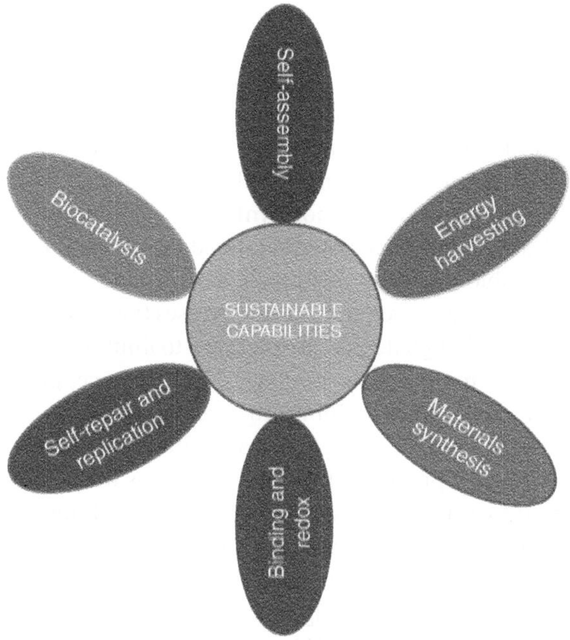

Figure 16.1 A schematic representation of sustainable capabilities in techno-biological applications.

16.2.2 Self-Repair and Replication, Energy Harvesting, and CO_2 Fixation and Chemosynthesis

Photosynthesis remains one of nature's marvels, where the radiant energy is trapped without human intervention, serving as an excellent drive toward sustainable energy infrastructure. It is this process that typically demonstrates the potency of inexpensive self-assembling catalysts on a massive scale. It was inferred that globally photosynthesis facilitates the storage of ~4000 exajoule (EJ) of radiant energy in the form of biomass within a period of one year. The energy stored in the system contributes to a mean instantaneous energy storage rate of ~130 terawatts (TW). On the other hand, according to the global energy consumption in 2012, ~550 EJ per year (~18 TW) of energy was exhausted. It is quite evident that autotrophs have the remarkable ability to trap solar energy and atmospheric CO_2 into energy-rich compounds such as starch, sucrose, cellulose, and oils (Salimijazi et al. 2019).

With the surge in the demand for natural resources, the amount of energy generated by photosynthesis will eventually remain larger to meet our future needs. The total energy use across the world is expected to increase to ~800 EJ per year

(~30 TW) by 2040. If the same trend continues till 2100, there will be an increase in global energy consumption at the rate of ~1400 EJ per year (~45 TW) (Plunkett 2011). However, impending energy-efficiency standards and technologies pertain to reducing this number. Furthermore, all forms of sustainable energy that are being consumed now or in the future nevertheless need to be derived from photosynthesis (wind, solar, nuclear, hydroelectric, and geothermal power). This approach typically relieves the pressure on the photosynthetic machineries in ratifying all energy demands, despite their huge potential. In addition, these biological systems possess a remarkable ability to self-repair. For instance, a tree trunk remains far less resistant to breakage even from severe bruises than the steel frames, this property bestow them with high degrees to tolerance to limit the lifespan inturn promulgating as modern renewable technologies. For instance, giant *Sequoia* (*Sequoiadendron giganteum*) trees exhibit an average lifespan of 3000 years. In contrast, photovoltaic cells showcase functional lifespans likely to be measured in decades. This self-repair property of biological systems would certainly offer the inherent potential to radically increase the shelf life of RE investments. Much as with conventional energy sources such as nuclear and fossil fuel-fired power plants, extensions in lifespan increase the value that can be extracted from RE investments, hence greatly raising their competitiveness with conventional energy installations (Adesina et al. 2017).

Furthermore, it is anticipated that the biological route (biomass) of capturing and storage of solar energy equally attributes to cost-effectiveness and is more reliable than an abiotic alternative as it is one of the approaches in replenishing the oldest agriculture technology. For instance, it was assumed that for a lower heating value of ~19 kJ/g, the cost involved in biological route of energy capture and storage ranged between \$2 and \$4 GJ^{-1} for loblolly pine and switchgrass, respectively. On cost grounds, the mean extraction cost of crude oil accounts to ~\$6 GJ^{-1}, but the production cost of shale oil worked out to \$14 GJ^{-1} (Rodriguez et al. 2014). It was noted that there must be a proper point of comparison for biological to abiotic solar energy conversion with photovoltaic-driven H_2 production (Blankenship et al. 2011). Alongside, Rodriguez et al. (2014) have opined on various designing strategies that could significantly aid the reduction of the production cost of solar-hydrogen generators at an approximate range of \$2 kg^{-1} or \$6 GJ^{-1}.

The proficiency and cost-effectiveness in the deployment and self-repair remain two main constraints involved in the biofuel production from biologically fixed carbon. However, the most successful solar energy storage and capture schemes are yet to be explored. It was estimated that biomass-driven energy provides ~4.9% of all US energy consumption, which is ~10 times the energy supplied by solar and other sustainable energy sources (Dunn 2016). At the global level, the energy provided by biomass accounted for ~10% of all energies, larger than all sustainable energy sources including nuclear energy (Hoeven 2012). Hitherto, the longevity, and reliability of biology open up the possibility of democratizing energy generation, permitting rural communities in the developing and developed worlds for the supply of clean, cheap, and merchandisable energy in revitalizing rustic economies all across the world (Lester et al. 2015).

16.3 Challenges and Motivations

The challenge in biological applications relies on leveraging the cost-effectiveness of energy capture and transforming the fixed carbon into value-added products with much ease as the process involved in the effective transformation of refining crude oil into petroleum products. Interestingly, the bio-based approach serves as a rich repository of biocatalysts, the enzymes with an inherent potential of transforming the carbon fixed in biomass into a wide array of infrastructure-compatible fuels and fuel precursors. These enzymes are effective in handling at room temperature and pressure within a limited space, they are efficient enough in materializing mass conversions and in restricting CO_2 emission on a par with thermochemical processes including Fischer–Tropsch (Fülöp and Ecker 2020).

Moreover, the enzymes are deployed at large in the production of biofuels owing to their improved performance characteristics. Alongside, these biocatalysts have been integrated into microorganisms such as bacteria and yeasts enabling the conversion of plant biomass into biofuels. This approach has also been encoded into autotrophic microorganisms like algae and diatoms to facilitate direct conversion into products obtained from carbon fixation, conversion into bio-enriched fuels have been further contrived into plants. Bio-based strategy voids the demand for large-scale conversion plants thereby reducing the costs associated with logistics and augmenting the diffuse rate capturing of solar energy (Bangaru et al. 2022).

The deployment of microorganisms in the refining or conversion process confers unprecedented governance over the redox state of the metals. For instance, electroactive microorganisms recruit unique extracellular electron transfer (EET) protein complexes to mobilize electrons in their metabolic processes and to counter a wide range of substrates such as metal ions and electrodes. These electroactive microbes are classified into metal reducers and oxidizers. The presence of EET confers the metal-reducing microbe, *Shewanella oneidensis* MR-1 the ability to respire array of substrates such as Fe, Mn, U, Tc, Cr, Np, Pu, iodate, selenite, tellurite, and vanadate more than O_2. Acquisition of this property provides metal-reducing microbial factories to be utilized for wider applications such as bioremediation and radionuclide cleanup in contaminated sites thereby significantly reducing the mobility of toxic metals in the environment. Furthermore, the microbes' ability to reduce solid surfaces such as electrodes identifies them as a pivotal component in microbial fuel cells (MFCs) that can generate electrical energy from wastes and sediments (Korth and Harnisch 2019; Yee et al. 2020; Thapa et al. 2022).

Electroactive microorganisms on the other hand are capable of oxidizing substrates such as organic acids and/or hydrocarbons using anode as an electron acceptor through different metabolic pathways. In this line, in *S. oneidensis* MR-1, the EET is empowered by a wider network of *c*-type cytochrome electron transfer proteins. This includes the *mtr* EET operon, which encodes a set of specialized outer membrane cytochromes that help connect the electron transport chain (ETC) in the inner membrane of the cell to the external environment (Ikeda et al. 2021). This is the mechanism responsible for sustaining most of the electron flux from metabolic pathways in the cytoplasm to the external electron acceptors beyond the outer membrane (Zerfa et al. 2019). There are several metal-reducing electroactive microbes that adopt similar mechanisms. There is a growing body of evidence suggesting that a variant of the EET system has been tuned to take up electrons into the inner membrane and metabolism to be used by carbon-fixing neutrophilic metal-oxidizing microbes (Adesina et al. 2017).

It is noteworthy to mention that most of the electroactive microbes are capable of transferring electrons over long distances at much higher rates by using conductive pili (also known as bacterial nanowires). There are some electroactive microbes that adopt this solid-matrix conduction to transfer electrons even over short distance (centimeter length) scales (Walker et al. 2018). According to ancient folklore, these metal-oxidizing microbes pioneered the oxygenation process through photosynthesis even before the advent of mankind preparing the earth's biosphere suitable for living. Alongside, the hydrothermal activity has resulted in a significant flux of the reduced iron leached into the oceans. The iron-oxidizing electroactive microbes would take up this reduced Fe as an electron and energy source to drive the carbon-fixing metabolism. Moreover, *S. oneidensis* MR-1, one of the electroactive bacteria could use oxidized Fe as a terminal electron acceptor during anaerobic respiration, facilitating the recycling by subduction, fuelling global Fe cycle, and accumulating the Banded Iron Formations in the pattern of rocks. It was estimated that the crucial production of Fe-based cycling ecosystems could have been as high as 6 gigatons per year of organic carbon with ~8 times lower than the primary production of photosynthesis-based ecosystems. In accordance, the total power flux through a Fe-based cycling ecosystem remains higher at 10 TW to support a viable ecosystem (Canfield et al. 2006; Ward et al. 2019).

On the other hand, there are contemporary electroactive metal-oxidizing organisms bestowed with the carbon fixation potential by photosynthesis, suggestive of an electron transfer mechanism involved in the reduction of NADPH (from an average of ~−100 to ~−320 mV for the reduction). Although there is an effective transition from low energy to high energy versus standard hydrogen electrode

(SHE) in the reduction process, the exact mechanism involved in the "uphill pathway" and the voltage transformation needs further studies (George et al. 2020).

It is noteworthy to mention that some of the iron-oxidizing microbes are already deployed for industrial applications. For instance, *Acidithiobacillus ferrooxidans* remains one of the key components in the bioleaching of copper at the Escondida Mine, Atacama Desert, Chile. This bacterium could effectively oxidize Fe(II) to produce Fe(III), which further oxidizes Cu(I) to Cu(II), in turn reforming Fe(II) and leaching Cu(II). This bioleaching process is thought to be the world's largest bioreactor operating at a volume of ~one trillion liters (Rouchalova et al. 2020).

16.4 Biological Solution

Despite innumerous methods available for the physicochemical remediation, uranium excavating sites cleanup, decommissioning nuclear-power sites, waste-management and storage facilities, and accident sites, the microbe-mediated approach remains the most sought-after approach owing to its modest charge, low ecological impact, and long-term remediation. Microbe-mediated approach works on several mechanisms in de-mobilizing the metals, among which bio-reduction, biomineralization, and bio-sorption have a major role. There are two main mechanisms typically used for mobilization of metals such as bio-oxidation and bio-demineralization. These approaches are used not only in bio-remediation but also in the recovery of energy-critical elements, production of battery electrodes, and successful hybrid photosynthesis.

Furthermore, over and above the environmentally viable metals such as Iron and Manganese, there are electroactive microbes such as *S. oneidensis* MR-1, *Geobacter metallireducens*, and *Geobacter sulfurreducens* which could efficiently bio-reduce and precipitate radionuclides such as Uranium (VI) and clear off the waste products at the nuclear-associated sites through their EET machinery.

Biomineralization method also helps precipitate microbial-mediated ligands such as sulfide, phosphate, carbonates, and hydroxides in response to the localized alkaline conditions prevailing at the surface of the cell. For instance, the biomineralization of uranium into uranyl phosphate using *Citrobacter* sp., and autunite using *Rahnella* sp., *Bacillus* sp., and *Aeromonas* sp. have been well documented. In addition, sulfate-reducing bacteria capable of precipitating sulfide were bio-mineralized using *Desulfovibrio* sp. and *Desulfotomaculum* sp. In addition, metals are also removed through sorption or electrostatic interaction to negatively charged groups such as $-COOH$, $-NH_2$, $-OH$, $-[PO_4]^3-$, and $-SH$ groups present on the surface of the cell. However, the bio-sorption method has not been widely considered a strategy in bioremediation of nuclear-associated sites.

Microorganisms are also widely used to mobilize Uranium, attributing it to an eco-friendly approach pertaining to extraction from the ore. Pal et al. (2010) have recently demonstrated the use of *A. ferrooxidans* for the direct bio-oxidation and bioleaching of uranium and similarly, Rashidi et al. (2012) have also demonstrated the use of *A. ferrooxidans* with Fe cycling for uranium bioleaching. Energy storage on the other hand at a large scale remains one of the key features of the sustainable energy infrastructure. It is accomplished by the shift in time of the availability of renewable sources, reducing the congestion time, altering the accessibility of renewables, and overcoming the bottleneck at the payload. The energy generated from a few kWh to multiple MWh shall be efficiently stored. This has a substantial impact not only on the production of CO_2 but also on the pliability and dissemination integrity of the grid. One of the attractive solutions for this problem is the advent of batteries as they conglomerate energy storage potential with high power density and recyclability, however, challenges pertaining to their widespread adoption such as duration, greenhouse gas footprint, and key ingredient supply constraints.

16.5 Life Cycle Assessments (LCA)

Life cycle assessment (LCA) remains one of the powerful support tools that evaluates the environmental impacts of either a product or process and the steps involved from materials extraction to waste disposal (Guinée et al. 2011). Currently, the involvement of LCA has found a surmountable increase in addressing the environmental impacts of energy technologies (Haque 2020; Lan and Yao 2022). Its prime objective of decision making and framing policies associated with energy is heralded with some unresolved queries concerned with the outcome of indeterminate conventions in this fast-evolving sector. The inquiries arise over the fact that LCA is neither spatially explicit nor temporally resolved (thus representing a snapshot in space and time or with general data that do not represent a location and particular time). There are a few cases where the life cycle results were impactful in the evolution of the energy sector. This has improved our understanding about the energy systems influencing policy making and decisions. On the other hand, the assessments are challenged by a fast-paced, economically powerful, and highly innovative sector, in particular the high regional and temporal variability of operations. Such variability (Miller 2022) has been noted to contribute to unresolved problems in LCA (Kosior and Crescenzi 2020). Notably, the electric grid comprises highly diverse generation technologies – from nuclear to renewables and fossil fuels – resulting in different regional mixes. Such differences are highly influential on LCA results (Gutowski 2018).

LCA revolves around three related unresolved problems and highlights the importance for the advancement of LCAs in electricity generation: spatial variation, time horizons, and local technical uniqueness (Turconi et al. 2013). First, LCAs that are not involved in capturing spatial variation oversee the important factors influencing the results such as logistics, population distribution, and local environmental conditions such as meteorology, topography, and hydrology. Second, the time horizons and their selection can highly influence the results when discounting the amplifying short-term costs and benefits or neglecting the longer-term effects. In addition, the choice of the temporal resolution applied for each analysis can also influence the accuracy of results. Moreover, there exists some limitation on the temporal variation over time, as the impacts associated with the dynamic supply and demand of the electric sector remain variable across scales. Third, the local technical uniqueness of operations and their varying results across technologies used at the extraction, production, distribution, and end-of-life phases (resource demands, energy use, and waste management). It was noted that the direct link with the attributes to pedigree matrices served as qualitative tools to assess uncertainty in LCA, as they already include temporal and geographic correlation as two categories to evaluate data quality (Bamber et al. 2020). Furthermore, the International Organization for Standardization's (ISO) LCA standards refer to the spatial and temporal attribute without specific guidance, suggesting that the development needs to be thoroughly vetted to date.

16.6 Composite Sustainability Indices (CSI)

Composite sustainability index (CSI) is a measure that shows the percentage improvement of asset sustainability on a par with the baseline. In terms of sustainability, there is a growing importance of CSI as a policy and as a tool useful for policy makers upon making a sustainable decision. Composite indicators (CIs) congregate multifaceted issues into one index, thus providing comprehensive information. However, there remains a controversy that CIs are too subjective, as their results undesirably depend on the normalization method, a specific weighting scheme, and the sub-indicators aggregation method (Ordouei 2021).

Sustainable development indicators are recognized as useful tools for the assessment and anticipation of production performance and trends, being able to provide primary warning to prevent economic, societal, and environmental damage and support decision-making (Kubiszewski et al. 2022). The indicators are divided into two groups viz. content indicators, describing the state of the system and performance indicators responsible for measuring the behavior of the system (Rinne et al. 2013). In total, the CI can be defined as an aggregation of different indicators according to a well-developed and predetermined methodology (Gasparatos et al. 2008). CIs can be divided into several different categories depending on the various methods selected during their formulation (Niemeijer 2002): (i) data driven, when data availability is the central issue concerning the development of the CIs and high-quality data must be provided, (ii) theory driven, when selecting the best possible indicators for CIs construction is done from a theoretical point of view, and data availability is only one of the many aspects considered, and (iii) policy driven, when the indicators are selected, especially for the monitoring of a certain policy.

Nardo et al. (2008) described a framework for the construction of a CI, which includes the selection of relevant indicators and data, imputation of missing data, normalization of the selected indicators, weighting, and aggregation. Besides, all the steps are important for improving the quality of the final CI among which the weighting and congregation seem to have the greatest impact. Zhou et al. (2007) proposed a mathematical-programming approach for the construction of CIs using multiple criteria decision analysis (MCDA) for aggregation of the selected indicators and two sets of weights (the most and least favorable) that are generated from the data themselves. The recommended approach does not demand prior knowledge with respect to the weights of the selected indicators. Cherchye (2007) used data envelopment analysis (DEA) in their construction with the aim of neutralizing the recurring sources of criticism about CIs. The application of such a method would help skip the normalization stage as it is invariant to measurement units and the weights were generated by a "flexible benefit of the doubt" method. Sikdar (2009) simplified the comparison process of aggregated metrics by expressing them as the geometric mean of the ratios of the individual metrics for pairwise comparison. The generated aggregate metric remains sensitive to the individual metrics and weighting factors. Hatefi and Torabi (2010) proposed a common weight MCDA–DEA approach for constructing CIs. Zhou et al. (2010) proposed a multiplicative optimization approach for constructing CIs, using the weighted product (WP) method. In their approach, the weights were generated by resolving a series of multiplicative DEA-type models which were transformed into equivalent linear programs. The developed model enables the incorporation of additional relevant information about the weights.

16.7 Construction of a CSI

The goal is to determine the more suitable approaches for creating a CSI by applying sensitivity analysis to the results acquired. A CSI is created using the evaluated technique by gradually averaging the chosen environmental, societal, and economic indicators. During the pre-modeling stage, the pertinent indicators spanning many facets of sustainability are chosen, and their favorable or unfavorable effects on the growth of sustainability are assessed. The gathering of data, including indicators, benchmarks, and weights, comes next. Normalization is a step in the modeling process that turns the indicators into dimensionless forms. Prior to the aggregation process, weights are applied to the various indicators and sustainability subindices in order to gauge their significance for the company's sustainable growth. The indicators are initially sorted into categories by using a step-by-step process, and then grouped into sustainability subindices, which are then combined into a CSI. The ultimate output of the methodological process offers a series of subindices for several sustainability dimensions in relation to the performance of the organization, thereby offering a level of sustainability implementation.

16.7.1 Selection, Judgment, and Data Collection

According to the described methodology, the first step during the integrated sustainability performance assessment is the selection of a suitable set of indicators. In our approach, the indicators were selected based on global reporting initiative (GRI) guidelines. These guidelines provide a set of core and additional indicators. The core indicators are those indicators, which are identified as interesting to most stakeholders and assumed to be material, unless deemed otherwise, on the basis of the GRI reporting principles. However, the final selection should always be performed in close connection with the industry. Any additional indicators are optional, for example, some additional indicators can be included if the company is interested in evaluating its influence on sustainable development. Likewise, some core indicators can also be excluded, as they are not being measured or are seen as unimportant. The environmental indicators are selected based on these three aspects such as Material, Energy and Emissions, and Effluents and Waste. On the other hand, the societal indicators are chosen from the next three groups such as Employment, Occupational Health and Safety, and Diversity and Equal Opportunities.

From the perspective of economic performance, eight economic indicators are provided. The indicators are then grouped according to how they affect sustainable development in the following stage. These categories are divided between those whose rising values have a beneficial effect on sustainable development, such as paper recycling, and those whose rising values have a negative effect, such as freshwater usage. For each normalization technique used, the normalization equation will vary depending on which group the indicator is assigned to. The step that takes the greatest time is data gathering because there are many values that must be obtained from the company on its environmental, societal, and economic performance. Additionally, the standards must be established and used in the normalization process (distance to reference). The "benefit of the doubt" approach is used for determining the weights.

The benchmark values for each of the chosen indicators were established using information from the best available techniques (BAT), internal measurements and standards, local regulatory requirements, GRI reports for certain production sectors, and other pertinent publications. If the benchmark values are specified as a range, the tightest limit within the range is used. When establishing the standards, common technologies, raw materials, production methods, and energy types employed during production are all taken into consideration. For the BAP weighting calculations, expert opinions on the significance of various indicators and sustainability subindices must also be gathered. For the purpose of incorporating the chosen indicators into a CSI, normalization is integrated. The following normalization techniques are used: minimum–maximum, separation from a reference, and the proportion of annual differences over a period of years. Minimum-Maximum (Min-max) normalization. According to this technique, the equation normalizes any indicator that has a beneficial impact on sustainable development.

The greatest value will be attained in cases where an indicator has a detrimental effect on sustainable development, taking into account the nature of the indicator. It should be noted that as this transformation is time-dependent, the time period being studied may need to be adjusted if fresh data become available. Some indicators' minimum and maximum values may vary as a result of this modification, which will then have an impact on the normalized values. To preserve comparability between the old and new data, the CSI for the old data needs to be recalculated.

16.8 The Concept of Biorefinery and their Applications

A biorefinery is a specialized establishment that employs equipment and supplies to transform biomass into fuels, power, and valuable chemicals. The petrochemical sector, which produces a variety of petroleum products, including plastic and petrol, is located next to the biorefinery. Utilizing various biomasses and intermediate products helps a biorefinery maximize the value of the biomass feedstocks. Low-volume but high-value chemical compounds and high-volume but low-value fluid or liquid transportation fuel, such as biodiesel or bioenergy, are two possible examples of biorefinery products. Energy-efficient technology that produces electricity and captures the heat (CHP technology) can produce electricity for its own usage and sell excess electricity to the community due to its high efficiency.

Profitability is increased by high-value goods, energy needs are met by fuels produced in large quantities, and power generation lowers energy costs and reduces greenhouse gas emissions from conventional power plant installations.

However, the biorefinery has a limited capacity for producing chemicals and fuel, which could get more difficult. The majority of the energy and chemical needs of interdependent civilizations and trading organizations are met by consuming enormous amounts of fossil fuels. Fossil fuel production and use result in poisons, greenhouse gases, and hazardous compounds that harm the ecosystem. The environment becomes increasingly dependent as the need for energy and chemicals rises. Along with the expansion of the world's population, trash production is also steadily increasing. Every year, our nation produces over 250 million tonnes of municipal solid waste (MSW). Overall, 35% of MSW is recycled and composted, of which 13% is used to generate power the remaining 53% of MSW is buried as landfills.

Due to the scarcity of fossil fuels and the rising demand for energy and chemicals, it is critical to develop alternate alternatives to produce these materials. In this procedure, the environmental hazards of burning fossil fuels have been addressed, and biomass has been identified as a potential source of chemicals and energy in the future. There are many different biomass sources, but they can also be gathered from many different waste sources, including municipal solid trash, agricultural waste, and industrial contamination from pulp and paper mills. Biomass waste valorization also plays a significant role in waste recycling in a comprehensive waste management plan. In comparison to using fossil fuels, a biorefinery that employs renewable biomass as a feedstock for the creation of chemicals is more environmentally friendly. A biorefinery might promote economic growth while lowering environmental air pollution.

16.9 Biorefinery Technology

The goal of an integrated biorefinery is to build on the biomass conversion process in order to extract the most value possible from the sustainable biomass feedstock. To do this, a variety of different technologies is combined. In an effort to save the world, novel bio-refining techniques are being investigated. According to IEA Bioenergy Task 42, "sustainable processing of biomass into a marketable commodity and energy spectrum" refers to the combining or integrating of various technologies for the conversion of biomass into a variety of products (such as food, feed, chemicals, materials, petroleum, coal, heat, and/or electricity). The idea focuses on the process through which different petroleum products are refined and converted into useful fuels.

The actual technical specifications of a biorefinery can vary from case to case since they are susceptible to uncontrollable conditions, such as when farmers utilize different farming techniques and when the climate changes. Which feedstock is available for processing depends on environmental and socioeconomic conditions. A biorefinery can utilize a variety of plant materials, including switchgrass, sugarcane, wheat, corn, wood, agricultural waste, straw, surplus food, freshwater biomass, and the biomass component of municipal and other sources of waste (MSW). The main product categories in a biorefinery are chemicals, biofuels, electricity and heat, materials, food and feed, minerals, and CO_2 (biorefinery, etc.).

Based on the various primary qualities that they possess, biorefineries can be divided into different categories. Various feedstocks for biofuel production include perennial grasses, starch crops (e.g. wheat and maize), sugar crops (e.g. beetroot and cane), lignocellulose crops (e.g. controlled forest, short-growing coppice, and switchgrass), lignocellulose residues (e.g. stover and straw), oil crops (e.g. palm and oilseed rape), and inorganic matter (e.g. stover and straw) (e.g. industrial, commercial, and post-consumer).

Biorefineries can process feedstocks on a variety of platforms. These platforms include lignin, oils (plant-based or algal), organic grass solutions, pyrolytic liquids, starch, sucrose, or cellulose carbon carbohydrates, as well as biogas, which is composed of single carbon molecules like methane and carbon dioxide. These fundamental building blocks can be modified to create various products by merging biological, thermal, and chemical processes. When a biorefinery's feedstock, platform, and product are known, it is possible to evaluate the company critically. The development of biorefinery practices enables the development of technological distinctions, enhances understanding of the evolution of global biorefineries, and allows comparison of biorefinery systems. Examples of biorefineries that fall under this category include C6 sugar refineries that produce ethanol and animal feed from staple crops, syngas refineries that produce FT-diesel and naphtha from cellulosic biomass, and C6 and C6/C5 sugar and syngas refineries that produce ethanol, FT-diesel, and furfural from cellulose and hemicellulose crops.

16.9.1 Pathways

Biomass can be transformed into a liquid or gaseous form, which can then be used as an energy source, as well as used as fuel for heating. Biofuels can be produced from biomass using a variety of techniques. There is a contrast between biological and thermochemical conversion. Common conversion techniques used in a variety of sectors include anaerobic digestion, saccharification, and hydrolysis. Gasification, pyrolysis, liquefaction, gasification, and combustion are five different thermochemical conversion subcategories.

The aerobic breakdown of organic carbon reduces the organic nonwoody content as all oxygen is eliminated. Methane and carbon dioxide are among the less volatile, easier-to-sell compounds. However, only a small percentage of the biomass is used in the biochemical conversion process, which is time-consuming. More often than not, thermochemical conversion methods are regarded as superior due to their adaptability and effectiveness.

Concentrating lignocellulosic biomass into various products is how most second-generation biofuels are made, composed primarily of cellulose, hemicellulose, and lignin. Hemicellulose decomposes at a temperature range of 200–380 °C, while cellulose decomposes at a temperature range of 320–400 °C. At a temperature of 400 °C, cellulose is the most stable substitute that disintegrates.

In chemical-reaction thermochemistry, three key variables must be considered: temperature, heating rate, and residence time. Around 80% of the energy produced globally today comes from combustion, which is currently the main source. Due to their high cost and poor efficacy, a number of alternative pollution management techniques, including gasification and pyrolysis, are still in the research and development stage. There are two main pathways by which the biomass is converted into an extended range of products, they are thermochemical and biochemical pathways.

16.9.2 Thermochemical Pathway

There are two main pathways for thermochemical processing: one creates biomass by heating the biomass with controlled oxygen amounts at high temperatures and pressure (a mixture of carbon monoxide and hydrogen). Gasification is the name of the process. Many industrial chemicals are produced when solid biomass is gasified. The biofuels unit will discuss how gasification affects various liquid fuels.

The second technical method works without the aid of the atmosphere and includes heating the biomass at high temperatures. Pyrolysis is a process that is widely used. The reaction time must be quick since glue must be simple to use. If not, the carbon industry (char) will take the top spot. Organic oil is the principal end product of this procedure, which is referred to as rapid pyrolysis.

The thermochemical conversion seeks to convert as little biomass as possible to steam during the chemical-making phase. The Fischer–Tropsch procedure is an illustration of a thermochemical conversion. A significant use of chemical transformation is not the thermochemical conversion of biomass. The output of thermal energy serves as the primary motivator for this five-way conversion route, which are the Gasification, Liquefaction, Pyrolysis, Torrefaction/ Carbonization, and Combustion. The biomass is initially converted into syngas in the thermochemical pathway, which is then converted into ethanol by synthesis or another process.

16.9.3 Combustion

Combustion was the first way to use living things for the generation of energy, especially because humans were the ones who discovered fire. People have learned how to live, heat their homes, and prepare food by burning wood. Combustion of biomass is an exothermic reaction that relies on oxygen. Two important stable molecules, hydrogen and carbon dioxide, oxidize the biomass in this situation. At the moment, more than 90% of energy is consumed by the heat the reaction produces.

Heat and electricity are the two main sources of energy created from biomass. In rural areas, biomass also offers heat and a sustainable source of fuel for cooking. Additionally, biomass is burned for regional and industrial heating. In regions with chilly weather, wood-burning fireplaces and pellet stoves are frequently used. For current environmental practices, it is crucial to use biomass as a source of electricity. The two most frequent processes are the steam turbine that generates power and biomass combustion in boilers. In a boiler, biomass is utilized as an alternative to fossil fuel, usually for heating. The latter strategy is more successful than the first one at reducing carbon dioxide emissions from a high-emission fossil-fuel plant.

16.9.4 Carbonization

Carbonization is advised for the productivity of biomass as a secure and effective solid fuel, similar to torrefaction. In torrefaction, the biomass is continually heated between 200 and 300 °C with little to no oxygen interaction. This procedure alters the chemical structure of the biomass hydrocarbon, lowering its oxygen content while raising its carbon content. Additionally, torrefaction increases hygroscopic biomass and biomass density. Thus, these characteristics raise the timber's worth in the marketplace for the production and transportation of power. Other carbonization techniques' main objective is to produce carbon-rich, dependable products under various circumstances.

16.9.5 Pyrolysis

In contrast to combustion, pyrolysis takes place in an atmosphere devoid of oxygen, even when partial combustion is used to heat the reaction. It is capable of converting biomass into gases, liquids, and solids swiftly and efficiently. The pyrolysis procedure results in the breakdown of certain biomass. Rapid pyrolysis yields only bio-oil, but slow pyrolysis yields reliable charcoal and gas as end products. Pyrolysis is appropriate to convert biomass into liquid fuels. While combustion emits heat, this reaction is not endothermic.

16.9.6 Gasification

Fossil fuels that are solid, liquid, or gaseous are transformed into usable gases. For gasification reactions, either water or steam is required as a medium. The gaseous atmosphere is composed of either air, oxygen, or both. Rather than using biogasification to create biogasoline, natural gas is more frequently produced from the emissions of fossil fuels. Fuel is converted through gasification from one fuel type to another. The switch from one approach to another occurs for a variety of reasons.

In order to extract sulfur from the fuel gas and prevent it from causing global warming, lower the overall carbon to hydrogen (C/H) ratio in the gasoline, and improve the fuel's hydrofluorocarbon (HFC) and perfluorocarbon (PFC) emissions, noncombustible organic content such as nitrogen and water must be removed from the fuel. A fuel is more likely to be in its gaseous condition if it has a higher proportion of hydrogen. By introducing air to the material through gasification or pyrolysis, the relative hydrogen content in the substance is created.

A product rich in hydrogen was produced as a result of indirect exposure to high temperature and pressure. Steam reforming is another step in the process. The oxygen concentration in natural gas is lower than the amount of oxygen in cellulosic biomass. Gasification produces a more useful product while reducing the overall carbon footprint.

Some natural gas is gasified in order to be used as an energy source and to create ammonia. Reforming of natural gas (a mixture of H_2 and CO) aids in the production of steam. The biogas's CO is indirectly hydrogenated by the smog, resulting in the production of methanol. However, because these systems rely on natural gas, more carbon dioxide is created as a result than with other systems. Various manufacturing processes can employ biomass in place of fossil fuels. By converting liquid transportation fuels, biomass is gasified, creating a stable source of carbon dioxide and hydrogen. Methane, which can be used as an energy source, may also be produced by the process.

16.9.7 Liquefaction

The conversion of solid biomass into liquid fuels can be accomplished through a number of techniques, such as pyrolysis, gasification, and hydrothermal processes. By heating the biomass at higher temperatures (300–350 °C) and under high pressure, it can be transformed into an oily liquid.

16.9.8 Biochemical Pathway

Biochemical conversion calls for the breakdown of the biomass to release the sugars that can be refined into biofuels and other bioproducts with the aid of bacteria and catalysts. The following are potential sources for fuel mixtures and other bioproducts, including renewable diesel, ethanol and other alcohols, and renewable petrol. One of the major difficulties in destroying the intricate structures of cellulosic biomass is converting the biomass biochemically. The Bioenergy Technologies Office investigates more efficient and cost-effective ways to process the sugars so that people can obtain these advantageous sugars.

The crucial problem is to more efficiently and effectively convert sugars into biofuels. The Bureau has created novel approaches and technology to help us reach our goal.

16.9.9 Step by Step Chemical Conversion

The hemicellulose and cellulose are transformed by biochemical conversion into an intermediate stream of sugar using biocatalysts in addition to heat and other chemicals. Such sugars are either chemically catalyzed to produce valuable molecules or are a necessary intermediary in the production of advanced biofuels and other biochemicals. The following necessary steps make up the entire process. The characteristics needed for biochemical reactions are present in the feedstocks that are chosen. Feedstock handling systems need to be effective in order for plants to operate quickly and efficiently. Pretreatment: Heat is applied to the biomass to degrade the fibrous cell walls and facilitate the hydrolysis of cellulose and hemicellulose (see subsequent step). In the pretreated content, the sugars are day-by-day separated from cellulose and hemicellulose following hydrolysis. In order to create new chemicals from sugars, such as fuels or the components of other chemicals, bacteria are introduced.

Instead of chemical conversion, fuels and other valuable products can be made from sugars utilizing chemical catalysts. Recycling: The process of removing items' leftover contaminants, sediments, and oils. Fuels and other products are sent to refineries for distribution. Manufacturing facilities may receive other items and intermediary products for use in a variety of consumer products. The majority of the organic material that is still present is lignin, which may be burned to produce natural gas and fuel.

16.10 Circular Economy

A circular economy (CE) recognizes the limited character of the earth and seeks to decouple economic expansion from the depletion of natural resources and environmental deterioration. However, just stopping material cycles, as is frequently proposed in writings related to CE, is insufficient to stop the breaching of Earth system boundaries because the intensity of the produced material and energy flows is still undisputed. In fact, the discussion of circularity is mostly focused on material cycles, ignoring the fact that these cycles will require the extensive development of RE resources in order to be powered in a sustainable way. Only in theory, with infinite energy available, are fully closed material cycles conceivable. A CE may be extremely energy intensive, which raises the question of how much energy it is permitted for a CE to appropriate within the limitations of the Earth system. In addition, the effort to recover materials (i.e. necessary energy) grows nonlinearly with improved recycling yield.

Fossil energy sources, which are essentially sun energy that has been stored for more than two centuries, have been the primary source of energy for the linear industrial economy. Burning fossil fuels depletes these resources at an alarming rate (considering that they have been built up over millions of years) and is also one of the main causes of global environmental disruption, such as the start of a climate crisis, despite their small contribution to Earth's energy budget (0.003% in 2016). While the current fission reactor fleet's use of nuclear energy does little to address the climate crisis, it rapidly depletes the uranium supply, poses catastrophic risks to the health of people and the environment in the event of accidents, increases the likelihood that nuclear weapons will be developed and used, and generates radioactive waste that must be managed for decades. Since sustainable development is defined as "development that meets the needs of the present without compromising the ability of future generations to meet their own needs," both fossil and nuclear energy resources are incompatible with this idea. Therefore, in the future, a sustainable CE can only be powered by RE fluxes.

Solar irradiance, terrestrial heat, and tides are the three incoming RE fluxes that fuel the Earth system. The last two, at 0.03% and 0.002%, respectively, make up a very little portion of the energy budget for the planet. Therefore, in a closed Earth system, solar irradiance serves as the primary energy source for the movement of both natural and man-made elements. Infrared emittance, which is regulated by the Earth's surface temperature, is the primary mechanism by which the Earth system counteracts these energy imports. These energy fluxes, which are around five orders of magnitude more than the current human demand for technical energy (i.e. energy used to power technical activities, such as electric energy), permit all material cycles within the Earth system, whether they be natural or artificial.

These energy flows have been fully utilized by the Earth system throughout its history (for example, to drive the water cycle). The appropriation of RE fluxes by humans has become a driving force in the Earth system, for instance through the demand for chemical energy (i.e. energy used to provide mankind with food and biogenic materials, such as feed, fibers, and lumber). So the question is, how much of the theoretical RE potential can we safely use without going outside the bounds of the Earth system?

16.11 Limitations

Organizations frequently encounter obstacles when integrating sustainability into their plans and initiatives. Doppelt (2003) recognized seven different categories of these restrictions and offered remedies to lessen the likelihood that they would occur (Doppelt 2003). Patriarchal thinking, a compartmentalized approach to environmental and socioeconomic problems, a lack of a clear vision for sustainability, a muddled understanding of cause and effect, a lack of information, insufficient learning opportunities, and a failure to institutionalize sustainability are some of these problems.

The challenges that businesses and project teams have when implementing sustainability are described in Doppelt's seven sustainability errors. The likelihood of those mistakes occurring is decreased by remembering them. It is crucial to translate the sustainability strategy into quantifiable goals in order to be able to assess the advantages of integrating sustainability into projects and organizations (Edwards 2009; Epstein and Roy 2001). The project's contribution to sustainable development is not always obvious since it can be challenging for organizations to link project performance to higher-level objectives (Boswell et al. 2005). According to Gilbert Silvius, the way projects are managed, measured, and reported do not align with the sustainability concept, hence the impact of sustainability is not yet understood in project management. Therefore, there is a growing need to practically implement the concept in the management of projects (Maltzman and Shirley 2012).

16.12 Conclusions

As the most affordable and extensively used technology for capturing and storing solar energy, biofuels represent today's most prominent application of biology in sustainable energy. However, the continued development of conventional biofuels has raised serious concerns about their environmental effects, leading many to wonder whether biology has any place in sustainable energy. We hope that these issues are no more insurmountable for non-biological energy technologies than they are for biofuels, as opposed to seeing them as reasons to abandon biofuels. The improvements in genetic engineering and our growing comprehension of fundamental biology give us hope that these problems can be solved.

Furthermore, we are confident that biology will soon be employed for more than only the manufacture of biofuels due to its amazing capacity to regulate processes such as carbon fixation, the redox status of metals, cell multiplication, and self-assembly. Applications for biology are anticipated in traditionally abiotic technologies including nuclear power, battery energy storage, the manufacture and recycling of energy-critical components, and the development of new materials for lighter vehicle construction. Additionally, the incorporation of biology into cutting-edge technologies like enhanced and hybrid photosynthesis would be crucial.

In the development of applied biology technologies, there are two paradigms: first, the accumulation of small-scale innovations on top of naturally occurring organisms, and second, the application of larger innovations fueled by the resolution of fundamental scientific issues.

By drawing an analogy with advancements in computers during the last 25 years, the incremental development paradigm's promise may be demonstrated. Computers from 25 years ago and today have a lot in common (metaphorically speaking): they both have von Neumann architectures, use electronics that run on transistors created by CMOS technology, could (or do) run a variant of the Unix operating system, have graphical user interfaces based on high-speed rendering, and can communicate via Transmission Control Protocol/Internet Protocol (TCP/IP). However, concurrent technological advancements in battery technology, thin, high-resolution displays, encryption, and wireless chipsets have transformed the computer from a specialized tool to one that is startlingly commonplace and has a significant impact on the user.

Although future organisms used in the synthesis of biofuels will essentially share the same DNA as current organisms, compounded incremental improvements in the fuel synthesis enzymes and genes for enhanced tolerance to fuel and environmental conditions could produce an organism that is radically different from its original wild counterpart, one that has a transformative impact on the market for liquid fuels. We believe that the creation of organisms utilized in bioremediation, the manufacture of sophisticated materials, and the mining of energy-critical elements might follow a similar path.

It will also be necessary to solve issues at the nexus of applied and basic research for biology to be applied to sustainable energy. The development of sustainable energy applications involving biology will necessitate continued basic research into the fundamental concepts of biology, from evolutionary mechanisms and light-independent autotrophic metabolisms to carbon-fixation pathways and basic characteristics of biological materials. In a similar way to how the invention of the aeroplane necessitated advances in the basic science of aerodynamics.

References

Adesina, O., Anzai, I.A., Avalos, J.L., and Barstow, B. (2017). Embracing biological solutions to the sustainable energy challenge. *Chem* 2 (1): 20–51. https://doi.org/10.1016/j.chempr.2016.12.009.

Bamber, N., Turner, I., Arulnathan, V. et al. (2020). Comparing sources and analysis of uncertainty in consequential and attributional life cycle assessment: review of current practice and recommendations. *International Journal of Life Cycle Assessment* 25: 168–180. https://doi.org/10.1007/s11367-019-01663-1.

Bangaru, A., Sree, K.A., Kruthiventi, C. et al. (2022). Role of enzymes in biofuel production: recent developments and challenges. In: *Bio-Clean Energy Technologies*, Clean Energy Production Technologies, vol. 1 (ed. P. Chowdhary, N. Khanna, S. Pandit, and R. Kumar), 81–112. Singapore. https://doi.org/10.1007/978-981-16-8090-8_4: Springer.

Blankenship, R.E., Tiede, D.M., Barber, J. et al. (2011). Comparing photosynthetic and photovoltaic efficiencies and recognizing the potential for improvement. *Science* 332: 805–809. https://doi.org/10.1126/science.1200165.

Boswell, J., Wallace, B., and Boswell, P. (2005). Project sustainability management: translating words into action. *Civil Engineering: Magazine of the South African Institution of Civil Engineering* 13 (8): 12.

Brenner, M.P., Bildsten, L., Dyson, F., et al. (2006). Engineering microorganisms for energy production. Report of the JASON Program Office, the MITRE Corporation, June 23, 2006, *JSTR-05-300*.

Canfield, D.E., Rosing, M.T., and Bjerrum, C. (2006). Early anaerobic metabolisms. *Philosophical Transactions of the Royal Society of London. Series B, Biological Sciences* 361: 1819–1836.

Cherchye, L., Moesen, W., Rogge, N., and van Puyenbroeck, T. (2007). An introduction to 'benefit of the doubt' composite indicators. *Social Indicators Research* 82: 111–145.

Church, G.M., Elowitz, M.B., Smolke, C.D. et al. (2014). Realizing the potential of synthetic biology. *Nature Reviews. Molecular Cell Biology* 15: 289–294.

Deutch, J.M., Forsberg, C.W., Kadak, A.C., Kazimi, M.S. (2009). Update of the MIT 2003 future of nuclear power. *Report for Massachusetts Institute of Technology* (Volume 23): 856–860.

Doppelt, B. (2003). Overcoming the seven sustainability blunders, 2003. *The Systems Thinker* 14 (5): 2–6.

Dunn, D.R. (ed.) (2016). *Monthly Energy Review – September 2016*. US Energy Information Administration.

Dunn, B., Kamath, H., and Tarascon, J.-M. (2011). Electrical energy storage for the grid: a battery of choices. *Science* 334: 928–935.

Edwards, M.G. (2009). An integrative metatheory for organisational learning and sustainability in turbulent times. *The Learning Organization* 16 (3): 189–207.

Emerson, D., Fleming, E.J., and McBeth, J.M. (2010). Iron-oxidizing bacteria: an environmental and genomic perspective. *Annual Review of Microbiology* 64: 561–583.

Epstein, M.J. and Roy, M.-J. (2001). Sustainability in action: identifying and measuring the key performance drivers. *Long Range Planning* 34 (5): 585–604.

Fülöp, L. and Ecker, J. (2020). An overview of biomass conversion: exploring new opportunities. *PeerJ* 8: e9586. https://doi.org/10.7717/peerj.9586.

Gasparatos, A., El-Haram, M., and Horner, M. (2008). A critical review of reductionist approaches for assessing the progress towards sustainability. *Environmental Impact Assessment Review.* 28 (4–5): 286–311. https://doi.org/10.1016/j.eiar.2007.09.002.

George, D.M., Vincent, A.S., and Mackey, H.R. (2020). An overview of anoxygenic phototrophic bacteria and their applications in environmental biotechnology for sustainable Resource recovery. *Biotechnology Reports (Amsterdam, Netherlands)* 28: e00563. https://doi.org/10.1016/j.btre.2020.e00563.

Guinée, J.B., Heijungs, R., Huppes, G. et al. (2011). Life cycle assessment: past, present and future. *Environmental Science & Technology* 45 (1): 90–96.

Gutowski, T.G. (2018). A critique of life cycle assessment; where are the people? *Procedia CIRP* 69: 11–15. https://doi.org/10.1016/j.procir.2018.01.002.

Haque, N. (2020). Chapter 29: The life cycle assessment of various energy technologies. In: *Future Energy*, 3e (ed. T.M. Letcher), 633–647. ISBN 9780081028865. Elsevier.

Hatefi, S.M. and Torabi, S.A. (2010). A common weight MCDA–DEA approach to construct composite indicators. *Ecological Economics* 70 (1): 114–120. https://doi.org/10.1016/j.ecolecon.2010.08.014.

van der Hoeven, M. (ed.) (2012). *World Energy Outlook 2012*. International Energy Agency.

Ikeda, S., Takamatsu, Y., Tsuchiya, M. et al. (2021). *Shewanella oneidensis* MR-1 as a bacterial platform for electro-biotechnology. *Essays in Biochemistry* 65 (2): 355–364. https://doi.org/10.1042/EBC20200178.

Korth, B. and Harnisch, F. (2019). Spotlight on the energy harvest of electroactive microorganisms: the impact of the applied anode potential. *Frontiers in Microbiology* 10: 1352. https://doi.org/10.3389/fmicb.2019.01352.

Kosior, E. and Crescenzi, I. (2020). Chapter 16: Solutions to the plastic waste problem on land and in the oceans. In: *Plastic Waste and Recycling* (ed. R. Letcher), 415–446. Academic Press https://doi.org/10.1016/B978-0-12-817880-5.00016-5.

Kubiszewski, I., Mulder, K., Jarvis, D., and Costanza, R. (2022). Toward better measurement of sustainable development and wellbeing: a small number of SDG indicators reliably predict life satisfaction. *Sustainable Development* 30 (1): 139–148. https://doi.org/10.1002/sd.2234.

Lan, K. and Yao, Y. (2022). Dynamic life cycle assessment of energy technologies under different greenhouse gas concentration pathways. *Environmental Science & Technology* 56 (2): 1395–1404.

Lester, T.W., Little, M.G., and Jolley, G.J. (2015). Assessing the economic impact of alternative biomass uses: biofuels, wood pellets, and energy production. *J. Reg. Anal. Policy* 1: 36–46.

Lovins, A.B. (2005). More profit with less carbon. *Scientific American* 293: 74–83.

Maltzman, R. and Shirley, D. (2012). *Green Project Management*. CRC Press.

Miller, S.A. (2022). The capabilities and deficiencies of life cycle assessment to address the plastic problem. *Frontiers in Sustainability* 3: 1007060. https://doi.org/10.3389/frsus.2022.1007060.

Nardo, M., Saisana, M., Saltelli, A. et al. (2008). *Handbook on Constructing Composite Indicators: Methodology and User Guide*. OECD Publishing.

Niemeijer, D. (2002). Developing indicators for environmental policy: data-driven and theory-driven approaches examined by example. *Environmental Science and Policy* 5 (2): 91–103. https://doi.org/10.1016/S1462-9011(02)00026-6.

Ordouei, M.H. (2021). Chapter 11: Composite sustainability indices (CSI); a robust tool for the sustainability measurement of chemical processes from "early design" to "production" stages. In: *Methods in Sustainability Science* (ed. J. Ren), 169–196. https://doi.org/10.1016/B978-0-12-823987-2.00001-5. Elsevier.

Pal, S., Pradhan, D., Das, T. et al. (2010). Bioleaching of low-grade uranium ore using *Acidithiobacillus ferrooxidans*. *Indian Journal of Microbiology* 50 (1): 70–75.

Peralta-Yahya, P.P., Zhang, F., del Cardayre, S.B., and Keasling, J.D. (2012). Microbial engineering for the production of advanced biofuels. *Nature* 488: 320–328.

Plunkett, J.R. (2011). *International Energy Outlook and Projections.*, 1–453. Nova Science Publishers.

Rashidi, A., Safdari, J., Roosta-Azad, R., and Zokaei-Kadijani, S. (2012). Modeling of uranium bioleaching by *Acidithiobacillus ferrooxidans*. *Annals of Nuclear Energy* 43: 13–18. https://doi.org/10.1016/j.anucene.2011.12.020.

Rinne, J., Lyytimäki, J., and Kautto, P. (2013). From sustainability to well-being: lessons learned from the use of sustainable development indicators at national and EU level. *Ecological Indicators* 35: 35–42. https://doi.org/10.1016/j.ecolind.2012.09.023.

Rodriguez, C.A., Modestino, M.A., and Psaltis, D. (2014). Design and cost considerations for practical solar-hydrogen generators. *Energy & Environmental Science* 7: 3828–3835.

Rouchalova, D., Rouchalova, K., Janakova, I. et al. (2020). Bioleaching of iron, copper, lead and zinc from the sludge mining sediment at different particle sizes, pH and pulp density using *Acidithiobacillus ferrooxidans*. *Minerals*. 10 (11): 1013. https://doi.org/10.3390/min10111013.

Salimijazi, F., Parra, E., and Barstow, B. (2019). Electrical energy storage with engineered biological systems. *Journal of Biological Engineering* 13: 38. https://doi .org/10.1186/s13036-019-0162-7.

Shi, L., Rosso, K.M., Clarke, T.A. et al. (2012). Molecular underpinnings of Fe(III) oxide reduction by *Shewanella oneidensis* MR-1. *Frontiers in Microbiology* 3: 50.

Sikdar, S.K. (2009). On aggregating multiple indicators into a single metric for sustainability. *Clean Technologies Environ Policy* 11 (2): 157–161.

Thapa, B.S., Kim, T., Pandit, S. et al. (2022). Overview of electroactive microorganisms and electron transfer mechanisms in microbial electrochemistry. *Bioresource Technology* 347: 126579. https://doi.org/10.1016/j.biortech.2021.126579.

Tilman, D., Socolow, R., Foley, J.A. et al. (2009). Beneficial biofuels–the food, energy, and environment trilemma. *Science* 325: 270–271.

Turconi, R., Boldrin, A., and Astrup, T. (2013). Life cycle assessment (LCA) of electricity generation technologies: overview, comparability and limitations. *Renewable and Sustainable Energy Reviews* 28: 555–565. https://doi.org/10.1016/j .rser.2013.08.013.

Walker, D., Adhikari, R., Holmes, D. et al. (2018). Electrically conductive pili from pilin genes of phylogenetically diverse microorganisms. *The ISME Journal* 12: 48–58. https://doi.org/10.1038/ismej.2017.141.

Ward, L.M., Rasmussen, B., and Fischer, W.W. (2019). Primary productivity was limited by electron donors prior to the advent of oxygenic photosynthesis. *Journal of Geophysical Research – Biogeosciences* 124: 211–226. https://doi.org/10.1029/ 2018JG004679.

Woodward, A., Smith, K.R., Campbell Lendrum, D. et al. (2014). Climate change and health: on the latest IPCC report. *Lancet* 383: 1185–1189.

Yee, M.O., Deutzmann, J., Spormann, A., and Rotaru, A.-E. (2020). Cultivating electroactive microbes – from field to bench. *Nanotechnology* 31: 174003. https:// doi.org/10.1088/1361-6528/ab6ab5.

Zerfa, C., Asally, M., and Soyer, O.S. (2019). Interrogating metabolism as an electron flow system. *Current Opinion in Systems Biology* 13: 59–67. https://doi.org/10.1016/ j.coisb.2018.10.001.

Zhou, P., Ang, B.W., and Poh, K.L. (2007). A mathematical programming approach to constructing composite indicators. *Ecological Economic* 62: 291–297.

Zhou, P., Ang, B.W., and Zhou, D.Q. (2010). Weighting and aggregation in composite indicator construction: a multiplicative optimization approach. *Social Indicators Research* 96: 169–181.

Index

Photosynthesis-Assisted Energy Generation: From Fundamentals to Lab Scale and In-Field Applications, First Edition. Edited by Sathish-Kumar Kamaraj and Iryna Rusyn.

Printed and bound by CPI Group (UK) Ltd, Croydon, CR0 4YY
17/09/2026
14964308-0006